W9-CSP-372

SMART DEVICES: MODELING OF MATERIAL SYSTEMS

To learn more about AIP Conference Proceedings, including the
Conference Proceedings Series, please visit the webpage
http://proceedings.aip.org/proceedings

SMART DEVICES: MODELING OF MATERIAL SYSTEMS

An International Workshop

IIT Madras, Chennai, India   10 - 12 January 2008

EDITORS

Srinivasan M. Sivakumar
IIT Madras
Chennai, India

Vidyashankar Buravalla
India Science Lab, GM R&D
Bangalore, India

Arun R. Srinivasa
Texas A&M University
College Station, Texas, U.S.A.

SPONSORING ORGANIZATIONS
India Science Lab, GM R&D
Brakes India Limited
All India Council for Technical Education (AICTE)
Indian Institute of Technology Madras

Melville, New York, 2008
AIP CONFERENCE PROCEEDINGS ■ 1029

Phys
Sep/oe

Editors

Srinivasan M. Sivakumar
Dept. of Applied Mechanics
Indian Institute of Technology Madras
Chennai 600036
INDIA

E-mail: mssiva@iitm.ac.in

Vidyashankar Buravalla
India Science Lab, GM R&D
Bangalore 560 066
INDIA

E-mail: vidyashankar.buravalla@gm.com

Arun R. Srinivasa
Dept. of Mechanical Engineering
Texas A&M University
College Station, TX 77943
U.S.A.

E-mail: asrinivasa@tamu.edu

L.C. Catalog Card No. 2008929303

ISBN 978-0-7354-0553-0
ISSN 0094-243X

Printed in the United States of America

CONTENTS

INAUGURAL LECTURE

SESSION 1

SMART DEVICES—AN OVERALL PERSPECTIVE

SESSION A1

SHAPE MEMORY MODELING APPROACHES

SESSION A2

SHAPE MEMORY MODELING IN APPLICATIONS

vi

POSTER SESSION

Preface

Smart materials, adaptronics, multifunctional materials, intelligent materials, biomimetic materials - these words signify a new and exciting era in the development of integrated sensing and actuation systems; systems whose core materials and microstructures enable the integration of structure, sensing and actuation in a single device or component. The promise of radical changes in designs and the possibility of hitherto unimaginable range of functionalities and performance has been attracting a variety of scientists and engineers - ranging from physicists interested in a first principles description of new materials, metallurgists and materials scientists developing new ceramics, polymers and metal alloys, modelers developing simulations and engineers developing novel devices -to explore this area.

Nevertheless, the progress towards ubiquitous use of such devices has been rather slow. Many of the behaviors of shape memory alloys, magneto active and electro active solids and fluids have been explored for decades and processes have been developed for the manufacture of these materials. Yet, barring a few exceptions like the defence and aerospace industries, these materials have not seen widespread use. The SDMoMS international workshop that was organized at IIT Madras, in January, 2008 is meant to identify and address some of the issues that are at the heart of the rather slow adoption of these technologies.

The organizers and sponsors of the conference recognized that while there are many applications of adaptive materials, especially the shape memory alloys (SMAs), wherein, on a laboratory or a technology demonstration stage there have been significant achievements, not entirely unsurprisingly, there has been only limited success in transitioning these into full scale industrial applications. One of major delimiting factors for this is the lack of adequate predictive capability in terms of the materials response at a device or a system level. This can almost entirely be attributed to the fact that the response of these materials is coupled and highly nonlinear rendering conventional material models grossly inadequate. For instance, in SMAs, the underlying martensitic phase transformation responsible for shape memory and superelasticity involves interaction between thermal and mechanical energies calling for simultaneous consideration. Thus, the designers and engineers, lacking simple robust models for these materials, have been quite conservative in the use of these materials and have largely remained skeptical of the claims made regarding their full potential. A classic case to the point of maturity of a smart material in terms of application development is the contrast between the comfort of the designers with piezoelectrics for sensing and actuation purposes as opposed to the case of SMAs. Currently the most developed framework of simulation and design among the ``smart materials' is that of the piezoelectric materials in the linear regime. Here, depending upon the type of material and the nature of application, equivalent circuit representations (in both time and frequency domain) have been standardized (by the IEEE) and are readily available. This is of immense help to the controls systems design and to the modeling community. On the other hand, a systematic design and simulation methodology with SMAs and shape memory polymers (SMPs) is still relatively in its early stages. Most SMA material models are too complex for implementation into real time control systems.

The modeling community, in general, has focused predominantly in developing a variety of increasingly sophisticated and complex material models to understand and explain the essential behavior of such materials and has seen significant success in that direction. However, it has not paid much attention to the needs of the designers and engineers involved in designing smart devices and systems that require much more effort in terms of addressing issues hitherto ignored. In contrast, the design community has restricted itself to use simplistic highly 'linearized' models, thereby arriving at highly conservative solutions.

There are some essential features of designing with smart materials that have to be taken into account:

1. Owing to the hysteretic and nonlinear behavior, simulations need to play an essential role in the design process; it is not, in general, possible to interpolate reliably using the response from a few special cases.
2. In general, there is significant interplay between several different types of energies, viz. mechanical, electromagnetic and thermal.
3. The initial processing of the material (especially with SMAs) plays a major role in the subsequent response of the material. Thus, the models need to account for the loading history or the initial state of the material.
4. Either intentionally or otherwise, the material can undergo incomplete or partial hysteresis, which can affect both the reliability and the durability of these systems.
5. Since the materials undergo some level of functional changes from one cycle to another, a feedback control system is a necessary element of the design of these materials to compensate for the changes in their behavior.
6. Due to the nonlinearities and the time lag, simple PID controllers are no longer sufficient necessitating more sophisticated state space control designs

The 'smartness' of the materials is due to the underlying significant transformation(s) or changes in the microstructure. For piezoelectrics, it is the change in the configuration of a set of unit cells from a non-polarized to a polarized state with alignment of the polarized domains. For the shape memory alloys, it is the change in the shape of the crystal from a cubic to a tetragonal shape together with mutual accommodation. In shape memory polymers, it is the freezing of the networks into low entropy configurations through the formation of secondary bonds. This change in configuration also makes for a certain degree of ``instability'' of these materials and their sensitivity to external influences as compared to more conventional materials. Furthermore, the nonlinearity and the hysteresis can also be traced to this behavior. In this context, it becomes desirable or even necessary to integrate various scales of the material into a model, thereby, calling for a multiscale approach. Furthermore, to satisfactorily account for interaction between different types of energies, a multiphysics approach is also called for. These pose significant challenges to the modeling community.

One of the major objectives of the SDMoMS '08 workshop was to provide a forum for the exchange of ideas between the modeling community and the design community. This facilitates discussions and deliberations on specific issues like those mentioned above leading to a clear perception of the needs of the design community in terms of modeling tools necessary for achieving greater maturity and success in device development. This could potentially lead to several initiatives towards developing models that specifically address the issues faced by the designers. On the one hand, the design community can become aware of the fact that simulations of the response of these materials play a key role in their design. At the same time, the modeling community whose ultimate clients are the designers may become aware of the vast gulf separating their current simulation endeavors (which are focused towards increasingly detailed ``universal'' models with complex constitutive relations capable of incorporating a wide range of effects under a single umbrella) and the needs of the design community (who need capability to develop control systems and predictive capabilities for application specific geometry and loading). In this context, the modeling efforts, from a designer's perspective, have been misdirected thereby calling for a 'course correction'.

Thus, the papers presented in this workshop deal with the current state of the art in the design and modeling of smart material systems, the challenges faced by designers in the widespread incorporation of these materials into useful products. Further, the possibility of the development of a common modeling framework for smart materials is also explored so as to enable designers to exercise choice of materials from an array of smart materials, gather material property data, compare different smart materials and simulate the response of these materials in a relatively straightforward manner before arriving at an optimal design.

The plenary lecture by Prof. K. R. Rajagopal, Forsyth Chair, and Professor, Departments of mechanical, civil, and biomedical engineering at Texas A&M University and an authority in the area of constitutive modeling, is particularly germane to the issue.

(1) He emphasizes the role played by thermodynamic considerations (especially the second law of thermodynamics) in the development of models for smart material behavior. This is particularly important since many of these materials can be used for energy transducers (or heat engines) and ad-hoc models may easily violate the impossibility of perpetual motion.

(2) He points out the role of the changing natural states of the material. In other words, for many materials (notably ferroelectrics, SMAs and SMPs), their multifunctionality, as noted earlier, comes from the fact that they have more than one equilibrium or natural configuration.

Subsequent deliberations of the workshop attendees fall into the following headings:
1. Shape memory materials
2. Magnetic solids and fluids
3. Polymers for smart applications, and
4. Ferroelectrics

The first series of lectures and discussions focused on the continuum based modeling of the response of smart materials

In the first session that focused on modeling strategies of shape memory materials, more specially, on the prescription of free energy functions and their forms for modeling. Prof.Arun Srinivasa presented a simple macroscopic thermodynamically consistent phenomenological way of developing constitutive equations for SMP wires using a prescription their Gibbs' free energy. Prof.Srikanth Vedantam, highlighted the need for a two scale modeling approach – the interatomic scale and the continuum scale in order to construct a free energy function for SMAs in the continuum framework,. He demonstrated through a shape memory alloy example, how the free energy could be constructed using this approach. Dr. D.R. Mahapatra presented a mathematical framework to the structure of the free energy due to multiple martensitic variants in SMAs and demonstrated how several microstructural properties and kinetics can be obtained based on the free energy. Prof. P.Thamburaja presented his work on a finite deformation 3-D model based on a single crystal plasticity type approach to describe the martensitic transformations, martensitic reorientation and detwinning. Finite element simulations were carried out using ABAQUS and were compared with the experimental results. Dr.B.Vidyashankar described how a thermodynamically consistent phenomenological model of shape memory alloy wires could be constructed without the mathematical inconsistencies in the material functions found in the literature. He presented the model simulations for the hysteretic behavior of SMAs under arbitrary thermo-mechanical loading conditions. Prof.Hui Hui Dai, showed how a geometrical size effect arises due to the instability phenomena that accompanies phase transitions in a slender circular cylinder. He presented an analytical solution to the model developed for this.

The next set of talks focused on the issues related to the design and modeling of specific devices.

Dr.Dayananda described how an SMA based carbon fiber reinforced smart landing gear was modeled, characterized and validated for a semi rigid radio controlled airship. He highlighted some of the needs that arise in modeling in the course of this design process. Dr. L. Chandrasekaran briefly explained how he designed a safety mechanism application using an SMA trigger. Several models have been proposed in the past describing the fundamental behavior of MR fluids. However, the development of MR devices have faced obstacles that had to be overcome at the design state, argued Dr.John Ulicny. He presented the strengths and limitations of various model proposed for MR fluids including a model that accounted for the details of the dispersed phase through a Molecular dynamics-like approach in the MR Fluids. Dr. M.Mahendran presented his work on magnetic shape memory alloy (NiMnGa) polymer composites for vibration energy absorption. Prof.Gopalakrishnan explained how ANN could be used to model the nonlinear response of the magnetostrictive materials.

The next set of talks focused on the fact that in smart polymers, processing and deployment are closely coupled and that desired response characteristics can be obtained by a suitable processing of the material.

Prof.Ashwini Agrawal demonstrated that improvements in mechanical response in PH sensitive copolymers can be obtained by varying the architecture and the polymerization procedure. Prof. Bishak Bhattacharya described how a single link rotating flexible IPMC manipulator can control large deflection vibrations and at the same time be used effectively for a manipulator application. Prof.Abhijit Deshpande listed out various performance parameters that may be necessary to model the shape memory polymers for smart applications.

The session on ferroelectric devices highlighted the need for better characterization, modeling and utilization of the nonlinear phenomena in ferroelectrics so that they can be used more effectively.

Dr. Marc Kamlah summarized the experimental studies that are carried out at his laboratory on the nonlinear behavior of piezoceramics under uni-axial and multi-axial loading conditions to capture the ferroelectric and ferroelastic hysteretic behavior and also the room temperature creep effects. Prof.Andreas Menzel, in turn, presented a constitutive model developed to describe the nonlinear effects of piezoelectric ceramics due to micro-cracking with fatigue phenomena and domain switching. Grain boundary effects play a major role in the domain switching phenomena that contribute to nonlinear effects in the ferroelectric material. Prof. S.M.Sivakumar described how these grain boundary effects can be incorporated into a micromechanically motivated thermodynamically consistent model and embedded into a three dimensional finite element frame work for modeling simple applications.

Prof. Binu Mukherjee elaborated on the current practices on characterization of piezoceramics, electrostrictive ceramics and single crystals. He also emphasized the need for carrying out tests under a variety of external environments like large applied electric fields, stresses, wide range of frequencies and temperatures. Dr. D.D.Ebenezer presented how the effect of internal losses in the piezo ceramics is modeled using complex material properties and linearized equations under steady state conditions.

Drs. S.Sen and A.Sen described their study on the effect of varying compositions in PZT wafers and deposition parameters in PZT thin films.

Mr.R.Shanker presented in his poster, his experimental study on extracting the dynamic characteristics of a frame structure using piezo-electric ceramic (PZT) transducers. Mr.R.Maranganti, in his poster presentation, showed the different length scales at which the gradient effects manifest for different materials such as semiconductor, metallic, amorphous and polymeric materials are analyzed.

We hope that these papers will serve as a starting point for engineers and scientists of different disciplines to further embark on a much needed cooperative venture to enable the ubiquitous use of ``smart technologies'' in the design of devices and components.

Srinivasan M. Sivakumar
Professor
Dept. of Applied Mechanics,
Indian Institute of Technolgy Madras
Chennai 600036.
INDIA

Vidyashankar Buravalla
Staff Researcher
India Science Lab, GM R&D,
Bangalore 560 066
INDIA

Arun R. Srinivasa
Associate Professor
Dept. of Mechanical Engg,
Texas A&M University,
College Station, TX 77943
USA

Acknowledgments

We express our heartfelt thanks to ISL, GM R&D for giving an almost unconditional support for conducting this workshop. Their support (moral, physical and financial) was the backbone for the success of this workshop. Mere words cannot express our gratitude towards Prof. K.R. Rajagopal who gave the vital moral support during the testing times while arranging this workshop. We also thank all the others sponsors – Brakes India Limited, AICTE through Center for Continuing Education, Indian Institute of Technology Madras (IITM), and some of the academic departments at IITM for making this workshop a possibility. All the members of the organizing committee and faculty of Applied Mechanics department, IITM pitched in with help when needed. There were many volunteers who were working like the busy bees to make sure all arrangements were in order leading to the success of this workshop. Several departments at IITM facilitated the smooth conduct of the workshop. The video graphing team covered all the proceedings of the workshop. Our sincere thanks to them all.

We thank Maya Filkop and Kristen Girardi of AIP Special Publications and Proceedings Division for readily agreeing to publish the proceedings of this workshop.

Lastly, but importantly, we thank from the bottom of our heart, all the invited speakers and the participants for making this workshop a fruitful one. Without their keen interest and cooperation, this proceedings volume would not have come out in time.

COMMITTEES

Advisory Committee

CHAIRMAN
M. S. Ananth, Director, Indian Inst. of Tech. Madras, India
MEMBERS
B. G. Prakash, Director, India Science Lab, GM R&D, India
Jan Aase, Director, VDRL, GM R&D, Warren, USA
A. R. Srinivasa, Mechanical Engg., Texas A&M Univ, USA
S. Krupanidhi, IISc, Bangalore, India

Organizing Committee

CHAIRMEN
K. Ramesh, Head, Department of Applied Mechanics, IIT Madras, India
P. D. Mangalgiri, Lab Group Manager, ISL, GM R&D, India
SECRETARY
Srinivasan M. Sivakumar, IIT Madras, India
JOINT SECRETARY
B. Vidyashankar, India Science Lab, GM R&D, India
MEMBERS
Nancy L. Johnson, GM R&D, USA
S. Gopalakrishnan, IISc, India
A. Arockiarajan, Nottingham Univ., U.K.
K. Vijayaraju, ADA, Bangalore
C. Lakshmana Rao, IITM, India
D. Roy Mahapatra, IISc, India
G. M. Kamath, NAL, Bangalore
K. Balasubramanian, IITM, India
K. Ravishankar, SERC, Chennai, India

SPONSORS

India Science Lab, General Motors Research and Development, Bangalore, INDIA
Brakes India Limited, Chennai, INDIA
All India Council for Technical Education (AICTE), INDIA
Indian Institute of Technology Madras, Chennai, INDIA
   - Departments of Applied Mechanics, Physics and Ocean Engineering

LIST OF PARTICIPANTS

INVITED SPEAKERS

Arockiarajan A., University of Nottingham, U.K.
Ashwini K Agrawal, IIT Delhi, India
Bhattacharya, Bishakh, IIT Kanpur, India
Buravalla, V., India Science Lab, GM R&D, India
Dai, Hui Hui, City University of Hong Kong, PRC
Dayananda G., National Aerospace Laboratories, India
Deshpande, Abhijit, IIT Madras, India
Ebenezer D. D., Naval Physical and Ocean Laboratories, India
Gopalakrishnan S., IISc Bangalore, India
Kamlah, Marc, Forschungszentrum Karlsruhe, Germany
Mahapatra, D Roy, IISc, Bangalore, India
Mahendran, M., TEC, Madurai, India
Mangalgiri, P.D., India Science Lab, GM R&D, India
Menzel, Andreas, University of Siegen, Germany
Mukherjee, Binu, Royal Military College of Canada, Canada
Rajagopal K. R., Texas A&M University, USA
Sen, A., Central Glass and Ceramic Research Institute, CSIR, India
Sen, S., Central Glass and Ceramic Research Institute, CSIR, India
Sivakumar, S.M., IIT Madras, India
Srinivasa, A, Texas A&M University, USA
Thamburaja, P., National University of Singapore, Singapore
Ulicny, John, GM R&D and Planning, Warren, USA
Upadhya, A. R., National Aerospace Laboratories, India
Vedantam, S., National University of Singapore, Singapore

OTHER DELEGATES

Anbuvelan, K., Bharath University, India
Ashesh Sha, IIT Kanpur, India
Atanu Banerjee, IIT Kanpur, India
Babu, V.S., PVP Siddhartha Instt Of Technology, India
Chandrashekara, C.V. IIT Delhi, India
Chanrasekaran, L.C., Qinetiq, UK.
Chandran.V.S., Amal Jyothi College Of Engg. , India
Gouda, B.J., Hirasugar Institute Of Technology, India
Gunasegarane, G.S., Pondicherry Engg. College, India
Jaget Babu N.R. College Of Engineering, India
Jayabal, K., IIT Madras, Chennai, India
John Alexis, S., Srikrishna College Of Engg. , India

OTHER DELEGATES (contd.)

Joselin, R., C.S.I. Institute Of Technology, India
Kanasogi, R.M., Malnad College Of Engg, India
Kantha Babu, SSN College Of Engineering, India
Kumar, R.S., Government Engineering College, India
Lawrence, Shimi, SCT College Of Engineering, India
Madhekar, S.N., College Of ENGG.PUNE, India
MALLIK, U.S., Siddaganaga Institute Of Technology, India
Manjunath, T.S., G.M. Institute Of Technology, India
Maranganti, R.,University of Houston, USA
Mini.R.S, College Of Engineering, India
Mohan Lal, D, Anna University, India
Murugarajan, A., Iitmadras, Chennai, India
Murugesh, M.C., G.M. Institute Of Technology, India
Nanda Kumar, N.S., Kongu Engg.College, India
Natarajan, N., Bannari Amman Institute Of Tech, India
Nathi Shyam Kumar, Kakatiya Institute Of Technology&Sciene, India
Pandit, D., IIT Madras, Chennai, India
Priyadarsini, R.S., College Of Engg., Trivandrum, India
RAJ KUMAR, E., M.R.R.Institute Of Tech & Science, India
Rajak, A.R.A., Secab Inst. Of Tech., India
Rekha, R., College Of Engineering, Munnar, India
Sanjay Kumar, S.M., Sri Krishna Institute Of Tech, India
Satish Babu, D., ADE, DRDO, India
Senthur Pandi.R, Thiagarajar College Engg, India
Shanker, R., IIT Delhi, India
Tamilarasan, Pallavan College Of Engg. , India
Thomas, A., College Of Engineering, Munnar, India
Vasantha Nathan, Mepco Schlenk Engg.,College, India
Venkat Raman, S., Accelrys KK, Bangalore, India
Vijay Kumar, G., PVP Siddhartha Instt Of Technology, India
Yeole, M.P., ISRO, Trivandurum, India

Row1: Srivatsan, Srikrishna, Jayavel, Gopalakrishnan, Sivakumar (sitting)
Row2: Arun Srinivasa, Binu Mukherjee, Roy Mahapatra, Sivaramakrishnan, Hui Hui Dai, Nancy Johnson, A.Menzel, M.Mahendran, Prakash Thamburaja, Srikanth Vedantam, Ebenezer, Jayabal, Rajasudha, Kandavel
Row3 : Ashwini Agrawal, Arockiarajan, delegate,delegate, Chandrasekaran, Ashish, John Ulicny, delegate
Row4: delegate, A Sen, delegate, Prabhakar, Paramanand Singh, delegate, Santhanam, Amol, Vidyashankar Buravalla, delegates

VOLUNTEERS, PARTICIPANTS and CONTRIBUTORS – SDMoMS '08

INAUGURAL LECTURE

Modeling of Entropy Producing Processes

K. R. Rajagopal

Department of Mechanical Engineering, Texas A&M University, College Station, TX-77843

Abstract: In all processes concerning real materials, entropy is produced. While undergoing such processes, the natural configuration of a body (the configuration that the body would take on the removal of all external stimuli) changes. It is also possible that the material symmetry of the body in these various natural configurations could be different. Usually, one requires that all processes meet the second law of thermodynamics namely that the rate of entropy production is non-negative. In this study we make a more stringent requirement, namely that the rate of entropy production is maximized. A framework is developed wherein knowing how the material stores energy, produces entropy, conducts heat, absorbs or emits radiation, etc., allows one to determine the constitutive equation for the stress and other relevant quantities.

INTRODUCTION

In all processes, a real body produces entropy, though one does find idealized bodies such as elastic materials that are incapable of producing entropy to be useful in approximately modeling the response of some real bodies within a special class of processes. The response of bodies to external stimuli is characterized by the many ways in which bodies store energy, how they release this energy that is stored, ways in which they produce entropy, how they conduct heat, how they emit and absorb radiation, the structures for their latent heat and latent energy (the difference in the internal energy associated with the different phases of the body), how much of the working that is supplied is converted into heat, and in general other pertinent information concerning the response of bodies. For instance, a particular body might be able to store the energy that is supplied to the body in such a manner that all of it can be recovered in a purely mechanical process (such bodies are usually called elastic). Crystalline bodies with dislocations are capable of storing energy due to the rearrangement of the dislocation structure that cannot be recovered in a purely mechanical process, but the energy being recovered in a thermodynamic process (annealing). The external stimuli could be thermal, chemical, electrical and magnetic fields. Here, for the sake of simplicity, attention will be restricted to purely mechanical and thermal stimuli.

Given external stimuli, it is possible that different natural configurations might be achieved dependent on how the external stimuli are removed. The natural configuration achieved might be different based on whether the external loading is removed instantaneously or very slowly. In an elastic body the manner in which the external load is removed is irrelevant and one attains the same stress free

CP1029, *Smart Devices: Modeling of Material Systems, An International Workshop*
edited by S. M. Sivakumar, V. Buravalla, and A. R. Srinivasa

configuration. It is important to recognize that a traction free configuration does not necessarily lead to a stress free configuration and thus all equilibrium solutions for elastic bodies cannot be obtained by requiring that the stored energy be a global minimum.

While in an elastic body the natural configuration does not change, a real body that has undergone classical "plastic deformation" has an infinity of natural configuration associated with its deformation while a body that is undergoing solid to solid phase transition has a finite number of natural configurations (see Rajagopal (19995) and Rajagopal and Srinivasa (2004a), (2004b) for a detailed discussion of the role of natural configurations in thermomechanics). The question then arises as to the criterion which one could use for determining how the natural configuration changes. It turns out that natural configurations change whenever entropy production takes place. Eckart (1948) seems to have been the first to recognize the important role that natural configurations play in specifying the response of materials.

Eckart's work implies that classical plasticity ought to be viewed as infinity of response functions from infinity of evolving natural configurations. While Eckart's work made a significant advance concerning the notion of natural configurations, he did not recognize the role of the changing material symmetry associated with these natural configurations and other related issues (see Rajagopal (1995) for a discussion of the same).

We shall find that if one requires that the thermodynamic processes proceed according to the rate of entropy production being maximized one can determine the manner in which the natural configurations evolve.

Let the current configuration of the body is denoted by κ_t, and further suppose that on the removal of the external stimuli the body attains the configuration $\kappa_{p(t)}$ (we shall suppress the t in the suffix and denote the configuration by κ_p for the sake of convenience) the preferred natural configuration of the body amongst the several natural configurations that are available to the body. As mentioned earlier, different natural configurations can be attained based on the class of allowable thermodynamic processes.

We have to decide on the set of properties that define the state of each material point that belongs to the body. For a certain class of bodies, based on the process class that they are going to be subject to, it might be appropriate to think of the deformation gradient, temperature, stress, velocity gradient, the various temporal and spatial derivatives of the above quantities as well as several other quantities to define the state of a particle. An important point to bear in mind is that the set of natural configurations that a body can attain will also be a part of the specification of the state variables. Constitutive relations are in their most general form relationships between the various state variables. It may not be possible to express a variable such as the stress as a function of variables such as the strain and temperature. Once the state space associated with the body can be defined we can discuss the processes that take the particle from one state to another. These processes cannot be arbitrary; they have to be such that the basic balance equations and the second law of thermodynamics are met.

4

At times a more stringent requirement introduced by Onsager (see Onsager 1931, Prigogine 1967, Glansdorff and Prigogine 1971) is required, namely that the rate of entropy production be minimized. Onsager's requirement, which is often referred to as Onsager's Principle does not have the same universality as the second law and is expected to hold for only special materials in special processes. Recently, Rajagopal and Srinivasa in a series of papers require that the rate of entropy production be maximal[1] to choose a subset of constitutive relations from those in which the rate of entropy production is non-negative. While Ziegler (1963) had earlier appealed to such a requirement, he did not use it to obtain constitutive relations in the manner of Rajagopal and Srinivasa (see Rajagopal and Srinivasa (2004a) for a discussion of the differences between their approach and that of Ziegler (1963)). At this juncture it is important to clear up a possible misunderstanding that might arise as Onsager requires the rate of entropy production to be minimized while Rajagopal and Srinivasa require that the rate of entropy production be maximized. There is no contradiction between these two requirements as they concern different issues. The appropriate form of the rate of entropy production is picked from amongst a class of rates of entropy productions that are non-negative by requiring that the rate of entropy production be maximal, and once it is so chosen it will attain a minimum with respect to time during the process. In fact, Rajagopal and Srinivasa (2005) also provide a way for generalizing Onsager's "principle" to non-linear phenomenological law.

KINEMATICS AND BASIC EQUATIONS

Let κ be a reference placer that maps the abstract body onto its configuration $\kappa(B)$ in a three dimensional Euclidean space. Let $\kappa_t(B)$ denote the configuration of the body B, at time t. By the motion χ of the body, we mean a one to one mapping at each instant of time t, that associates a particle $\mathbf{X}_\kappa \in \kappa(B)$ with a particle $\mathbf{x} \in \kappa(B)$, i.e.,

$$\mathbf{x} = \chi_\kappa(\mathbf{X}_\kappa, t). \tag{1}$$

A property φ can be defined through

$$\varphi = \varphi_\kappa(\mathbf{X}_\kappa, t) = \varphi_{\kappa_t}(\mathbf{x}, t) \tag{2}$$

We shall use the following notation to represent the derivatives based on referential and current configurations :

$$\nabla \varphi = \frac{\partial \varphi_\kappa}{\partial \mathbf{X}_\kappa}, \quad \text{grad}\,\varphi = \frac{\partial \varphi_\kappa}{\partial \mathbf{x}}, \quad \frac{d\varphi}{dt} = \frac{\partial \varphi_\kappa}{\partial t}, \quad \frac{\partial \varphi}{\partial t} = \frac{\partial \varphi_{\kappa_t}}{\partial t}. \tag{3}$$

The gradient of the motion (usually called the deformation gradient) is defined through

---

[1] A few words concerning the rationale for the maximization are warranted. In an isolated system, the entropy of the system tends to a maximum and the system attains equilibrium. The quickest way for the system to attain the maximal value of entropy is to undergo processes that produce the maximal rate of entropy. Though this might seem a reasonable expectation for closed (isolated) systems we shall also require this of open systems.

$$F_\kappa = \frac{\partial \chi_\kappa}{\partial \mathbf{X}_\kappa}.$$ (4)

The velocity $\mathbf{v}$ and the velocity gradient $\mathbf{L}$ are defined through

$$\mathbf{v} = \frac{\partial \chi_\kappa}{\partial t},$$ (5)

$$\mathbf{L} = \text{grad } \mathbf{v}.$$ (6)

We define the Cauchy-Green Stretch tensors $\mathbf{B}_\kappa$ and $\mathbf{C}_\kappa$ through

$$\mathbf{B}_\kappa = F_\kappa F_\kappa^T, \mathbf{C}_\kappa = F_\kappa^T F_\kappa,$$ (7)

and the Green-St.Venant strain $\mathbf{E}_\kappa$ and Almansi-Hamel strain $\mathbf{e}_\kappa$ through

$$\mathbf{E}_\kappa = \frac{1}{2}(\mathbf{C}_\kappa - 1), \mathbf{e}_\kappa = \frac{1}{2}(1 - \mathbf{B}_\kappa^{-1}).$$ (8)

It immediately follows that

$$\mathbf{L} = \dot{F}_\kappa F_\kappa^{-1}.$$ (9)

The symmetric part of the velocity gradient $\mathbf{D}$ is defined through

$$\mathbf{D} = \frac{1}{2}(\mathbf{L} + \mathbf{L}^T).$$ (10)

Any process undergone by the body has to meet the balance of mass, linear and angular momentum, and energy which are given by

$$\frac{\partial \rho}{\partial t} + div(\rho v) = 0,$$ (11)

$$\rho\{\frac{\partial v}{\partial t} + [grad v]v\} = div\mathbf{T} + \rho\mathbf{b},$$ (12)

$$\mathbf{T}^T = \mathbf{T},$$ (13)

$$\rho\frac{d\varepsilon}{dt} = \mathbf{T}.\mathbf{L} + div\mathbf{q} + \rho r,$$ (14)

where ρ denotes the density, $\mathbf{v}$ the velocity, $\mathbf{T}$ the Cauchy stress, $\mathbf{b}$ the specific body force, ε the specific internal energy, $\mathbf{q}$ the heat flux and r the radiant heating.

Finally, we record the second law of thermodynamics which the body has to meet in every process:

$$\frac{\partial \eta}{\partial t} + div\left(\frac{\mathbf{q}}{\theta}\right) = \frac{\rho r}{\theta} + \rho\xi, \quad \xi \geq 0,$$ (15)

where η is the entropy, θ the temperature, $\mathbf{q}$ the heat flux, and ξ is the rate of entropy production. The second law expressed as above is different from the usual expression where the last term in the above equation is omitted and the equality is replaced by an inequality. The above approach was used by Green and Nagdhi (1977)

6

and more recently by Rajagopal and Srinivasa (see the review articles by Rajagopal and Srinivasa (2004a), (2004b)) to study the thermodynamic response of bodies.

On combining the balance of energy with the above equation we obtain

$$\mathbf{T} \mathbf{L} - \rho \dot{\varepsilon} + \rho \theta \dot{\eta} - \frac{\mathbf{q} \cdot \frac{\partial \theta}{\partial x}}{\theta} = \rho \theta \xi = \zeta \geq 0, \qquad (16)$$

where we refer to ζ as the rate of dissipation (usually the rate of dissipation refers to the product of the density, temperature and the rate of entropy production associated with working being converted to heat, i.e., energy in its thermal form, but here we shall use it to mean the product of the density, temperature and the rate of entropy production associated with all forms of entropy production).

On introducing the specific Helmholtz potential

$$\psi = \varepsilon - \theta \eta, \qquad (17)$$

we can rewrite the above equation as

$$\mathbf{T} \mathbf{L} - \rho \dot{\psi} - \rho \eta \dot{\theta} - \frac{\mathbf{q} \cdot \frac{\partial \theta}{\partial x}}{\theta} = \rho \xi. \qquad (18)$$

If one further assumes that the rate of entropy production ξ can be expressed additively (it is not necessary to make this assumption and one can deal with more complicated forms of entropy production) as

$$\xi = \xi_c + \xi_d, \qquad (19)$$

where ξ_c is the rate of entropy production due to conduction and ξ_d the rate of entropy production due for the various other entropy producing processes, and if we further assume

$$\xi_c := \frac{\mathbf{q} \cdot \frac{\partial \theta}{\partial \mathbf{x}}}{\theta} \geq 0, \qquad (20)$$

then we are left with

$$\mathbf{T} \mathbf{L} - \rho \dot{\psi} - \rho \eta \dot{\theta} = \zeta_d \geq 0, \quad \zeta_d := \rho \theta \xi_d. \qquad (21)$$

Depending on the problem under consideration we will have to assume appropriate forms for the specific Helmholtz potential ψ and the rate of entropy production ξ_d, and the entropy η. We are making assumptions concerning two scalar functions instead of six scalar functions for the components of the stress as is usually done.

Let us consider, for the sake of illustration isothermal[2] processes. In this case, equation (18) will further simplify to

$$\mathbf{T} \mathbf{L} - \rho \dot{\psi} = \zeta_d \geq 0. \qquad (22)$$

---

[2] A fully thermodynamic theory that allows for phase transformations, conduction, radiation, etc., can be found in the paper by Rao and Rajagopal (2002), and Kannan et al (2002) where they consider the crystallization of polymeric melts and Rajagopal and Srinivasa (2004) where they consider solid to solid phase transformations.

We need to maximize the rate of entropy production subject to the second law expressed in the form (22) (or in the more general case (16)) enforced as a constraint. If any other constraints such as those of incompressibility are to be enforced, then one has to incorporate them into the function which is being maximized by introducing the appropriate Lagrange multipliers. It is important to bear in mind that the above requirement of the maximal rate of entropy production is not a "principle" of thermodynamics, only a plausible and reasonable assumption. Ziegler (1963) appealed to such an assumption, but from a different perspective (see Rajagopal and Srinivasa (2004a) for a detailed discussion of the distinction between Ziegler's approach and that advocated here).

In the case of isothermal processes, we maximize the rate of entropy production ξ subject to (22) as a constraint, i.e., we maximize

$$\Phi = \xi + \lambda_1 \left(\mathbf{T} \Box \mathbf{L} - \rho \dot{\psi} \right). \tag{23}$$

If in addition the body under consideration is incompressible, then we have to also take into consideration that

$$tr\mathbf{D} = div\mathbf{v} = 0, \tag{24}$$

and maximize

$$\Phi = \xi + \lambda_1 \left(\mathbf{T} \Box \mathbf{L} - \rho \dot{\psi} \right) + \lambda_2 tr\mathbf{D}. \tag{25}$$

EXAMPLE: VISCOELASTIC FLUIDS

In order to illustrate the efficacy of the framework we consider its applicability to viscoelastic fluids. Suppose the fluid stores energy like a neo-Hookean spring and the rate of entropy production is that of a viscous fluid that takes into account the change in the natural configuration. Then

$$\psi = \frac{\mu}{2} \left(I_{\mathbf{B}_{k_p}} - 3 \right), \tag{26}$$

and

$$\xi = \eta \mathbf{D} \Box \mathbf{B}_{k_p} \mathbf{D}, \tag{27}$$

where μ and η are constants.

A straightforward application of the procedure leads to (see Rajagopal and Srinivasa (2000) for details) the following representation for the Cauchy stress and the evolution of the natural configuration for the fluid:

$$\mathbf{T} = -p\mathbf{1} + \mu \mathbf{B}_{k_p}, \tag{28}$$

$$-\frac{1}{2} \overset{\triangledown}{\mathbf{B}}_{k_p} = \frac{\mu}{\eta} \left[\mathbf{B}_{k_p} - \lambda\mathbf{1} \right], \tag{29}$$

where

$$\lambda = \frac{3}{\left(tr\mathbf{B}_{k_p}^{-1} \right)}. \tag{30}$$

The above model is a generalization of the non-linear Maxwell model. If one assumes that the elastic response is that of a linearized elastic solid, then one obtains the three dimensional generalization of the one dimensional model developed by Maxwell (1866).

The above thermodynamic framework is able to describe a plethora of disparate phenomena exhibited by materials: viscoelasticity, traditional inelastic response, twinning, solid to solid transitions in shape memory alloys (see Rajagopal and Srinivasa (2004a), (2004b)), crystallization of polymers (Rao and Rajagopal (2002)), and shape memory polymers (Barot et al. (2008)). The procedure seems to have a lot of promise, but it is important to recognize that the procedure will have its limitations, it cannot be expected to describe all types of responses exhibited by materials.

ACKNOWLEDGEMENT

Rajagopal thanks the National Science Foundation for the support of this work.

REFERENCES

1. G. Barot, I. J. Rao and K. R. Rajagopal, A thermodynamic framework for the modeling of crystallizable shape memory polymers, International Journal of Engineering Science, In Press.
2. R. Clausius, On the motive power of heat and the laws which can be deduced from it for the thoery of heat. In The second law of thermodynamics (ed. W. F. Magie). New York: American Book Company (1899).
3. de Groot, S. R. & Mazur, P., Non-equilibrium thermodynamics. Interscience (1962).
4. C. Eckart, The thermodynamics of irreversible processes IV, the theory of elasticity and anelasticity. Physical Review 73, 373-382 (1948).
5. Glansdorff, P. & Prigogine, I. Thermodynamic theory of structure, stability and fluctuations. Wiley (1971).
6. E. Green and P. M. Naghdi, On thermodynamics and the nature of the second law, Proc. Roy. Soc. Lond. A 357 (1977), 253–270.
7. K. Kannan, I. J. Rao and K. R. Rajagopal, A thermomechanical framework for the glass transition phenomenon in certain polymers and its application to fiber spinning, J. Rheology, 46 (2002), 977–999.
8. J. C. Maxwell, On the dynamical theory of gases, Phil. Trans. Roy. Soc. London A157 26-78 (1866).
9. Onsager, L. Reciprocal relations in irreversible thermodynamics. I. Phys. Rev. 37, 405–426 (1931).
10. Prigogine, I. Introduction to thermodynamics of irreversible processes, 3rd edn. Interscience (1967).
11. M. Planck, Second law of thermodynamics. In Introduction to theoretical physics. 4. Theory of heat (translated by H. L. Brose). London: M. Planck, Second law of thermodynamics. In Introduction to theoretical physics. 4. Theory of heat (translated Macmillan (1932).
12. J. Rao and K. R. Rajagopal, A thermodynamic Framework for the study of Crystallization in Polymers, Z. angew. Math. Phys. 53 (2002) 365-406.
13. R. Rajagopal, Multiple configurations in continuum mechanics. Reports of the Institute for Computational and Applied Mechanics, University of Pittsburgh (6), (1995).
14. R. Rajagopal and A. R. Srinivasa, On the thermomechanics of materials that have multiple natural configurations, Part I: Viscoelasticity and Classical Plasticity, Z. angew. Math. Phys. 55 , 861–893 (2004a).

15. R. Rajagopal and A. R. Srinivasa, On the thermomechanics of materials that have multiple natural configurations, Part II: Twinning and Solid to Solid Phase Transitions, Z. angew. Math. Phys. 55, 1074–1093 (2004b).

16. Thomson, On a universal tendency in nature to the dissipation of mechanical energy. Phil. Mag. 4, 304–306 (1852).

17. H. Ziegler, Some extremum principles in irreversible thermodynamics. In Progress in solid mechanics (ed. I. N. Sneddon & R. Hill), vol. 4. New York: North Holland (1963).

SESSION 1

SMART DEVICES—AN OVERALL PERSPECTIVE

Challenges In Modeling Of Smart Materials And Devices

Prakash D Mangalgiri

Lab Group Manager (Smart Systems Modeling), India Science Lab, General Motors R&D, Bangalore 560066

Abstract. Over the last few decades the interest in Smart Materials – materials which respond to some external stimuli with manifestation of a change in mechanical parameters – has grown tremendously and various engineering applications have been demonstrated using the "smart" function. With the traditional base of the sensor technologies in vogue, the sensing function in a smart system has received considerable attention and is relatively well developed. However, the "actuation" function has posed major challenges. Materials such as piezos, Shape-memory alloys, MR and ER fluids, magnetostrictive materials and some electro-active polymers have all been in focus of the engineering community for building Smart Actuators. Several prototypes have been built to demonstrate the actuator applications. However, converting these ideas into products has remained a major challenge. Issues of reliability of the functional performance, degradation with time, response to off-design conditions, hysterisis and nonlinear responses, scaling from small-scale and accelerated experiments and the cost of testing several parameters are some of the concerns that have surfaced in translating prototypes to products. With the advances in computational capabilities and materials science, building math-models for the behavior of these materials and the actuators built out of them is being seen as a fruitful direction in achieving predictive capability so much needed to help sort out many of the issues mentioned above.

General Motors R&D has taken a significant step in this direction with a consorted effort to create math models of smart materials and actuators. Several applications in the automobile sector are on the anvil. Initial focus of the work is on materials like SMAs, MR Fluids and electro-active polymers. This paper discusses some of the issues in building smart actuators and the work being pursued in creating math-models to help resolve them. The challenges posed by the complex behavior of such materials in creating the models are highlighted and illustrated with a case study of SMA based devices.

Keywords: Smart materials, modeling of smart materials, shape memory alloys, smart actuators, smart devices, modeling of smart devices.

INTRODUCTION

In the last two decades or so, the "Smart" technology has grown steadily and has attained a fair amount of maturity in several aspects. The growth in tools (software) and technology (hardware) in electronics and controls is enabling engineers to build active systems to improve performance, safety and reliability. Various "smart" materials – materials with unconventional properties - such as, Piezos, Electro-active polymers and Shape Memory materials are being investigated and many new smart materials are being developed to have exotic properties. Among the three key

CP1029, *Smart Devices: Modeling of Material Systems, An International Workshop*
edited by S. M. Sivakumar, V. Buravalla, and A. R. Srinivasa
©2008 American Institute of Physics 978-0-7354-0553-0/08/$23.00

functional areas in a smart system (see Fig 1), namely, sensing, data analysis to control and actuation, the sensing function and the control function appear to have gained significant maturity benefiting from the backbone of traditional sensors and controls

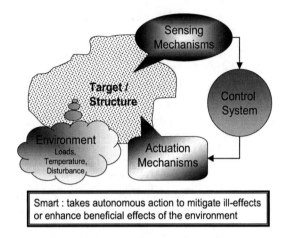

FIGURE 1. Schematic of a smart system

technologies. On the other hand, the actuation function – building actuators using smart materials to create useful forces, displacements or other physical parameters – has posed several challenges and remains a key development area for application of smart technology in many areas of engineering. While several laboratory demonstration devices and prototypes have been built, only a few have been translated into useful commercial products. This paper is focused on describing and discussing some of the challenges in developing smart materials based actuating devices as products.

Recognizing the importance and the potential benefits in the use of smart materials for building various actuation devices, GM-R&D identified this as one of the key focus areas for automotive research in this decade. In order to emphasize the role of the materials in the traditional mechatronics context, GM-R&D coined a new term – mechaMAtronics – motivating the researchers to pay increasing attention to the study of smart materials for use in building smart devices. As a result of these efforts, the newly established India Science Lab of GM-R&D is now actively engaged in the study of smart materials and modeling their behavior.

In this paper, after an initial overview of smart materials, we describe various issues in meeting the challenge and then illustrate them by taking up an example case study of one important class of smart materials, namely, the Shape Memory Alloys (SMAs).

SMART MATERIALS, THE GROWING INTEREST AND ISSUES

There are varieties of materials which have one or more special functional properties which can be used for creating actuation parameters (forces, displacements, shapes, etc). Some of these which have attracted a lot of attention of researchers are highlighted in the list given below.

TABLE 1. Smart materials of interest. The materials which are currently under investigations for actuators are underlined.

Class	Material	Actuation through	Trigger
• Metallic	- Shape Memory Alloys (e.g., Ni-Ti)	Shape or Stiffness change	Heat or Magnetic field
• Ceramics	- Piezoelectric (e.g. PZT) - Ferroelectric and anti-ferroelectric - Electrostrictive (e.g. PMN)	Mech strains	Electric potential
	- Magnetostrictive (e.g. Terfenol)	Mech strains	Magnetic field
• Polymeric	- Shape Memory Polymers	Shape change	
	- Polymeric Piezo Films of PVDF	Mech strains	Electric potential
	- Polymer Gels (Thermoresponsive, pH-sensitive, Magnetic) - Electro-active Polymers - ER and MR Fluids		

The actuators based on such smart materials are expected to perform with higher reliability and efficiency, achieve novel functionalities, provide flexibility and modularity through possible articulated shapes, reduce weight and cost and yet be

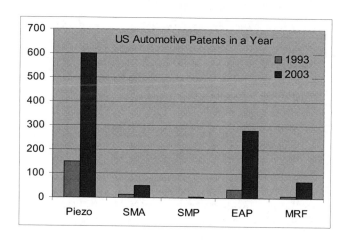

FIGURE 2. Number of US patents in smart materials

simple enough in terms of their construction. As such the interest in smart materials is growing significantly, as is evident from the rise in number of patents during the decade 1993 – 2003 as seen in the chart in Fig 2, [1].

Automotive companies like General Motors are planning to integrate smart materials into vehicles in near future [2]. Actuators and sensors made from smart materials could improve vehicle performance and fuel economy, and enable new comfort and convenience features. Some of the applications under consideration include active vehicle surfaces, such as air dams that adjust to govern airflow; an adaptive interior grab handles and latches that automatically present themselves from a folded position to make access easier [2].

Several schemes have been devised and demonstrated at lab-level for getting requisite functionality for many of the applications of the smart materials. However, in order to achieve the advantages which are claimed above, several issues need to be sorted out before they can be employed in a practical environment. These can briefly be stated as below.

- Important parameters for an actuator are the block-force and/or the displacement that can be obtained and the response time in which such performance can be obtained. Can we get these parameters in sufficient magnitude for practical structures?
- Presently there is an overwhelming reliance on testing to validate the design and estimate the performance. Every stage of design and design change need to be validated through elaborate tests. How do we reduce testing?
  - Can we do scaling from one set of requirements to the other? Can we carry over experience from testing of one configuration/concept to other configuration? And How?
  - Can we estimate effect of various changes from time to time
- How do these materials/ devices perform over long time under practical service conditions?
  - Certification, warranty, reliability, life

Due to the issues mentioned above, many applications which have been demonstrated in the laboratories have not come to fruition in the practical world. Even where successful prototypes have been build, translation to successful products has been comparatively rare. MR fluid based devices may be taken here as a case in point. Many devices such as clutches, brakes, dampers, etc have been successfully built in laboratories, but the successful applications of an MR fluid based device in production are only a few such as MR based active suspension used for ride control in the suspension systems of automobiles [3]. This has been in use for about decade, but the smart materials community has not been able to translate the acquired knowledge and experience into another application. Issues such as segregation of particles and thermal degradation due to heat generation still cause concern to the designers and engineers and modeling of MR fluid flows under the flow field and the magnetic field provide challenging problems for the researchers.

For several of the issues mentioned above, modeling can provide an effective means of resolution. Certainly, math-modeling of the complex behavior of the smart materials and their devices can go a long way in achieving a predictive capability giving some credible means to address these issues. We shall now illustrate this through a case study of Shape Memory Alloys for which our group working at the India Science Lab has been engaged in creating useful math-models.

CASE : SHAPE MEMORY ALLOY DEVICES

Issues

Development and design of smart devices for automotive applications using Shape Memory Alloys (SMA) is gaining increasing attention because of the high block-force and displacements that they can provide for mechanical applications. The most popular SMAs are Ni-Ti or Ni-Ti-Cu alloys and unlike conventional materials their underlying mechanical behavior is significantly different and complex involving solid state phase transformations under stress and/or temperature, leading to two main characteristics – the shape memory effect and the superelastic effect. These effects are exploited to build actuators (mostly by using SMA wires) which can be triggered by temperature and stress. Present technology of these materials is considered to be mature enough to have sufficiently high transition temperatures for practical applications (so as to avoid accidental triggering by ambient conditions) and achieve consistent properties at least in the wire form. The material behavior as an actuator is dominated by the solid-state phase transformations leading to nonlinear path-dependent load-displacement relationship accompanied by hysteresis over cyclic loading. While several actuation schemes can be constructed and indeed have been demonstrated in laboratory experiments, the lack of predictive capability (due to the complex material behavior) about reliable operation over a period of time has hindered their translation into practical products or applications. Following questions arise when trying to utilize the thermo-mechanical response of an SMA for an actuation device.

- How do we predict the state of the material at a given point in an arbitrary loading path? Note that the behavior is path dependent.
- How do we relate the state of the material to the mechanical response – force vs displacement or stress vs strain?
- Will the functionality (thermo-mechanical actuation) be affected by stray loads or temperature perturbations, from vibrations, jerks, ambient fluctuations?
- If the above factors affect the behavior, then how does one ensure reliability?
- How does the material degradation - mechanical fatigue as well as functional degradation such as loss in recovery force or repeatability - take place over a period of time and over a number of operational cycles? How does one ensure durability?

The issue is proposed to be addressed by building reliable math-models for the SMA behavior (validated through lab level experimental data) which can then be used with numerical techniques such as FEM to predict the actuator behavior under various conditions.

The SMA Behavior

The typical SMA behavior is dominated by the behavioral changes brought about by the phase transformation; see Fig 3a as an illustration and Fig 3b for phase diagram in the stress-temperature field. This behavior is well documented, see for example, Refs [4,5], and is schematically illustrated in Fig 3 for pictorial recap.

Two important aspects of modeling this behavior are:

- The Evolution Kinetics, i.e., how does the martensitic phase fraction (ξ) evolve, what is the criteria for evolution, how do the stress induced (ξ_S) and

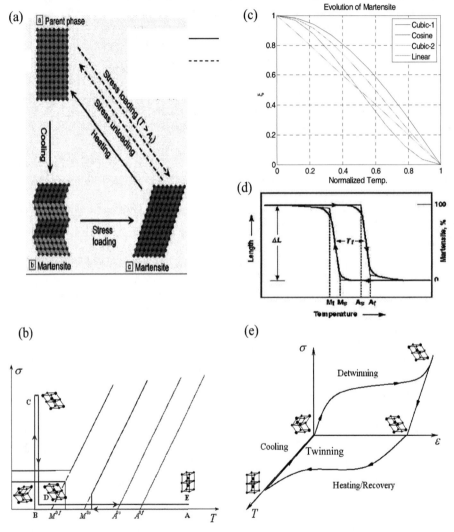

FIGURE 3. SMA behavior. (a) Solid State Phase transformations under stress and temperature (b) Phase diagram in the Stress-temperature plane; a typical thermo-mechanical cycle for actuation is shown (c) Evolution of martensitic phase fraction with temperature, typically proposed math-models (d) Hysteresis under cyclic thermal loading (e) A typical thermo-mechanical cycle of actuation using Shape Memory effect shown in stress-strain-temperature regime.

temperature induced (ξ_T) martensitic variants evolve. Relate ξ, ξ_S, ξ_T with the loading parameters such as stress and temperature.

- The Constitutive Relationship, i.e., the relations between the stress and strain at any given state of the material described by ξ, ξ_S, ξ_T and the loading parameters such as stress and temperature.

Two basic approaches have been adopted by the researchers in order to account for the complexity of the SMA behavior.

(i)   Thermodynamics based models use some form of free energy (Gibbs, Helmholtz, Landau, etc.,) composed of two parts, viz., temperature dependent chemical part dealing with the entropies of volume fraction of the individual phases and the mechanical part dealing with the stress/strain field due to external loading and the interaction between the various phases. Using this approach evolution of phase boundaries can be determined. Some of the models evolved using such approaches are given in Refs [6-15].

(ii)   Phenomenological models separate out the two aspects of modeling mentioned above. The phase evolution and transformation conditions are incorporated using empirically determined phase diagram. Subsequently, a constitutive relationship that uses the phase fraction derived out of the explicit evolution kinetic is used to describe the thermo-mechanical behavior. This leads to simplified models facilitating their use as design tools. Several investigators have proposed such models and these have evolved significantly over the years, see, for example, Refs [16-22].

Motivated by the need to produce design aids accounting for the behavior of SMAs, we (at ISL) have focused on phenomenological models, taking the cue from widely used Brinson model [18]. By incorporating a few refinements, see Buravalla and Khandelwal [23] and Khandelwal and Buravalla [24], we have been able to arrive at models which can be effectively used to predict the state of the material undergoing an arbitrary loading path in the stress-temperature field, can account for the path dependency, can cater to partial and full loops in the stress-temperature (or strain-temperature, or stress-strain) regime thus accounting for hysteresis and thus predict the consequent load-displacement behavior. Some of the typical results are shown in Figs. 4-5. Noting that for materials like SMA, the difference in approaches for stress-temperature and strain-temperature frameworks would be non-trivial, these models have been refined to work effectively in both the regimes. Further, the current work has focused on evolving a Distance Based Approach – using a criterion based on the distance of a point on the load path from the transformation finish boundary - to facilitate a robust simulation of several perturbations in the load.

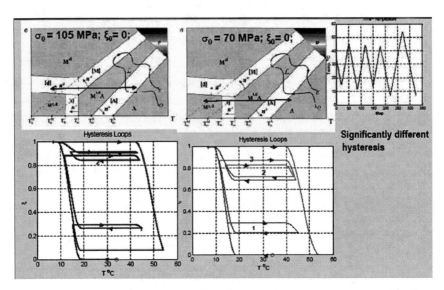

FIGURE 4. The phase diagram in Stress-temperature plane. Arbitrary path in this regime can be taken. Typical solutions for temperature cycling at two different stress levels are shown. Significant differences in hysteresis can be seen.

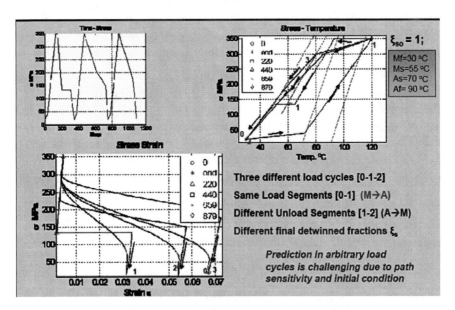

FIGURE 5. Arbitrary load cycles and the corresponding path in the phase diagram. Note that the loading path in each cycle is same, but the unloading paths are slightly different. This makes significant difference in the state of the material reached.

The Device Behavior

An SMA based actuating device can be schematically represented as in Fig 6a and the corresponding components of the model are shown in Fig 6b. In order to trigger

(a)

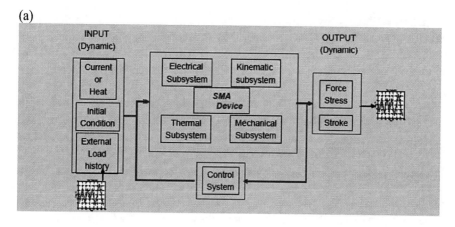

(b)

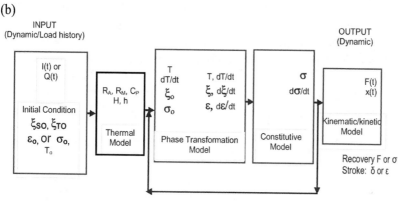

FIGURE 6. Device level modeling. (a) A typical device, schematic (b) Modeling modules for the device

the temperature-based changes in SMA, the SMA element (say, a wire) needs to be heated and cooled. While heating is generally affected by passing an electric current or by direct heating by a heat source, cooling is generally uncontrolled. Also, at the other end, a mechanical system may be provided to manipulate the forces and displacements generated by the SMA element into the required useful form. Thus, in order to arrive at a device level model, we must incorporate such a thermal/electrical system as well as a mechanical (kinematic) system and integrate with the SMA element behavior. The model then should incorporate

(i)   Shape memory and superelastic effects of SMA under arbitrary stress-temperature or strain-temperature loading, complete and partial load cycles and inner hysteresis loops

(ii)   Spatial variation in SMA (as a result of differential heating or cooling or stress)

(iii)   Temperature as a function of time – heating and cooling cycles

(iv)   Mechanical system with controls and feedback.

The last one – the mechanical system – is similar to a conventional mechanical systems and should not pose any specific problems related to SMA. The thermal system model when developed for a given state of the material again follows the

21

conventional methods. However, as the SMA material state changes, the thermal and electrical properties of SMA may change and accounting for this forms an essential challenge in developing a thermal model. As a simplified engineering approach, it is usual to create separate models for the thermal and the mechanical behavior assuming they are uncoupled and then use an iterative procedure to arrive at the solution to the combined thermo-mechanical behavior. While such an approach neglects the coupling of the thermal and mechanical effects, it can provide useful results in terms of the design parameters and provide reasonable estimates of the performance. Such an approach has been used, for example in Refs [25-26] using numerical techniques coupling SMA behavior with FE analysis of a host structure through an iterative procedure. Spatial variation in SMA (as a result of differential heating or cooling or stress) behavior can pose problems and one way to address them is by using suitable forms FEM for SMA and then coupling it with the analysis of the host structure.

As an example, we consider a simple case of a cantilever beam actuator, see Fig 7. Typically, an SMA wire may be connected at an offset distance (*a*) provided by rigid connectors at a number of points to the cantilever beam which is expected to actuate under some load. A key issue, then, is to provide an FE framework for the combined problem of the actuated element (beam) and the actuating element (the SMA)

As a first-cut analysis, suitable simplifying assumptions can be made - that the host

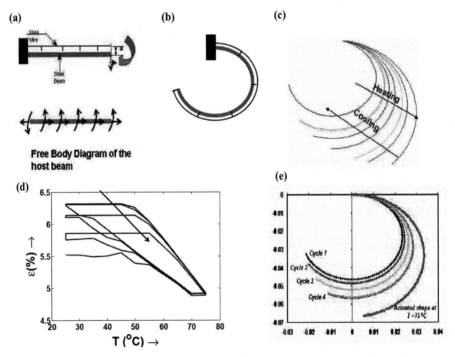

FIGURE 7. An SMA actuated Beam Actuator. (a) The Cantilever beam with an SMA wire at an offset "a". Also, the free body diagram.(b) The deformed shape (c) Various shapes that can be achieved, as heating and cooling progresses (d) The cyclic loading path in the strain-temperature plane as the beam is actuated cyclically (e) the non-repeatable shapes as the various cycles of actuation take place. Ref [27].

structure (the beam) is thermally isolated from SMA and the structural response of SMA and the host are coupled using only kinematic constraints at anchor points. The spatial variations of stress-temperature-strain in SMA wire can be ignored so that a Lumped Parameter Model (LPM) is used for SMA. The host structure can then be analyzed through FEM and an iterative technique then can be developed to couple the SMA forces and displacements with those of the host. Figure 7 depicts some of the results that were obtained using such an approach. Further refinements can then be incorporated using FE meshing for the SMA wire to take into account the spatial variations in SMA and developing techniques to couple it with FE analysis of the host.

Future directions in SMA

On the material level modeling, the current efforts based on phenomenological models need to be strengthened with incorporating degradation models to address durability issues with more confidence. Further, the dependence on extensive material data needs to be reduced through integration with thermodynamics based models. For the translation of these into the FE-based analysis at device level, these material models need to be integrated with standard General Purpose FE software so that a designer can address the problem in an integrated way. Coupled problems of thermal and kinematic behavior need to be solved to check out the criticality of the coupling. Spatial variations in SMA will be addressed through a more general application of FE. At the device level and further at the system level (system in which an SMA device used as an active component), issues of control need to be addressed. With these tools in hand it is expected that the predictive capability achieved will sort out several issues mentioned earlier and will go a long way in creating in a credible translation of the technology into practical products.

CONCLUDING REMARKS

Use of Smart Materials such as Shape Memory Alloys, MR Fluids, Piezos, Electroactive Polymers, etc is being explored vigorously for building smart actuating devices which can produce useful mechanical forces and /or displacements which can be used for several applications in various sectors. In automotive sector alone, the potential for such application is considered to be significant and several prototypes or lab-level models have been built. However, issues of reliability and durability need to be addressed before we can see these translated into commercial products. Creating math-models for the behavior of these materials and the actuating devices based on them would go a long way in achieving certain predictive capability, reduce reliance on testing, produce design guidelines, provide scaling tools to translate experience on one device into another and thus create credibility about these new form of devices. These issues and the modeling route have been discussed with the help of a case study on Shape Memory Alloys and devices. It is expected that the developments in such areas would lead to several applications to be put in practice in the automotive sector.

ACKNOWLEDGEMENTS

The paper is based on the current work being carried out at the India Science Lab of GM-R&D, Bangalore by the Smart Systems Modeling group. The author is thankful to many members of the group especially, Vidyashankar Buravalla, Ashish Khandelwal and Shivaram AC for providng useful material to write the paper. Many useful discussions with these colleagues as well as other colleagues from GM-R&D, especially, Nancy Johnson and John Ulicny are also gratefully acknowledged.

REFERENCES

1. Diann Brei, Jonathan Luntz, John Shaw, Nancy L. Johnson, Alan L. Browne, Paul W. Alexander, and Nilesh D. Mankame, "General Motors and the University of Michigan smart materials and structures collaborative research laboratory" in Industrial and Commercial Applications of Smart Structures Technologies 2007, L. Porter Davis; B. K. Henderson; M. Brett McMickell, eds. Proceedings of the SPIE, Volume 6527, SPIE Pub. (2007).
2. "GM plans to integrate smart materials by 2010", Automotive Engineer, 32 (4), 44-44, 2/3p, (2007)
3. "Active suspension no longer 'high end'", Automotive Engineer, 32 (4), 44-44, 1/4p, (2007)
4. V. Birman, Review of mechanics of shape memory alloy structures Appl. Mech. Reviews 50, 629-645 (1997)
5. D. Bernardini and T. J. Pence, "Shape-memory materials, modeling", in Encyclopedia of Smart Materials, volume 2, M. Schwartz ed., John Wiley and Sons Inc., New York, 2002, pp. 964-980.
6. D.C. Lagoudas, J.G. Boyd and Z Bo, J. Eng. Mater. Tr ASME 116, 337–347 (1994).
7. J.G. Boyd and D. Lagoudas, J. Intell. Mater. Syst. Struct., 333–346 (1994).
8. J.G. Boyd and D. Lagoudas, Int. J. Plast. 12, 805–841 (1996).
9. J.G. Boyd and D. Lagoudas, Int. J. Plast. 12, 843–873 (1996).
10. K.R. Rajagopal and A.R. Srinivasa, Z. Angew. Maths. Phys 50, 459–496 (1999).
11. I.V. Chenchiah and M.S. Siva, Mech. Res. Comm. 26, 301–307 (1999).
12. S. Zhang and G.P. McCormick, Acta Mater. 48, 3091–3101 (2000).
13. J.A. Shaw and S. Kyriakides, Int. J. Sol. Struct. 39, 1275–1305 (2002).
14. R.D. Reynolds, "A nonlinear thermodynamic model for phase transitions in shape memory alloy wires", Ph. D. Thesis, Rice University, 2003.
15. F. Auricchio and L. Petrini, Int. J. Num. Meth. Engg. 61, 807–836 (2004).
16. K. Tanaka, Res. Mechanica 18, 251–263 (1986).
17. C. Liang and C.A. Rogers, J. Intell. Mater. Syst. Struct. 2, 207–234 (1990).
18. L.C. Brinson, J. Intell. Mater. Syst. Struct. 4 (2), 229–242 (1993).
19. L.C. Brinson, R. Lammering,. Int. J. Sol. Struct. 4 (2), 229–242 (1993).
20. X. Wu and T.J. Pence, J. Intell. Mater. Syst. Struct. 9, 335–354 (1998).
21. H. Prahlad and I. Chopra, J. Intell. Mater. Syst. Struct. 11 (4), 272–282 (2001).
22. H.M. Elahinia and M. Ahmadian, "On the shortcomings of shape memory alloy phenomenological models" in: Proc. IMECE04, ASME International Mechanical Engineering Congress and Exposition. ASME, Anaheim, CA, USA, 2004, pp. 13–20.
23. V. R. Buravalla and A. Khandelwal, Int. J. Sol. Struct. 44, 4369–4381 (2007)
24. A. Khandelwal and V. R. Buravalla J. Intell. Mater. Syst. Struct.. (2007)
25. M Cho and S Kim, Int. J. Sol. Struct. 42 1759-1776 (2005)
26. M Cho and S Kim, Smart Mater. Struct 16 372-381 (2006)

SESSION A1

SHAPE MEMORY MODELING APPROACHES

A coupled thermo-mechanical theory for polycrystalline SMAs

P. Thamburaja* and N. Nikabdullah[†]

*Department of Mechanical Engineering, National University of Singapore, Singapore 117576
[†]Department of Mechanical Engineering, Universiti Kebangsaan Malaysia, Bangi, Malaysia 43600

Abstract. In this work we develop a finite-deformation-based and thermo-mechanically-coupled constitutive model for polycrystalline shape-memory alloys (SMAs) capable of undergoing austenite ↔ martensite phase transformations. The theory is developed in the classical isotropic metal-plasticity setting using fundamental thermodynamic laws. The constitutive model is then implemented in the ABAQUS/Explicit [1] finite-element program by writing a user-material subroutine. The results from the constitutive model and numerical procedure are then compared to representative physical experiments conducted on a polycrystalline rod Ti-Ni undergoing superelasticity. The constitutive model and the numerical simulations are able to reproduce the stress-strain responses from these physical experiments to good accuracy. Experimental strain-temperature-cycling and shape-memory effect responses have also shown to be qualitatively well-reproduced by the developed constitutive model. Finally, we perform coupled thermo-mechanical finite-element simulations to demonstrate the effect of deformation rate on the overall behavior of SMAs experiencing austenite ↔ martensite phase transformations.

Keywords: Phase transformation, Constitutive model, Plasticity, Finite-elements

1. INTRODUCTION

In the last few decades, there have been tremendous efforts in the development of smart and functional materials e.g. shape-memory alloys, piezoelectric, electro-active polymers/gels, magnetostrictive materials etc. Among these different classes of materials, the polycrystalline shape-memory alloy (SMA) is one of the most researched smart material. Examples of SMAs include the Ti-Ni (or Nitinol), Cu-Al-Ni, Cu-Zn-Al, Ti-Ni-Cu and Au-Cd alloy systems. SMAs are primarily used in development of MEMS devices, micro-actuators, micro-clampers, damping devices and also biomedical devices such as stents and orthodontic wires.

Under suitable thermo-mechanical loading conditions, SMAs have an ability to undergo a *diffusionless* and *reversible* phase transformation between two phases : (1) the *austenite* phase which is stable at temperatures above the *austenite finish temperature*, θ_{af} and low stresses, and (2) the *martensite* phase which is stable at temperatures below the *martensite finish temperature*, θ_{mf} and high stresses. The stress-free cooling

CP1029, *Smart Devices: Modeling of Material Systems, An International Workshop*
edited by S. M. Sivakumar, V. Buravalla, and A. R. Srinivasa
©2008 American Institute of Physics 978-0-7354-0553-0/08/$23.00

of a fully-austenitic material will cause the martensitic phase to start nucleating from the austenitic phase at the *martensite start temperature*, θ_{ms}. The transformation from austenite to martensite will result in the *release* of latent heat. Further cooling to a temperature below θ_{mf} will result in the specimen existing in the fully-martensitic state. The stress-free heating of this fully-martensitic specimen will cause the austenite phase to start nucleating at the *austenite start temperature*, θ_{as}. The transformation from martensite to austenite is accompanied by the *absorption* of latent heat. Further heating to a temperature above θ_{af} will cause the specimen to transform back to the fully-austenitic state[1].

At temperatures above θ_{af}, cyclic uniaxial extension/compression of an SMA specimen under isothermal ambient conditions between zero and a finite strain (but small) causes a stress-induced transformation from austenite to martensite and back. This phenomenon is termed as *superelastic* behavior as there is very little permanent deformation experienced by the sample after such a strain cycle. However, the sample would have undergone *hysteresis* under this strain cycle due to motion of sharp interfaces between the austenite and martensite phases. Superelastic behavior typically results in a flag-type stress-strain response. Another manifestation of superelasticity is the *thermal cycling* at constant stress which can be explained as follows : An austenitic SMA is initially pre-stressed to a level which is not sufficient to cause austenite → martensite transformation. With the pre-stress maintained, a reduction in temperature will cause a phase transformation from austenite → martensite to occur at a critical temperature. An increase in temperature with the SMA now being fully-martensitic (with the pre-stress still maintained) will cause a reverse transformation from martensite → austenite to occur at another critical temperature. The thermal cycling behavior is also hysteretic and it results in a flag-type strain-temperature response.

SMAs can also exhibit the *shape-memory effect* by transformation (or simply the shape-memory effect). It occurs as follows : (1) At a temperature between θ_{ms} and θ_{as}, the sample is deformed under isothermal ambient conditions to cause austenite to martensite phase transformation; (2) upon reverse loading to zero stress, the deformation induced due to the prior austenite → martensite transformation is not recovered and a residual/permanent strain exists in the sample; (3) under stress-free conditions, the increase of the sample's temperature to above θ_{af} will cause the residual strain to be fully recovered.

Since the overall behavior of SMAs is very sensitive to temperature, it is of paramount importance to develop constitutive models which are capable of studying the austenite ↔ martensite phase transformations under thermo-mechanically-coupled settings. The developed constitutive models need to also be three-dimensional in nature since most practical applications result in the components experiencing multi-axial stress states.

Recently, there has been extensive worldwide activity in the the modeling of austenite ↔ martensite transformations in SMAs. Some one-dimensional constitutive models which describe the basic features of austenite ↔ martensite transformations include the models of Liang and Rogers[3], Abeyaratne and Knowles[4] and Bekker and

---

[1] For more information regarding the material science aspects of SMAs, please refer to Otsuka and Wayman[2].

28

Brinson[5]. Some three-dimensional constitutive equations which describe the behavior of SMAs include the works of Sun and Hwang[6], Boyd and Lagoudas[7], Patoor et al.[8], Auricchio et al.[9], Lu and Weng[10], Lim and McDowell[11], Thamburaja and Anand[12], and Jung et al.[13]. However, these aforementioned models were developed under isothermal conditions which do not take into the coupled interaction between the thermal and mechanical fields during phase transformations. The three-dimensional constitutive equations of Anand and Gurtin[14] and Thamburaja and Anand[15] were developed in a thermo-mechanically-coupled framework but they are numerically very expensive to implement since these models employ crystal-plasticity-based theories.

To aid in the design process of SMA components experiencing austenite ↔ martensite phase transitions, it is still very useful to develop *phenomenologically*-based constitutive models which are three-dimensional and thermo-mechanically-coupled since : (a) they can model the actual behavior of SMAs to reasonably good accuracy; (b) they are computationally inexpensive to implement; and (c) they have a structure which is theoretically less complicated compared to crystal-mechanics-based constitutive models. With these objectives in perspective, we will develop a three-dimensional, thermo-mechanically-coupled and phenomenologically-based constitutive model to describe the austenite ↔ martensite phase transformations exhibited by SMAs in this paper.

The structure of this paper is as follows : In Section 2, we develop a three-dimensional, finite-deformation and coupled thermo-mechanically-based constitutive model for SMAs capable of undergoing austenite ↔ martensite phase transformations. The constitutive model is developed in a thermodynamically-consistent framework using classical rate-independent and isotropic metal-plasticity. A robust time-integration procedure based on the constitutive model is developed and implemented in the ABAQUS/Explicit[1] finite-element program by writing a user-material subroutine. In Section 3, we verify our constitutive model and its corresponding numerical algorithm with respect to physical experiments conducted on a polycrystalline rod Ti-Ni. With the aid of finite-element simulations, we show that our constitutive model is able to *quantitatively* and *qualitatively* reproduce the basic features exhibited by SMAs i.e. superelasticity, shape-memory effect etc. We finally conclude in Section 4.

2. CONSTITUTIVE MODEL

Here we develop a constitutive model for shape-memory alloys capable of undergoing austenite-martensite phase transformation using the well-established theory of rate-independent isotropic metal plasticity. The list of the governing variables in the constitutive model are[2] : (i) The Helmholtz free energy per unit *reference* volume, ψ. (ii) The

---

[2] *Notation* : The terms Div, ∇ and ∇^2 denote the *referential* divergence, gradient and laplacian, respectively. All the tensorial variables in this work are second-order tensors unless stated otherwise. For a second-order tensor $\mathbf{B}$, $\mathbf{B}^\top$ denotes the transpose of the tensor $\mathbf{B}$, $\mathbf{B}^{-1}$ denotes the inverse of the tensor $\mathbf{B}$ and $\left(\mathbf{B}^{-1}\right)^\top = \mathbf{B}^{-\top}$. We also write trace $\mathbf{B}$ and det $\mathbf{B}$ for the *trace* and *determinant* of a second-order tensor $\mathbf{B}$, respectively. Also sym $\mathbf{B} = (1/2)(\mathbf{B} + \mathbf{B}^\top)$ denotes the symmetric portion of tensor $\mathbf{B}$ whereas skew $\mathbf{B} = (1/2)(\mathbf{B} - \mathbf{B}^\top)$ denotes the skew portion of tensor $\mathbf{B}$. The deviatoric portion of tensor $\mathbf{B}$ is denoted by $\mathbf{B}_0 \equiv \mathbf{B} - (1/3)(\text{trace}\,\mathbf{B})\mathbf{1}$. The scalar product of two second-order tensors $\mathbf{A}$ and $\mathbf{B}$ is denoted by

Cauchy stress tensor, $\mathbf{T}$. (iii) The deformation gradient tensor, $\mathbf{F} = \nabla \mathbf{y}$ with $\det \mathbf{F} > 0$ and $\mathbf{y}$ denoting the position vector of the material point in the *current* configuration. With $\mathbf{x}$ denoting the position vector of the material point in the *reference* configuration, the deformation gradient $\mathbf{F}$ maps the line element $\mathbf{dx}$ in the *reference* configuration to the line element $\mathbf{dy}$ in the *current* configuration viz. $\mathbf{dy} = \mathbf{F(x)dx}$. (iv) The First Piola-Kirchoff stress tensor, $\mathbf{S} = (\det \mathbf{F})\mathbf{TF}^{-\top}$. (v) The inelastic deformation gradient tensor, $\mathbf{F}^p = \nabla \mathbf{z}$ with $\det \mathbf{F}^p > 0$ and $\mathbf{z}$ denoting the position vector of the material point in the *relaxed* configuration. It represents the cumulative effect of austenite $\leftrightarrow$ martensite phase transformations in the RVE. The inelastic deformation gradient $\mathbf{F}^p$ maps the line element $\mathbf{dx}$ in the *reference* configuration to the line element $\mathbf{dz}$ in the *relaxed* configuration viz. $\mathbf{dz} = \mathbf{F}^p(\mathbf{x})\mathbf{dx}$. (vi) The elastic deformation gradient tensor, $\mathbf{F}^e$ with $\det \mathbf{F}^e > 0$. From the Kroner-Lee decomposition (Kroner [16], Lee [17]), the elastic deformation gradient is given by

$$\mathbf{F}^e = \mathbf{FF}^{p-1}. \tag{1}$$

The elastic deformation gradient $\mathbf{F}^e$ maps the line element $\mathbf{dz}$ in the *relaxed* configuration to the line element $\mathbf{dy}$ in the *current* configuration viz. $\mathbf{dy} = \mathbf{F}^e(\mathbf{x})\mathbf{dz}$. Thus, it describes the elastic stretches as well as the lattice rotations from the *relaxed* configuration to the *current* configuration. (vii) The total martensite volume fraction, ξ with $0 \leq \xi \leq 1$. (viii) The critical resistance to austenite-martensite phase transformation, f_c with $f_c > 0$. The critical resistance has units of energy per unit volume.

To construct the constitutive equations we focus on a continuum body occupying a region $\mathcal{R}$ in the *reference* configuration with $\mathbf{n}$ the outward unit normal on its boundary denoted by $\partial \mathcal{R}$. We also denote ∂A and ∂V as the *referential* area and volume integral, respectively. In our constitutive model, the fully-austenitic state is chosen as the *reference* configuration.

• **Kinematics and kinetics**

Since the total deformation gradient, $\mathbf{F} = \mathbf{F}^e \mathbf{F}^p$ the expression for the total velocity gradient tensor is then given by

$$\mathbf{L} \equiv \dot{\mathbf{F}} \mathbf{F}^{-1} = \mathbf{L}^e + \mathbf{F}^e \mathbf{L}^p \mathbf{F}^{e-1} \tag{2}$$

where $\mathbf{L}^e \equiv \dot{\mathbf{F}}^e \mathbf{F}^{e-1}$ and $\mathbf{L}^p \equiv \dot{\mathbf{F}}^p \mathbf{F}^{p-1}$ represents the elastic and inelastic velocity gradient tensors, respectively. We further decompose $\mathbf{L}^p = \mathbf{D}^p + \mathbf{W}^p$ where $\mathbf{D}^p = \text{sym}\,\mathbf{L}^p$ and $\mathbf{W}^p = \text{skew}\,\mathbf{L}^p$ denote the inelastic stretching rate and inelastic spinning rate tensors, respectively. Following the work of Anand and Gurtin[18] on isotropic solids, we set the inelastic flow to be *irrotational* :

$$\mathbf{W}^p = \mathbf{0} \longrightarrow \mathbf{L}^p = \mathbf{D}^p.$$

$\mathbf{A} \cdot \mathbf{B} = \text{trace}\,(\mathbf{B}^{\top}\mathbf{A})$. The scalar product of two vectors $\mathbf{u}$ and $\mathbf{v}$ is also denoted by $\mathbf{u} \cdot \mathbf{v}$. The tensor product of two vectors $\mathbf{u}$ and $\mathbf{v}$ is denoted by $\mathbf{u} \otimes \mathbf{v}$. The magnitude of vector $\mathbf{u}$ and tensor $\mathbf{B}$ is denoted by $|\mathbf{u}|$ and $|\mathbf{B}|$, respectively. The second-order identity tensor is denoted by $\mathbf{1}$.

Assuming that austenite-martensite phase transformations are accompanied by no changes in volume (Sun and Hwang,[6]), we take the inelastic stretching rate[3] to be purely *deviatoric* and *symmetric*, and augment it with a term to describe the tension-compression asymmetry (Orgeas and Favier,[22]) :

$$\mathbf{D}^p = k(1 + a\phi)^b \left\{ \dot{\xi}_1 \mathbf{N}_1 + \dot{\xi}_2 \mathbf{N}_2 \right\}. \tag{3}$$

The flow direction tensors $\mathbf{N}_\beta$ with $\beta = 1, 2$ are restricted by $\mathbf{N}_\beta = \mathbf{N}_\beta^\top$ and trace $\mathbf{N}_\beta = 0$. Here $\dot{\xi}_1 \geq 0$ and $\dot{\xi}_2 \leq 0$ denote the *transformation rates*. We define *forward transformation* to occur when $\dot{\xi}_1 > 0$, and *reverse transformation* to occur when $\dot{\xi}_2 < 0$. The total rate of change of martensite volume fraction is then given by

$$\dot{\xi} = \dot{\xi}_1 + \dot{\xi}_2. \tag{4}$$

A transformation from austenite to martensite occurs when $\dot{\xi} > 0$. Conversely, a transformation from martensite to austenite occurs when $\dot{\xi} < 0$. No phase transformation occurs when $\dot{\xi} = 0$. With $k > 0$ being a constant of proportionality, we will enforce $|\mathbf{N}_\beta| = \varepsilon_T$ for each β where $\varepsilon_T > 0$ denotes the transformation strain due to austenite-martensite phase transformation (to be determined experimentally). The scalar ϕ with $-1 \leq \phi \leq 1$ represents the J_3 measure with $a \geq 0$ and $b \geq 0$ denoting dimensionless constitutive parameters to be calibrated from experiments. The amount of tension-compression asymmetry exhibited in the material is measured by the J_3 parameter. The functional form for ϕ will be described later.

From microscopic considerations, the reverse transformation is the crystallographic recovery of the deformation which is induced during forward transformation i.e. the reverse transformation is restricted by the forward transformation history. Therefore, with t denoting time and the direction $\mathbf{N}_1$ to be determined, we generalize the work of Boyd and Lagoudas[7] to finite-deformations and take the flow direction $\mathbf{N}_2$ to be defined by :

$$\mathbf{N}_2 = \varepsilon_T \left[\frac{\int \mathbf{D}^p \, dt}{|\int \mathbf{D}^p \, dt|} \right]. $$

With $\mathbf{C}^e \equiv \mathbf{F}^{e\top}\mathbf{F}^e$, we define the measure of the elastic strain tensor as

$$\mathbf{E}^e \equiv \frac{1}{2} \left(\mathbf{C}^e - \mathbf{1} \right) \longrightarrow \dot{\mathbf{E}}^e = \mathbf{F}^{e\top} (\text{sym} \, \mathbf{L}^e) \mathbf{F}^e. \tag{5}$$

• **Standard force balance**

The standard force balance is given by

$$\int_{\partial \mathscr{R}} \mathbf{S} \mathbf{n} \, \partial A + \int_{\mathscr{R}} \mathbf{b} \, \partial V = \mathbf{0} \tag{6}$$

---

[3] The present constitutive model is not intended to describe the inelastic deformation due to martensitic reorientation and detwinning. For recent efforts in the development of single-crystal constitutive models describing the martensitic reorientation and detwinning in SMAs, please refer to the papers of Thamburaja and co-workers ([19],[20],[21]).

31

with **b** being the macroscopic body force vector per unit volume. Inertial forces are also included in the body force **b**. Using the divergence law on equation (6) and localizing the result within $\mathcal{R}$ yields

$$\text{Div}\,\mathbf{S} + \mathbf{b} = \mathbf{o}. \tag{7}$$

• Standard moment balance

From the standard moment balance we have

$$\int_{\partial\mathcal{R}} \mathbf{y} \times \mathbf{Sn}\,\partial A + \int_{\mathcal{R}} \mathbf{y} \times \mathbf{b}\,\partial V = \mathbf{o}. \tag{8}$$

Applying the divergence law on equation (8) and localizing the result within $\mathcal{R}$ while using equation (7) yields

$$\mathbf{SF}^\top = \mathbf{FS}^\top. \tag{9}$$

Substituting $\mathbf{S} = (\det\mathbf{F})\mathbf{TF}^{-\top}$ into equation (9) yields $\mathbf{T} = \mathbf{T}^\top$ i.e. the Cauchy stress is symmetric.

• Balance of energy

The first law of thermodynamics (the balance of energy) is stated as

$$\int_{\partial\mathcal{R}} \mathbf{Sn}\cdot\dot{\mathbf{y}} - \mathbf{q}\cdot\mathbf{n}\,\partial A + \int_{\mathcal{R}} \mathbf{b}\cdot\dot{\mathbf{y}} + r\,\partial V = \frac{d}{dt}\int_{\mathcal{R}} \varepsilon\,\partial V \tag{10}$$

where ε is the internal energy per unit *reference* volume. Here $\mathbf{q}$ is the heat flux vector measured per unit *reference* area, and r is the heat supply per unit *reference* volume. Applying the divergence law on equation (10) and localizing the result within $\mathcal{R}$ while using equation (7) yields

$$\mathbf{S}\cdot\dot{\mathbf{F}} - \text{Div}\,\mathbf{q} + r = \dot{\varepsilon}. \tag{11}$$

Using $\mathbf{S} = (\det\mathbf{F})\mathbf{TF}^{-\top}$, $\mathbf{L}^p = \mathbf{D}^p$, equations (2), (5)$_2$ and (9) in equation (11) results in

$$\mathbf{T}^* \cdot \dot{\mathbf{E}}^e + \overline{\mathbf{T}} \cdot \mathbf{D}^p - \text{Div}\,\mathbf{q} + r = \dot{\varepsilon} \qquad \text{where} \tag{12}$$

$$\mathbf{T}^* = (\det\mathbf{F})\mathbf{F}^{e-1}\mathbf{TF}^{e-\top} \quad \text{and} \quad \overline{\mathbf{T}} = \mathbf{C}^e\mathbf{T}^*. \tag{13}$$

Here $\mathbf{T}^*$ and $\overline{\mathbf{T}}$ are *frame-invariant*[4] measures of stresses with $\mathbf{T}^*$ being symmetric whereas $\overline{\mathbf{T}}$ is generally not symmetric.

---

[4] *Frame-invariance* (Anand and Gurtin,[18]): Consider the transformations of the form $\mathbf{x} \rightarrow \mathbf{x}$, $\mathbf{z} \rightarrow \mathbf{z}$ and $\mathbf{y} \rightarrow \mathbf{Q}(t)\mathbf{y} + \mathbf{a}(t)$ where t denotes time, $\mathbf{Q}(t)$ is a proper orthogonal rotation tensor and $\mathbf{a}(t)$ a translational vector. The *reference* and *relaxed* configurations are independent of the choice of such changes in frame. Under changes in frame of the form given above, the variables : (a) $\mathbf{F} \rightarrow \mathbf{QF}$, (b) $\mathbf{F}^p \rightarrow \mathbf{F}^p$, (c) $\mathbf{L}^p \rightarrow \mathbf{L}^p$ and $\mathbf{D}^p \rightarrow \mathbf{D}^p$, (d) $\mathbf{F}^e \rightarrow \mathbf{QF}^e$, (e) $\mathbf{C}^e \rightarrow \mathbf{C}^e$ since $\mathbf{F}^e \rightarrow \mathbf{QF}^e$, (f) $\mathbf{E}^e \rightarrow \mathbf{E}^e$ since $\mathbf{C}^e \rightarrow \mathbf{C}^e$, (g) $\mathbf{T}^* \rightarrow \mathbf{T}^*$ since $\det\mathbf{F} \rightarrow \det\mathbf{F}$, $\mathbf{F}^e \rightarrow \mathbf{QF}^e$ and $\mathbf{T} \rightarrow \mathbf{QTQ}^\top$, (h) $\overline{\mathbf{T}} \rightarrow \overline{\mathbf{T}}$ since $\mathbf{C}^e \rightarrow \mathbf{C}^e$ and $\mathbf{T}^* \rightarrow \mathbf{T}^*$, (i) $\mathbf{q} \rightarrow \mathbf{q}$ since the heat flux vector is referentially-defined.

• Dissipation inequality

The second law of thermodynamics is written as

$$\frac{d}{dt}\int_{\mathscr{R}} \eta\, \partial V \geq \int_{\partial\mathscr{R}} -\frac{\mathbf{q}}{\theta}\cdot\mathbf{n}\, \partial A + \int_{\mathscr{R}} \frac{r}{\theta}\, \partial V \tag{14}$$

with η representing the entropy per unit *reference* volume. Using the divergence law on equation (14) and localizing the result within $\mathscr{R}$ yields

$$\dot{\eta}\theta + \mathrm{Div}\,\mathbf{q} - \frac{\mathbf{q}}{\theta}\cdot\nabla\theta - r \geq 0. \tag{15}$$

The Helmholtz free energy per unit *reference* volume, ψ, is defined as

$$\psi = \varepsilon - \eta\theta \longrightarrow \dot{\psi} = \dot{\varepsilon} - \dot{\eta}\theta - \eta\dot{\theta}. \tag{16}$$

We use the functional expression for the free energy density of a shape-memory alloy as follows (cf. Helm and Haupt,[23]) :

$$\psi = \hat{\psi}(\mathbf{E}^e, \xi, \theta) \longrightarrow \dot{\psi} = \frac{\partial\psi}{\partial\mathbf{E}^e}\cdot\dot{\mathbf{E}}^e + \frac{\partial\psi}{\partial\xi}\dot{\xi} + \frac{\partial\psi}{\partial\theta}\dot{\theta}. \tag{17}$$

Substituting equations (16) and (17) into equation (12) yields

$$\left(\mathbf{T}^* - \frac{\partial\psi}{\partial\mathbf{E}^e}\right)\cdot\dot{\mathbf{E}}^e - \left(\eta + \frac{\partial\psi}{\partial\theta}\right)\dot{\theta} + \Gamma = \dot{\eta}\theta \qquad \text{where} \tag{18}$$

$$\Gamma \equiv \overline{\mathbf{T}}\cdot\mathbf{D}^p + (\mathrm{Div}\,\mathbf{c})\dot{\xi} - \frac{\partial\psi}{\partial\xi}\dot{\xi} - \mathrm{Div}\,\mathbf{q} + r. \tag{19}$$

Further substitution of equation (18) into inequality (15) results in the *dissipation inequality* :

$$\left(\mathbf{T}^* - \frac{\partial\psi}{\partial\mathbf{E}^e}\right)\cdot\dot{\mathbf{E}}^e - \left(\eta + \frac{\partial\psi}{\partial\theta}\right)\dot{\theta} + \Pi \geq 0 \qquad \text{where} \tag{20}$$

$$\Pi \equiv \overline{\mathbf{T}}\cdot\mathbf{D}^p - \frac{\partial\psi}{\partial\xi}\dot{\xi} - \frac{\mathbf{q}}{\theta}\cdot\nabla\theta. \tag{21}$$

From the principle of equipresence (Coleman and Noll,[24]), inequality (20) yields

$$\mathbf{T}^* = \frac{\partial\psi}{\partial\mathbf{E}^e} \quad \text{and} \quad \eta = -\frac{\partial\psi}{\partial\theta}. \tag{22}$$

Equations $(22)_1$ and $(22)_2$ are the constitutive equations for the stress measure $\mathbf{T}^*$ and the entropy, respectively.

• Phase transformation criteria and Fourier's law

Substituting equations (22) back into inequality (20) yields the *reduced dissipation inequality*:

$$\Pi \equiv \overline{\mathbf{T}} \cdot \mathbf{D}^p - \frac{\partial \psi}{\partial \xi} \dot{\xi} - \frac{\mathbf{q}}{\theta} \cdot \nabla \theta \geq 0. \tag{23}$$

Here Π represents the *total dissipation* and it is always non-negative. Recall that each $\mathbf{N}_\beta$ is defined to be *deviatoric* and *symmetric*. We substitute equations (3) and (4) into equation (23) and assume *each* dissipative mechanism to be *strongly dissipative* (Anand and Gurtin,[18]) :

$$f_1 \dot{\xi}_1 > 0 \quad \text{whenever} \quad \dot{\xi}_1 \neq 0 \tag{24}$$

where $f_1 \equiv \left(k(1+a\phi)^b \left[\operatorname{sym} \overline{\mathbf{T}}_0 \cdot \mathbf{N}_1\right] - \frac{\partial \psi}{\partial \xi} \right)$ denotes the driving force for *forward* transformation,

$$f_2 \dot{\xi}_2 > 0 \quad \text{whenever} \quad \dot{\xi}_2 \neq 0 \tag{25}$$

where $f_2 \equiv \left(k(1+a\phi)^b \left[\operatorname{sym} \overline{\mathbf{T}}_0 \cdot \mathbf{N}_2\right] - \frac{\partial \psi}{\partial \xi} \right)$ denotes the driving force for *reverse* transformation, and finally

$$-\frac{\mathbf{q}}{\theta} \cdot \nabla \theta > 0 \quad \text{whenever} \quad \nabla \theta \neq \mathbf{o}. \tag{26}$$

We will assume that inequalities (24), (25) and (26) are obeyed at all times so that the reduced dissipation inequality (23) will be concurrently satisfied.

To satisfy inequality (24) under the assumption of *rate-independence*, we choose an expression for f_1 as follows:

$$f_1 = f_c \left(\operatorname{sign}(\dot{\xi}_1)\right) \longrightarrow f_1 = f_c \quad \text{whenever} \quad \dot{\xi}_1 > 0. \tag{27}$$

Similarly, to satisfy inequality (25) under the assumption of *rate-independence*, we choose an expression for f_2 as follows:

$$f_2 = f_c \left(\operatorname{sign}(\dot{\xi}_2)\right) \longrightarrow f_2 = -f_c \quad \text{whenever} \quad \dot{\xi}_2 < 0. \tag{28}$$

Equations (27) and (28) are the criteria for *forward* and *reverse* phase transformations, respectively.

Finally, assuming the material obeys Fourier's law of heat conduction we enforce

$$\mathbf{q} = -k_{th} \nabla \theta \tag{29}$$

to satisfy inequality (26) where $k_{th} > 0$ denotes the constant coefficient of thermal conductivity.

• Free energy density

We take the free energy per unit *reference* volume, ψ to be in the separable form

$$\psi = \psi^e(\mathbf{E}^e, \theta) + \psi^\xi(\xi, \theta) + \psi^\theta(\theta) \quad \text{where} \tag{30}$$

34

$$\psi^e(\mathbf{E}^e, \theta) = \mu |\mathbf{E}_0^e|^2 + \kappa(\text{trace}\,\mathbf{E}^e)^2 - 3\kappa\alpha_{th}(\theta - \theta_o)(\text{trace}\,\mathbf{E}^e), \tag{31}$$

$$\psi^\xi(\xi, \theta) = \frac{\lambda_T}{\theta_T}(\theta - \theta_T)\xi + \frac{1}{2}h\xi^2 \quad \text{and} \quad \psi^\theta(\theta) = c(\theta - \theta_o) - c\theta\log(\theta/\theta_o). \tag{32}$$

Here ψ^e denotes the classical thermo-elastic free energy density with μ, κ and α_{th} denoting the shear modulus, bulk modulus and the coefficient of thermal expansion, respectively. The austenite $\leftrightarrow$ martensite phase transformation energy is denoted by ψ^ξ where λ_T and θ_T represents the *latent heat* released/absorbed (units of energy per unit volume) during the austenite $\leftrightarrow$ martensite phase transformation and the *phase equilibrium* temperature, respectively[5]. The energetic interaction coefficient, h has units of energy per unit volume. Finally, ψ^θ represents the purely thermal portion of the free energy with $c > 0$ being the specific heat per unit volume.

For simplicity, we will treat the material parameters $\{\mu, \kappa, \alpha_{th}, \lambda_T, \theta_T, h, c\}$ to be constants. Furthermore, we will also assume that there are no mismatches between the austenite phase and the martensite phase material parameters.

• Constitutive equation for the stress and entropy

Substituting the expression for the free energy density given in equation (30) into equation $(22)_1$ yields the constitutive equation for the stress measure $\mathbf{T}^*$:

$$\mathbf{T}^* = 2\mu\mathbf{E}_0^e + \kappa\left[\text{trace}\,\mathbf{E}^e - 3\alpha_{th}(\theta - \theta_o)\right]\mathbf{1}. \tag{33}$$

The constitutive equation for the entropy density is then given by substituting equation (30) into equation $(22)_2$:

$$\eta = c\log(\theta/\theta_o) + 3\kappa\alpha_{th}(\text{trace}\,\mathbf{E}^e) - (\lambda_T/\theta_T)\xi. \tag{34}$$

• Flow direction $\mathbf{N}_1$ and the J_3 parameter

Recall that $\overline{\mathbf{T}} = \mathbf{C}^e\mathbf{T}^*$. From equations $(5)_1$ and (33), we can see that the stress measure $\mathbf{T}^*$ is coaxial with $\mathbf{C}^e$ i.e. $\mathbf{T}^*$ and $\mathbf{C}^e$ share the same set of eigenvectors. Since $\mathbf{T}^*$ and $\mathbf{C}^e$ are symmetric with $\mathbf{T}^*$ and $\mathbf{C}^e$ being coaxial, we obtain

$$\mathbf{C}^e\mathbf{T}^* = \mathbf{T}^*\mathbf{C}^e = (\mathbf{C}^e\mathbf{T}^*)^\top \longrightarrow \overline{\mathbf{T}} \text{ is symmetric.}$$

Substituting equation (30) into the expressions for the driving force for phase transformation yield :

$$f_1 = k(1 + a\phi)^b(\overline{\mathbf{T}}_0 \cdot \mathbf{N}_1) - \frac{\lambda_T}{\theta_T}(\theta - \theta_T) - h\xi, \tag{35}$$

$$f_2 = k(1 + a\phi)^b(\overline{\mathbf{T}}_0 \cdot \mathbf{N}_2) - \frac{\lambda_T}{\theta_T}(\theta - \theta_T) - h\xi. \tag{36}$$

---

[5] In formulating the phase transformation free energy, we are guided by the one-dimensional model of Abeyaratne and Knowles[4]. The quantity (λ_T/θ_T) captures the Clausius-Clapeyron relationship in our model.

From the expression for the driving force shown in equations (35) and (36), we can see that the transformation between the austenite and martensite phase is affected by the *stress* and the *phase transformation energy*.

During forward transformation i.e. $\dot{\xi}_1 \neq 0$, substituting equation (35) into equation (27) results in

$$k(1+a\phi)^b \left(\overline{\mathbf{T}}_0 \cdot \mathbf{N}_1\right) = \frac{\lambda_T}{\theta_T}(\theta - \theta_T) + h\xi + f_c\left(\text{sign}(\dot{\xi}_1)\right). \tag{37}$$

To satisfy equation (37), we take

$$(\varepsilon_T)^2 \left[k(1+a\phi)^b\right]\overline{\mathbf{T}}_0 = \left\{\frac{\lambda_T}{\theta_T}(\theta - \theta_T) + h\xi + f_c\left(\text{sign}(\dot{\xi}_1)\right)\right\}\mathbf{N}_1 \tag{38}$$

since $|\mathbf{N}_1| = \varepsilon_T$. Taking the magnitude on both sides of equation (38) yields

$$\mathbf{N}_1 = \varepsilon_T \left[\frac{\overline{\mathbf{T}}_0}{|\overline{\mathbf{T}}_0|}\right] \longrightarrow f_1 = \bar{\sigma}\varepsilon_T(1+a\phi)^b - \frac{\lambda_T}{\theta_T}(\theta - \theta_T) - h\xi \tag{39}$$

where $\bar{\sigma} = k|\overline{\mathbf{T}}_0|$ represents an *equivalent stress*. By generalizing the work of Orgeas and Favier[22] to finite-deformations, we take the J_3 parameter to be given by

$$\phi = \sqrt{6}\left[\mathbf{N}_1 \cdot \mathbf{N}_1^2\right](\varepsilon_T)^{-3}.$$

• **Flow rule**

During forward transformation i.e. $\dot{\xi}_1 \neq 0$ and $\dot{\xi}_2 = 0$, we define

$$\varepsilon_T(1+a\phi)^b\dot{\xi}_1 \equiv \sqrt{2/3}\,|\mathbf{D}^p| \longrightarrow k = \sqrt{3/2} \quad \text{and} \quad \bar{\sigma} = \sqrt{3/2}\,|\overline{\mathbf{T}}_0|.$$

Therefore $\bar{\sigma}$ denotes the *equivalent tensile stress* or *Mises stress*. The final form for the inelastic stretching rate i.e the *flow rule* is then given by

$$\mathbf{L}^p = \mathbf{D}^p = \sqrt{3/2}(1+a\phi)^b\left\{\dot{\xi}_1\mathbf{N}_1 + \dot{\xi}_2\mathbf{N}_2\right\}. \tag{40}$$

• **Conditions on the driving forces and phase transformation rates**

The driving forces f_1 and f_2 are defined to be within the ranges :

$$f_1 \leq f_c \quad \text{for} \quad 0 \leq \xi < 1, \quad f_2 \geq -f_c \quad \text{for} \quad 0 < \xi \leq 1.$$

For $\xi = 1$, the driving force for forward transformation is defined for all values of f_1. For $\xi = 0$, the driving force for reverse transformation is defined for all values of f_2. In a *rate-independent* theory, the variables $\left\{f_1, \dot{\xi}_1\right\}$ and $\left\{f_2, \dot{\xi}_2\right\}$ must satisfy the following conditions :

- *Elastic range conditions* : If $f_1 \neq f_c$, then $\dot{\xi}_1 = 0$. If $f_2 \neq -f_c$, then $\dot{\xi}_2 = 0$.
- *Forward transformation* : If $0 \leq \xi < 1$ and $f_1 = f_c$, then

$$\dot{\xi}_1 \overline{(f_1 - f_c)} = 0. \tag{41}$$

- *Reverse transformation* : If $0 < \xi \leq 1$ and $f_2 = -f_c$, then

$$\dot{\xi}_2 \overline{(f_2 + f_c)} = 0. \tag{42}$$

- *End conditions* : If $\xi = 1$ and $f_1 = f_c$, then $\dot{\xi}_1 = 0$. If $\xi = 0$ and $f_2 = -f_c$, then $\dot{\xi}_2 = 0$.

Equations (41) and (42) are the *consistency conditions* for forward and reverse phase transformation, respectively. The consistency conditions are used to determine the transformation rates $\dot{\xi}_1$ and $\dot{\xi}_2$.

- **Balance of energy : revisited**

With $\mathbf{D}^p$ being symmetric and deviatoric, substituting equations (29), (30), (33) and (34) into equation (18) yields

$$\overline{\mathbf{T}}_0 \cdot \mathbf{D}^p - \frac{\lambda_T}{\theta_T}(\theta - \theta_T)\dot{\xi} + k_{th}(\nabla^2\theta) + r = \dot{\eta}\theta. \tag{43}$$

Taking the time-derivative of equation (34) results in

$$\dot{\eta} = \left(\frac{c}{\theta}\right)\dot{\theta} + 3\kappa\alpha_{th}(\text{trace}\,\dot{\mathbf{E}}^e) - \frac{\lambda_T}{\theta_T}\dot{\xi}. \tag{44}$$

The evolution equation for temperature is given by substituting equations (4), (40) and (44) into equation (43) :

$$c\dot{\theta} = k_{th}(\nabla^2\theta) + (\lambda_T/\theta_T)\theta\dot{\xi} - 3\kappa\alpha_{th}(\text{trace}\,\dot{\mathbf{E}}^e)\theta + f_1\dot{\xi}_1 + f_2\dot{\xi}_2 + r. \tag{45}$$

To summarize, the list of constitutive parameters/functions that needed to calibrated/specified are

$$\{\mu, \kappa, \alpha_{th}, \lambda_T, \theta_T, h, c, a, b, \varepsilon_T, f_c, k_{th}, r\}.$$

A time-integration procedure based on the isotropic-plasticity-based constitutive model for shape-memory alloys has been developed and implemented in the ABAQUS/Explicit[1] finite-element program by writing a user-material subroutine.

3. PHYSICAL EXPERIMENTS AND FINITE-ELEMENT SIMULATIONS

The material parameters for the constitutive model developed in the previous section was fitted to the stress-strain responses obtained from superelastic simple tension, simple

37

compression and tubular torsion experiments of Thamburaja and Anand[12] conducted on a polycrystalline rod Ti-Ni at an ambient temperature of 298 K. These experiments were conducted under very-low strain-rates to ensure that isothermal conditions prevail during the testing period. The critical transformation temperatures for the polycrystalline Ti-Ni material were determined to be $\theta_{ms} = 251.3$ K, $\theta_{mf} = 213.0$ K, $\theta_{as} = 260.3$ K and $\theta_{af} = 268.5$ K.

All the finite-element simulations in this work were conducted using a single ABAQUS C3D8R continuum-three-dimensional brick element shown in Figure 1a under unless stated otherwise. Each node on an ABAQUS C3D8R element has three displacement degrees of freedom. In the spirit of modeling elastic perfectly-plastic materials, we set the energetic interaction coefficient, $h = 0$ J/m^3. Finally the heat supply per unit volume, r, is ignored in all of our calculations by setting it to be zero. Furthermore, the material is fully-austenitic at the beginning of each simulation. The material parameters in the present constitutive model were determined by employing a similar methodology to that outlined in Thamburaja and Anand [12]. For more details regarding the material parameter fitting procedure, please refer to the aforementioned work. Using the material parameters listed out in Table 1[6] the *isothermal* stress-strain curves obtained from the simple tension, simple compression and simple shear finite-element simulations are plotted in Figures 1b, 1c and 1d, respectively, along with the corresponding stress-strain data obtained from the physical experiments. The fit from the numerical simulations are in good accord with the experimental stress-strain responses.

TABLE 1. Material parameters for the polycrystalline rod Ti-Ni

$\mu = 23.31$ GPa	$\kappa = 60.78$ GPa	$\alpha_{th} = 10 \times 10^{-6}$ K^{-1}	$\varepsilon_T = 0.048$
$a = 0.748$	$b = 0.136$	$c = 2.08$ MJ/Km^3	$k_{th} = 18$ W/mK
$\theta_T = 255.8$ K	$\lambda_T = 97$ MJ/m^3	$f_c = 7.8$ MJ/m^3	$h = 0$ J/m^3
$r = 0$ W/m^3			

The numerical stress-strain curves obtained from the simple tension and simple compression simulations conducted above are repeatedly plotted in Figure 1e for comparison. As shown by these stress-strain responses, the present constitutive theory is able to model the tension-compression asymmetry exhibited by polycrystalline rod Ti-Ni namely : (i) the stress level required to nucleate the martensitic phase from the parent austenitic phase is considerably higher in compression than in tension; (ii) the transformation strain measured in compression is smaller than that in tension; and (iii) the hysteresis loop generated in compression is wider (measured along the stress axis) than

---

[6] The austenitic phase values are chosen for the material parameters $\{\mu, \kappa, \alpha_{th}, c, k_{th}\}$. As mentioned previously, we have assumed no mismatches between the austenite and martensite phase material parameters for simplicity.

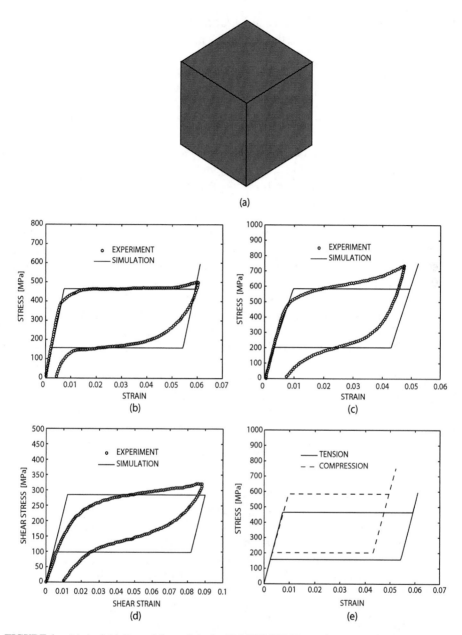

FIGURE 1. (a) An initially-undeformed single ABAQUS C3D8R continuum-three-dimensional brick element. Superelastic stress-strain curves in (b) simple tension, (c) simple compression and (d) simple shear (Thamburaja and Anand,[12]). The experimental data from these tests were used to fit the material parameters. The curve fits from the finite-element simulations are also shown. (d) Comparison of the stress-strain response from the tension and compression simulations to demonstrate the tension-compression asymmetry.

the hysteresis loop generated in tension. These features in the stress-strain responses are exhibited due to the influence of the J_3 parameter.

With the model calibrated, we perform a set of superelastic simulations in simple tension and simple compression under isothermal conditions at two other ambient temperatures : 288 K and 308 K. The stress-strain curves from these finite-element simulations are plotted in Figures 2a and 2b together with the simple tension and simple compression stress-strain curves obtained from the simulations conducted at an ambient temperature of 298 K (as also shown in Figures 1b and 1c). The stress-strain responses plotted in Figures 2a and 2b show that the stress required to induce austenite to martensite transformation or martensite to austenite transformation *increases* with increasing ambient/test temperature. This concurs very well with experimental findings (e.g. Thamburaja and Anand[15]).

Next we conduct a series of strain-temperature cycling simulations which can be described as follows : At a temperature of 320 K, the material is first pre-stressed to a desired stress level. With the pre-stress maintained, the temperature of the material is reduced to 220 K and then increased back again to 320 K. Depending on the pre-stress level, a transformation from the austenite to martensite phase will occur at a particular temperature as result of a sufficient reduction in temperature. At this point, a sufficient increase in temperature will then cause a transformation from the martensite to austenite phase to take place at another critical temperature.

In our finite-element simulations, we choose three different pre-stress levels under simple tension and simple compression conditions : 50 MPa, 100 MPa and 150 MPa. Figure 3a shows the strain versus temperature responses for the strain-temperature cycling simulations conducted under the aforementioned tensile stress levels. The strain versus temperature curves for the strain-temperature cycling finite-element simulations performed under the aforementioned compressive stress levels are plotted in Figure 3b. From the simulation results shown in Figures 3a and 3b, we can conclude the following trends : with increasing pre-stress level, a phase transformation from austenite to martensite or martensite to austenite occurs at a *higher* temperature. The results shown in Figures 3a and 3b also qualitatively reproduce the strain-temperature cycling experimental data shown in Thamburaja and Anand[15]. During the strain-temperature cycling, note that the transformation (actuation) strain obtained due to the tensile pre-stress is higher to that obtained using a compressive pre-stress. This is again due to the introduction of the J_3 parameter which takes different values under tensile or compressive stress states. Another point to note is that despite the amount of pre-stress and the sign of the pre-stress i.e. tensile or compressive, the temperature at which the martensite to austenite transformation occurs is always approximately 42 K higher than the temperature at which the austenite to martensite transformation takes place. To simulate the one-way austenite $\rightarrow$ martensite $\rightarrow$ austenite shape-memory effect, we perform the following finite-element calculations : With the temperature of the material initially at $\theta_1 = 265$ K where $\theta_{ms} < 265\,\text{K} < \theta_{af}$, we perform an *isothermal* simple tension and simple compression simulation to cause a transformation from austenite to martensite. When the material is fully-martensitic after the forward loading process, a reverse loading process to an unstressed state occurs through an elastic unloading of the material. With the applied stress in the material maintained at zero, the temperature of the material is then raised to 280 K. The stress-strain-temperature response from these finite-element simulations are plotted

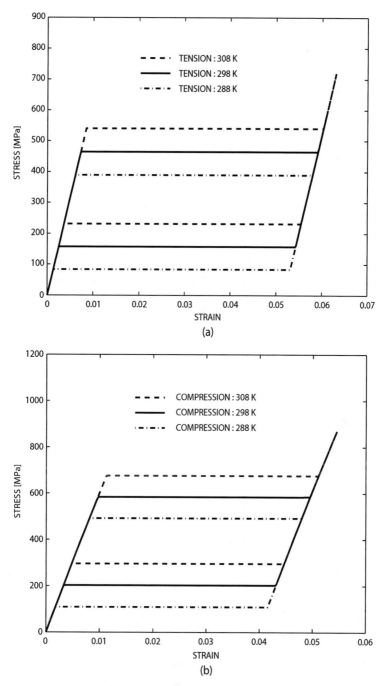

FIGURE 2. Simulated superelastic stress-strain responses in (a) simple tension and (b) simple compression at ambient temperatures of 288 K, 298 K and 308 K.

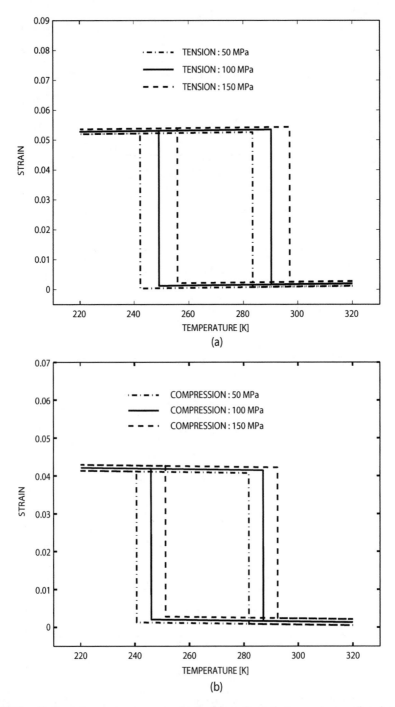

FIGURE 3.   Strain vs. temperature response obtained from the strain-temperature cycling simulations conducted under constant (a) tensile and (b) compressive stresses of 50 MPa, 100 MPa and 150 MPa. The thermal cycling is conducted between temperatures of 220 K and 320 K.

in Figure 4. Note that once the reverse loading process to an unstressed state has taken place, a residual strain will exist in the material as the temperature is not sufficiently high enough for reverse transformation from martensite to austenite to take place. However, as shown in Figure 4, the residual strain obtained from both these simulations during the deformation process at temperature θ_1 will be fully recovered once the temperature is raised above 276.6 K $> \theta_{af}$. Therefore, our constitutive model is able to qualitatively reproduce the one-way austenite $\rightarrow$ martensite $\rightarrow$ austenite shape-memory effect. To

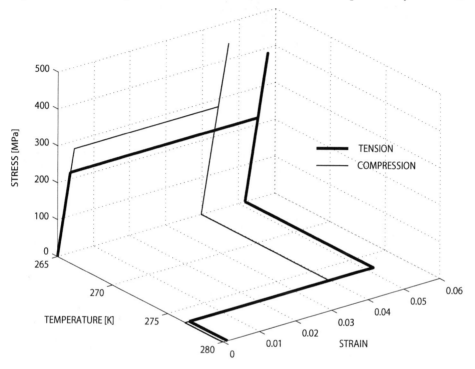

FIGURE 4. Stress-strain-temperature response obtained from the shape-memory effect simulations conducted in simple tension and simple compression. An isothermal stress-strain response from straining occurs at a temperature of 265 K. Following this, an increase in temperature to 280 K takes place.

investigate the non-isothermal behavior of shape-memory alloys during tensile superelastic deformation, we perform the following fully-coupled thermo-mechanical simulations : At an ambient temperature of 298 K, an initially-undeformed shape-memory alloy sheet with dimensions of 5mm by 20mm by 1mm is meshed using 200 ABAQUS C3D8RT elements as shown in Figure 5a. Each node on an ABAQUS C3D8RT element has three displacement and temperature degrees of freedom. The nodes on both ends of the specimen in the 1-3 plane act as grip sections, and their temperature is kept fixed at 298 K throughout the duration of the simulations i.e. the grips serve as a constant temperature bath. One grip section is constrained against motion along direction-2 and the other grip section is deformed along direction-2 under strain-rates of $1 \times 10^{-4} \, \text{s}^{-1}$ (Simulation A) and $5 \times 10^{-4} \, \text{s}^{-1}$ (Simulation B). Furthermore, heat convection from the outer surfaces of the sheet to the ambient environment (still air) is taken into account by setting the surface film heat transfer coefficient to be 12 W/m^{2}K. The stress-strain

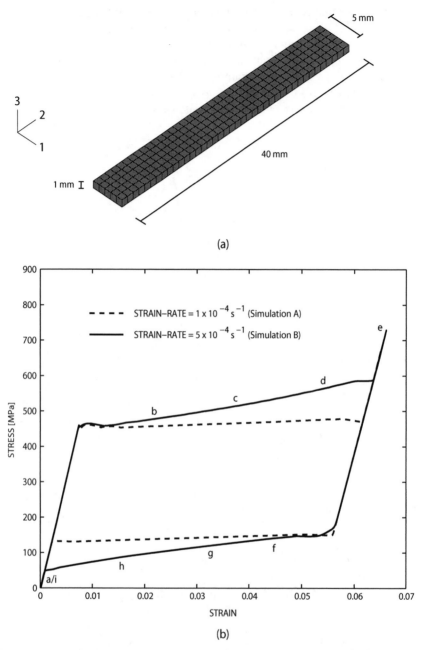

(a)

(b)

FIGURE 5. (a) Initially-undeformed finite-element mesh of an SMA sheet with dimensions of 5mm by 40mm by 1mm meshed using 200 ABAQUS C3D8RT elements. (b) Simulated tensile superelastic stress-strain response of the sheet shown in Figure 5a conducted under a strain-rate of $1 \times 10^{-4}\,\mathrm{s}^{-1}$ (Simulation A) and $5 \times 10^{-4}\,\mathrm{s}^{-1}$ (Simulation B). Both these simulations were performed using a fully-coupled thermo-mechanical analysis.

response from these two simulations using the initially-undeformed finite-element mesh shown in Figure 5a are plotted in Figure 5b. The contours of the martensite volume fraction in the sheet specimen obtained from Simulation B keyed to different points on its corresponding stress-strain curve is shown in Figure 6. Due to the boundary conditions imposed on the specimen as explained above, the austenite-martensite phase boundaries propagate from the grip sections to the specimen's center during the forward loading and reverse loading process. The contour plots presented in Figure 6 show the possibility of multiple austenite-martensite phase transformation fronts propagating in the specimen during superelastic deformation (cf. Shaw and Kyriakides[25]). Referring back to the stress-strain curves plotted in Figure 5b, Simulation B exhibits these following trends compared to Simulation A : (a) a wider hysteresis loop (measured along the stress axis), (b) a significantly larger hardening in the stress-strain response during the forward loading process; and (c) a significantly larger softening in the stress-strain response during the reverse loading process.

The causes for these observed trends are as follows : Recall that the energetic interaction coefficient h has been set to zero. Simulation A was conducted at a deformation rate which results in a nearly isothermal response i.e. the austenite $\leftrightarrow$ martensite phase transformations occur at nearly constant stress plateaus. However, Figure 7 show the contours of the temperature in the sheet specimen obtained from Simulation B keyed to different points on its corresponding stress-strain response shown in Figure 5b i.e. Simulation B was conducted at a strain-rate which results in a non-isothermal temperature field in the specimen during phase transformations. As shown in Figure 7, the temperature in the mid-section of the sheet increases by as much as 16 K above the ambient temperature (298 K) during the forward loading process. At a strain-rate of $5 \times 10^{-4}\,\mathrm{s}^{-1}$ the heat generated due to the *release* of the latent heat and mechanical dissipation is not conducted and convected out of the specimen quickly enough, and this results in the increase of the specimen temperature with respect to the ambient temperature. Thus, it is the increase in temperature which causes the hardening in the stress-strain response during the forward loading process as shown in Figure 5b.

Conversely, as also shown in Figure 7 the temperature in the mid-section of the sheet decreases to below the ambient temperature during the reverse loading process. At this deformation-rate i.e. $5 \times 10^{-4}\,\mathrm{s}^{-1}$ the heat loss in the specimen due to the *absorption* of the latent heat outweighs the amount of heat conducted and convected into the specimen, and hence causes a reduction in the specimen's mid-section temperature by as much as 14 K below the ambient temperature (298 K). Therefore, it is this decrease in temperature which causes the softening in the stress-strain response during the reverse loading process as shown in Figure 5b.

4. CONCLUDING REMARKS

A three-dimensional, finite-deformation-based and thermo-mechanically-coupled constitutive model to describe the austenite $\leftrightarrow$ martensite phase transformations in poly-crystalline SMAs has been developed. The theory was derived in a thermodynamically-consistent manner using the framework of classical isotropic metal plasticity. The constitutive model was also successfully implemented into the ABAQUS/Explicit[1] by

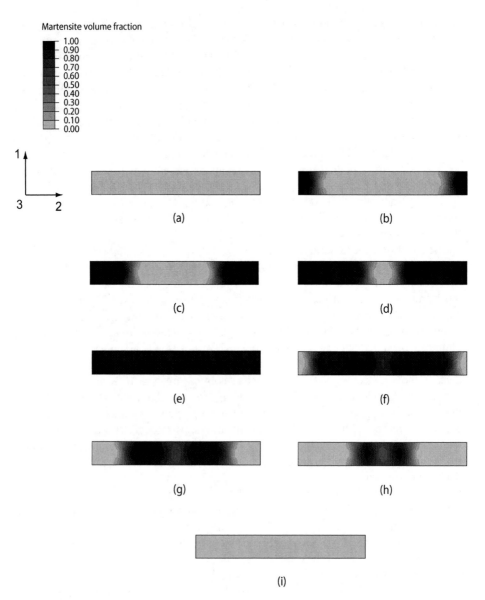

Martensite volume fraction

FIGURE 6. Contours of the martensite volume fraction keyed to various points of the stress-strain curve obtained from Simulation B as shown in Figure 5b. Due to the boundary conditions, both the forward and reverse phase transformations initiate from the ends and move towards the center of the specimen.

writing a robust time-integration procedure which was implemented as a user-material subroutine.

Representative superelastic stress-strain curves for experiments conducted in simple tension, simple compression and simple shear on a polycrystalline rod Ti-Ni were predicted to be in good accord by the constitutive model and the numerical simulations.

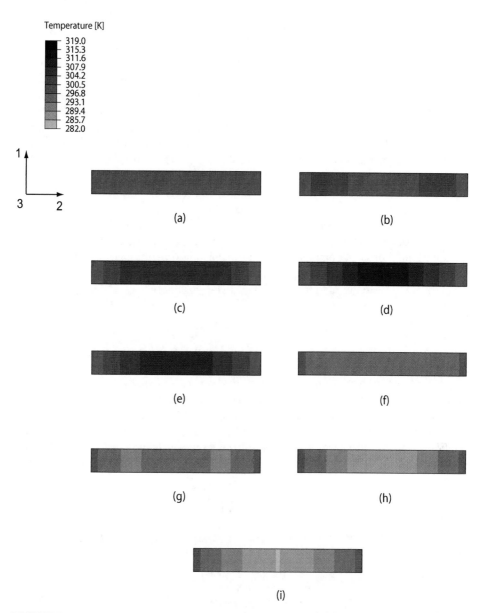

FIGURE 7. Contours of the temperature keyed to various points of the stress-strain curve obtained from Simulation B as shown in Figure 5b. During forward loading, the temperature in the specimen increases by approximately 16 K above the ambient temperature of 298 K due to the *release* of latent heat as a result of the austenite to martensite phase transformation. During reverse loading, the temperature in the specimen decreases by approximately 14 K below the ambient temperature of 298 K due to the *absorption* of latent heat as a result of the martensite to austenite phase transformation.

Furthermore, the constitutive model is also able to qualitatively reproduce the thermal cycling under constant stress and shape-memory effect experimental trends.

Finally, we also show the capability of our thermo-mechanically-coupled constitutive model in modeling the effect of strain-rate on the superelastic deformation of SMAs, namely : (a) the shape of the superelastic stress-strain responses depend *highly* on the deformation rate; (b) the apparent hardening in the stress-strain curves exhibited by SMAs during the austenite $\rightarrow$ martensite phase transformation is due to the inability of the released latent heat not being able to be conducted out of the specimen quickly enough, and (c) the apparent softening in the stress-strain curves exhibited by SMAs during the martensite $\rightarrow$ austenite phase transformation is due to the absorbed latent heat not being replenished quickly enough by the means of heat conduction into the specimen.

Some future work directions include : (1) The use of our present constitutive equations in modeling the behavior of shape-memory alloys under complicated multi-axial loading conditions e.g. combined tension-torsion, non-proportional tension-torsion conditions etc; (2) the modification of our locally-based constitutive model in order for the motion of the austenite-martensite interface to be accurately tracked during phase transformation.

5. ACKNOWLEDGEMENTS

PT would like to thank Profs. Srinivasan Sivakumar (IIT, Madras) and Arun Srinivasa (Texas A & M) for organizing the Smart Devices: Modeling of Materials (2008) conference. The financial support for this work was provided by the Ministry of Science, Technology and Innovation, Malaysia under Grant 03-01-02-SF0257. The ABAQUS finite-element software was made available under an academic license from HKS, Inc. Pawtucket, R.I.

REFERENCES

1. ABAQUS Reference Manuals, *Hibbit , Karlsson & Sorenson, Inc. Providence, R.I.* (2007)
2. K. Otsuka and C.M. Wayman, *Shape memory materials*, Cambridge University Press (1999)
3. C. Liang and C.A. Rogers, *J Intel Mat Syst Str*, **1**, 207-234 (1990)
4. R. Abeyaratne and J.K. Knowles, *J Mech Phys Solids*, **41**, 541-571 (1993)
5. A. Bekker and L.C. Brinson, *J Mech Phys Solids*, **45**, 949-988 (1997)
6. Q.P. Sun and K.C. Hwang, *J Mech Phys Solids*, **41**, 1-17 (1993)
7. J.G. Boyd and D.C. Lagoudas, *Int J Plasticity*, **12**, 805-842 (1996)
8. E. Patoor, A Eberhardt and M. Berveiller, *J Phys IV*, **6**, 277-292 (1996)
9. F. Auricchio, R.L. Taylor and J. Lubliner, *Comput Method Appl M*, **146**, 281-312 (1997)
10. Z.K. Lu and G.J. Weng, *J Mech Phys Solids*, **45**, 1905-1921 (1997)
11. T.J. Lim and D.L. McDowell, *J Eng Mater-T ASME*, **121**, 9-18 (1999)
12. P. Thamburaja and L. Anand, *J Mech Phys Solids*, **49**, 709-737 (2001)
13. Y. Jung, P. Papadopoulus and R.O. Ritchie, *Int J Numer Meth Eng*, **60**, 429-460 (2004)
14. L. Anand and M.E. Gurtin, *J Mech Phys Solids*, **51**, 1015-1058 (2003)
15. P. Thamburaja and L. Anand, *Acta Mater*, **51**, 325-338 (2003).
16. E. Kroner, *Arch Ration Mech An* , **4**, 273-334 (1960)
17. E.H. Lee, *J Appl Mech-T ASME*, **36**, 1-6 (1969)
18. L. Anand and M.E. Gurtin, *Int J Solids and Struct*, **40**, 1465-1487 (2003)

19. P. Thamburaja, *J Mech Phys Solids*, **53**, 825-856 (2005)
20. P. Thamburaja, H. Pan and F.S. Chau, *Acta Mater*, **53**, 3821-3831 (2005)
21. H. Pan, P. Thamburaja and F.S. Chau, *Int J Plasticity*, **23**, 711-732 (2007)
22. L. Orgeas and D. Favier, *Acta Mater*, **46**, 5579-5591 (1998)
23. D. Helm and P. Haupt, *Int J Solids and Struct*, **40**, 827-849 (2003)
24. B. Coleman and W. Noll, *Arch Ration Mech An*, **13**, 245-261 (1963)
25. J.A. Shaw and S. Kyriakides, *Acta Mater*, **45**, 683-700 (1997)

Discrete models for structural phase transitions

Srikanth Vedantam and S. Mohanraj

Department of Mechanical Engineering, National University of Singapore, Singapore 117576

Abstract. In this paper we study structural phase transitions from a microstructural viewpoint. The dynamic aspects of microstructure evolution is most easily studied in a discrete system. We begin by studying phase transitions in a discrete one dimensional system. The nearest neighbor interaction potential in this system is constitutively developed from a single oscillator model. The interaction potential has multiple local minima corresponding to the low temperature and high temperature phases. The local minima corresponding to the high temperature phase has a higher potential energy but a lower curvature than the global minimum of the potential energy of the low temperature phase. At low temperatures, the system is localized in the martensite minima but at higher temperatures the system is entropy-stabilized in the austenite phase.

Keywords: phase transformations, lattice models, hysteresis, free energy
PACS: 62.20.-x, 62.20.fg, 63.70.+h

INTRODUCTION

Most materials possess microstructure of some form or the other. It is well-known that the presence of fine length-scale sub-structures such as grains, twins, or second phase precipitates, strongly affects the constitutive response of materials. Historically, constitutive modeling approaches treated materials differing solely in their microstructure as different materials. Many currently technologically important materials such as shape memory alloys (SMAs) or nanocrystalline materials possess microstructure which evolves with either thermal or mechanical loading [1]. Structural phase transitions such as the martensitic transformations occurring in shape memory alloys generally involve microstructural features spanning multiple length scales. Since most applications involving use of SMAs are macroscopic [2], the presence of the smaller length scale microstructure must be taken into account for accurate prediction of the response of these materials. Any model designed to study these phenomena must incorporate information from the different length and time scales in some consistent manner. The classical continuum approach does not provide information regarding the role of the microstructure.

In order to describe the evolution of features at smaller length scales, some studies have augmented classical continuum theory with internal variables such as the volume fraction of the phases present [3]. Alternatively, higher gradients of strain have also been considered in order to provide the required length scale dependence [4]. These augmented theories still require specification of constitutive conditions in the form of kinetic relations to describe microstructure evolution either explicitly [3] or implicitly [5]. While the justification of these kinetic relations on the basis of empirical information has had some success [5], there is significant interest in determining them from an underlying physical description of the problem [6]. Generally discrete models such as molecular dynamics (MD) have been used to gather insight into the physical nature of

CP1029, *Smart Devices: Modeling of Material Systems, An International Workshop*
edited by S. M. Sivakumar, V. Buravalla, and A. R. Srinivasa
©2008 American Institute of Physics 978-0-7354-0553-0/08/$23.00

the smaller length scale problem. However, MD cannot describe the behavior of bulk samples at reasonable time scales due to the prohibitive computational cost. From this viewpoint, molecular dynamics coupled to continuum mechanics has been a preferred means of studying the multiscale problem [7].

There have been two approaches by which discrete and continuum scales have been coupled [8]. The first is the *concurrent* multiscale approach in which continuum and discrete descriptions of the body are simultaneously employed. In the regions of evolving microstructure, a discrete description is employed whereas in the remaining regions a continuum description is used. Such an approach requires transmission of information between the two descriptions as well as matching the length scales at the boundary. This "handshaking" problem has posed many difficulties so far [9],[10]. Moreover, the time scale is still limited by the MD time scales which are extremely small (on the order of a hundredth of the vibrational period of the atoms). Thus the computational costs of MD remain significant for this approach.

The second approach to multiscale modeling is the *sequential* approach (e.g. [6], [11]). In this, the problems at the different scales are solved independently and appropriate information is transmitted between the two models. The main advantages are that the physical insight at the different scales in obtained independently and can be used effectively. However, fundamental questions concerning the nature of the information required to be transmitted to the continuum scale remain unresolved. Also, the ability for a self-consistent approach in which the continuum information is transmitted back to the discrete scale is lost.

In both kinds of multiscale approaches, the discrete model poses the most significant challenges. Perhaps the most important challenge is in the specification of an adequate interatomic potential. Interatomic potentials for most materials are usually empirically specified. The lattice spacing and elastic moduli are used as constitutive inputs into the interatomic potentials. However, in the case of materials which undergo structural phase transitions, the ability to account for multiple stable lattice structures is essential. In addition, the latent heat of transformation found in first order transitions should be present. For this reason, we focus our study on the nature of the interatomic potential which permits description of a thermal first order phase transition. We being by studying a single oscillator model and extrapolate the model to a discrete one dimensional chain.

DISCRETE MODEL

As mentioned earlier, discrete mechanical models are useful in describing microstructural evolution and their effects on dynamic aspects of phase transitions. Two types of discrete models which have proved useful are MD and lattice dynamics (LD) approaches. For the solid-solid diffusionless phase transitions considered here, the LD approach is sufficient since there is no long range motion of the atoms. In the LD approach the nearest neighbors of every atoms are specified and considered to be fixed.

Determination of an appropriate interatomic potential for a given system is one of the major challenges of MD or LD simulations. This is particularly true for material undergoing phase transitions and in the following sections we systematically develop an appropriate interatomic potential for thermal structural phase transitions.

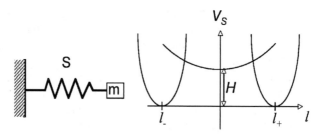

FIGURE 1. A schematic of the single oscillator model (left) and the spring potential V_S (right).

Properties of empirical interatomic potentials

Preliminaries

We begin development of our discrete constitutive model by studying a single oscillator. Consider a mass m connected to a rigid support by means of a spring S (Figure 1). The total energy of the system is

$$E = \frac{1}{2}m\dot{\ell}^2 + V_S(\ell) \tag{1}$$

where ℓ is the current length of the spring.

Mimicking a solid capable of undergoing structural phase transitions we assume the spring has two equilibrium states ℓ_+ and ℓ_-. Further, we seek a situation in which an intermediate state (located symmetrically between the + and - states) $\ell_o = \ell_+ - \delta = \ell_- + \delta$ is stable at high temperatures but not at lower temperatures. A form of the potential for such a situation is given by (Figure 1)

$$V(\ell) = \begin{cases} \frac{1}{2}k_m(\ell - \ell_-)^2, & \ell < \ell_o - \alpha \\ \frac{1}{2}k_a(\ell - \ell_o)^2 + H, & \ell_o - \alpha \le \ell \le \ell_o + \alpha \\ \frac{1}{2}k_m(\ell - \ell_+)^2, & \ell > \ell_o + \alpha \end{cases} \tag{2}$$

The side wells represent the low temperature phase of the material termed martensite and the central well is the high temperature phase of the material, termed austenite.

The equation of motion for this system is then given by

$$m\frac{d^2\ell}{dt^2} = \begin{cases} -k_m(\ell - \ell_-), & \ell < \ell_o - \alpha \\ -k_a(\ell - \ell_o), & \ell_o - \alpha \le \ell \le \ell_o + \alpha \\ -k_m(\ell - \ell_+), & \ell > \ell_o + \alpha \end{cases} \tag{3}$$

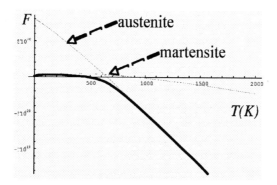

FIGURE 2. The free energy vs temperature of the single oscillator model (bold line). The light lines show the free energy of the pure austenite and martensite phases.

We will assume continuity of velocity at $\ell = \ell_o \pm \alpha$. The parameter α is determined by continuity of V:

$$\alpha = \frac{\delta - \sqrt{\tau\delta^2 + (2H/k_m)(1-\tau)}}{1-\tau}, \quad \tau = k_a/k_m \qquad (4)$$

We introduce the following non-dimensional quantities:

$$\bar{e} = \bar{\ell} - 1 = \frac{\ell - \ell_o}{\ell_o}, \quad \bar{\alpha} = \frac{\alpha}{\ell_o}, \quad \bar{\delta} = \frac{\delta}{\ell_o} \qquad (5)$$

Substituting the above into Eq. we obtain

$$\frac{d^2\bar{e}}{d\bar{t}^2} = \begin{cases} -\bar{k}_m(\bar{e}+\bar{\delta}), & \bar{e} < -\bar{\alpha} \\ -\bar{k}_a\bar{e}, & -\bar{\alpha} \leq \bar{e} \leq \bar{\alpha} \\ -\bar{k}_m(\bar{e}-\bar{\delta}), & \bar{e} > \bar{\alpha} \end{cases} \qquad (6)$$

where we have set $\bar{k}_{m,a} = (k_{m,a}\ell_o^2/\varepsilon_o)$ and $\bar{t} = t\sqrt{m\ell_o^2/\varepsilon_o}$ with $\varepsilon_o = k_B T_o$ (k_B is the Boltzmann constant and T_o is some reference temperature).

For convenience, we will drop the bars for the rest of the paper. In this non-dimensional setting it can be seen that the temperature of the single particle is given by its (twice) average kinetic energy

$$T = \langle (de/dt)^2 \rangle. \qquad (7)$$

Partition function and thermodynamic quantities

The partition function for the single degree of freedom system is given by

$$\mathscr{L} = \int_{-\infty}^{\infty} \int_{-\infty}^{\infty} \exp(-E/k_B T)\,d\dot{\ell}\,d\ell. \qquad (8)$$

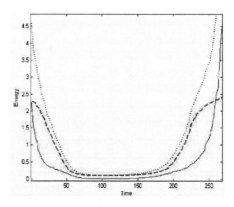

FIGURE 3. Typical plot of the average energy vs time for a cooling followed by heating cycle. The dotted line is the total energy. The dashed line represents the average potential energy and the solid line represents the average kinetic energy.

For the piecewise parabolic energy E assumed above, the partition function can be calculated explicitly.

The free energy of this system can then be calculated from the partition function using

$$F = -k_B T \ln \mathscr{Z}. \tag{9}$$

A plot of the free energy with temperature for $\tau \ll 1$ shows the transition between austenite and martensite phases (Figure 2). The transition is sharper for smaller values of τ. As a guide, the free energy curves of the pure martensite and austenite phases are also shown.

At low temperatures, the atom oscillates in the side wells and represents the martensite phase. At higher temperature, the vibrational energy increases and the atom traverses the three wells. However, since the curvature of the central well is smaller than that of the side wells, the time spent in the central well is longer and this represents the austenite phase. Even though the average potential energy in this phase is higher than the martensite side wells, the entropy is correspondingly higher and hence the free energy of the austenite phase is lower at higher temperatures.

In this single oscillator model, there are no dissipation mechanisms and hence there is no hysteresis. In real materials, hysteresis occurs through the dispersion of energy. This can happen only in a coupled system and we next consider the simplest such system: a one dimensional chain of atoms.

Phase transitions in one dimensional chains

We consider a chain of N atoms each connected to its nearest neighbors through nonlinear springs of the type given in the single oscillator model. Let the displacement of the i-th atom from a fixed point be given by u_i. The equation of motion for each atom

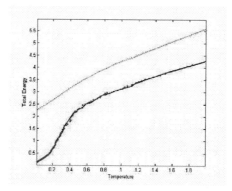

FIGURE 4. Heating and cooling curves for different *tau*. The transition is seen to be sharper for $\tau \ll 1$ (lower solid line and dots) than for $\tau = 1$ (the upper dotted curve). The upper curve is vertically offset upwards for clarity.

is

$$m\ddot{u}_i = -\frac{\partial V(u_{i+1} - u_i)}{\partial u_i} - \frac{\partial V(u_{i-1} - u_i)}{\partial u_i}. \tag{10}$$

The potential energy considered above arises purely from the nearest neighbor interactions. This allows a situation in which pairs of atoms could alternate between the two variants of martensite allowing fine twinning. In other words, there is no energy associated with twin or phase *boundaries* in the above description. In order to introduce domain wall energy we consider an addition potential energy contribution of the form

$$V_{nn} = \frac{k}{8}(u_{i+1} - 2u_i + u_{i-1})^2 \tag{11}$$

This contribution provides a gradient energy since it arises from the difference in strains between adjacent pairs of atoms. The equations of motion with the gradient contribution are now

$$m\ddot{u}_i = -\frac{\partial V(u_{i+1} - u_i)}{\partial u_i} - \frac{\partial V(u_{i-1} - u_i)}{\partial u_i} + \frac{k}{2}(u_{i+1} - 2u_i + u_{i-1}) \tag{12}$$

We solve the equations of motion numerically using the velocity verlet algorithm. The end atoms experience only the forces of the internal atoms. The atoms are initially given random velocities and positions around the austenite phase. The temperature is controlled through velocity scaling. The velocity scaling is performed over a few hundred steps. Averages of kinetic and potential energies are calculated over 10^4 steps. The velocities are first scaled down and after the atoms are localized to the side wells the velocities are then scaled up representing a cooling of austenite followed by heating of martensite after transition.

Figure 3 shows a typical plot of the average potential and kinetic energies in the cooling and heating cycle. It can be immediately noted that equipartition of the total energy into the kinetic and potential energies is violated due to the anharmonicity of the potential energy.

55

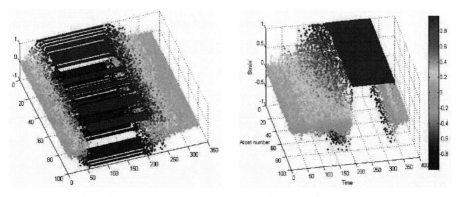

FIGURE 5. Strain distribution along the chain is shown with time during a cooling and heating curve. Zero strain represents the austenite phase whereas strains of ± 1 represent the two variants of martensite. The figure on the left is for $k = 0$. The absence of twin boundary energy allows the presence of fine scale twins. The figure on the right is for $k = 1$ and the high energy penalty for twin boundaries allows the formation of a single variant.

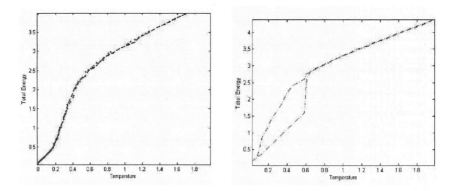

FIGURE 6. The total average energy(internal energy) vs temperature for a cooling and heating cycle for $k = 0$. The absence of non-nearest neighbor interaction does not allow dispersion of energy and hysteresis is negligible. The figure on the right shows the cooling and heating curves for $k = 1$. The presence of non-nearest neighbor interactions allows the dispersion of energy and this is manifested as hysteresis. The line to the right of the hysteresis loop is the heating curve whereas the line to the left is the cooling curve.

The effect of the curvature of the well corresponding to the high temperature phase is shown in Figure 4. The dotted line shows the total energy of the system (the internal energy) vs the temperature for heating and cooling for $\tau = 1$ (the curve is offset vertically upwards for clarity). The solid line and dots show the heating and cooling curves for $\tau \ll 1$. It can be seen that the transition is sharp when $\tau \ll 1$ and the model more closely represents a first order transition.

The presence of the gradient interactions penalizes the formation of twin boundaries as seen in Figure 5. Initially, the chain is in the austenite phase and the strain is close to zero. As the chain is cooled, martensite begins to form. The two variants of martensite are indicated with strains of ± 1. Subsequently, heating the chain causes a reverse transformation to austenite. In the simulation shown in the left of Figure 5, the boundary energy is set to zero (by setting $k = 0$) and this leads to the formation of fine twins. The resultant pattern formed is random, and depends on the initial conditions and the rate of cooling. However, the average length scale of the twins depends on the twin boundary energy. The simulation shown on the right of Fig 5 is performed with $k = 1$ and this high boundary energy causes the formation of only a single variant.

Another consequence of the gradient energy is the dispersion of energy which is manifested as hysteresis. The internal energy is plotted versus the temperature for $k = 0$ on the left and $k = 1$ on the right in Figure 6. It is seen that when $k = 0$, the hysteresis is negligibly small whereas when $k = 1$ a finite hysteresis is present. This is due to the dispersion of energy in the form of lattice waves. When an atom in the chain transforms from austenite to a variant of martensite, the increase in energy due to the formation of a phase boundary transforms to kinetic energy which is propagated along the chain. This dispersal of energy from a transformation event is not symmetric in the heating and cooling processes which results in hysteresis.

CONCLUSIONS

We have presented a discrete lattice model for thermally driven structural phase transitions. The key feature of the interatomic potential for a phase transforming material is that it has high curvature and low energy at the martensite interatomic separation and low curvature and high energy at the austenite interatomic separation distance. A calculation of a single uncoupled oscillator with this potential energy shows that the free energy of austenite is lower at higher temperatures. Phase transitions are thus driven by the higher entropy contribution for the austenite phase arising from the low curvature of the potential energy. In a one dimensional discrete chain, this potential demonstrates the sharp transition found in the first order phase martensitic phase transformations. A gradient term is introduced to penalize the presence of twin and phase boundaries. The gradient term disperses energy in the chain and this allows hysteresis in this system.

REFERENCES

1. K. Bhattacharya, *Microstructure of Martensite: Why it Forms and How it Gives Rise to the Shape Memory Effect*, Oxford University Press, London, 2003.
2. T. W. Duerig, K. N. Melton, D. Stockel and C. M. Wayman (Eds.), *Engineering Aspects of Shape Memory Alloys*, Butterworth-Heinemann, London, 1990.
3. P. Thamburaja, *J. Mech. Phys. Solids* **53**, 825–856 (2005).
4. R. Abeyaratne, *SIAM J. Appl. Math.* **51**, 1205–1221 (1991).
5. S. Vedantam, *Smart Mater. Struc.* **15**, 1172–1178 (2006).
6. R. Abeyaratne and S. Vedantam, *J. Mech. Phys. Solids* **51**, 1675–1700 (2003).
7. V. B. Shenoy, *Bull. Mater. Sci.*, **26**, 53–62 (2003).
8. R. E. Rudd and J. Q. Broughton, *Phys. Rev. B* **72**, 144104 (2005).
9. E. B. Tadmor, M. Ortiz and R. Phillips, *Phil. Mag. A* **73**, 1529–1563 (1996).
10. V. B. Shenoy, R. Miller, E. B. Tadmor, D. Rodney, R. Phillips and M. Ortiz, *J. Mech. Phys. Solids* **47**, 611–642 (1999).
11. L. Truskinovsky and A. Vainchtein, *Phil. Mag.* **85**, 4055–4065 (2005).

A Simple, Gibbs Potential Based Multinetwork Model for Shape Memory Polymers

Arun R. Srinivasa and P. Gosh

Dept. of Mechanical Engineering, Texas A&M University, College Station, Texas, USA 77843

Abstract. The aim of this paper is to demonstrate a Gibbs potential based approach for the development of the constitutive equations for a shape memory polymer where the shape retention and recovery upon heating is controlled by a glass transition process. The approach is motivated by the use of a simple spring-dashpot type analogy and the resulting equations are classified as *state equations* for the thermoelastic spring like response and a *kinetic equation* for the frictional dashpot. The governing equations are written in a state evolution form as a set of ordinary differential equations to be solved. The entire set of equations are developed starting from the basic conservation laws for a wire together with the laws of thermodynamics. The solution for a simple thermomechanical shape setting and recovery cycle is obtained using Matlab. It is shown that the results are in qualitative agreement with experiments performed on polyurethane.

Keywords: shape memory polymers, gibbs potential, multinetwork model, shape fixity

INTRODUCTION

The aim of this brief paper is to simulate the behavior of shape memory polymers (see e.g. Lendlein [3] for a survey of shape memory polymer technology) using a model with the barest minimum of constitutive parameters that will capture the behavior these polymers. These polymers are currently being investigated for a wide range of applications where their shape setting and shape recovery behavior is useful. Essentially, the canoical behavior of these polymers can be described as (1) they are elastomeric at temperatures above around 350 K and can be deformed by large amounts (2) If they are deformed to a particular shape, the shape fixed and the temperature lowered, they will retain their new shape, i.e., they can be "molded". (3) Subsequently, upon heating they will recover to their original shape. This process can be used to advantage in a variety of industrial processes. However, due to their relatively low modulii and large strains, care must be taken to ensure that the "spring back" during the molding process is accounted for in designing these materials and that the stresses and strains developed can be accurately characterized for providing design guidelines. A few theories of such shape setting behavior have been developed (see e.g. Tobushi [5]) but many of them are not amenable to subsequent incorporation into the control systems of a device. Moreover, a number of parameters have to specified as complex functions of temperature and thus it

CP1029, *Smart Devices: Modeling of Material Systems, An International Workshop*
edited by S. M. Sivakumar, V. Buravalla, and A. R. Srinivasa
©2008 American Institute of Physics 978-0-7354-0553-0/08/$23.00

becomes exceedingly difficult to evaluate these materials from a design and control point of view. The current paper shows that most of the features of a SMP can be obtained by the use of only ONE temperature dependent parameter (the yield stress). We point out that, we are NOT claiming that the other parameters are independent of temperature but only that their temperature dependence does NOT play a crucial role in the shape setting and recovery behavior. This allows us to develop a "minimalist theory of SMPs" for design purposes. Furthermore the equations are given in *state evolution form* rather than as integral equations (as is more common in the nonlinear viscoelasticity literature) with an eye to make it of practical use to control systems designers. It allows the constitutive behavior to be incorporated into control systems toolboxes such as SIMULINK.

We demonstrate that a systematic application of fundamental thermodynamics principles coupled with a suitable choice for the Gibbs' potential and a suitable kinetic equation for the inelastic variables will allow us to develop a complete simulation of the behavior of shape memory polymeric materials.

PRELIMINARIES

Consider a cable or wire (i.e., a one-dimensional continuum) made up a shape memory polymer, lying along the x axis of a coordinate system. Let us, for definiteness, assume that the temperature is high enough that the material is in its rubbery state. Let the mass density (per unit length in the reference configuration be (ρ_r)The position of a particle in this wire at time $t = 0$ is indicated by X. Assume that the cable elongates due to a combination of external stimuli so that, a any other time t, the particle which was at X is now at x. Thus, the motion of the continuum, its deformation gradient and its velocity are given by

$$x = \mathbb{F}(X,t), F(X,t) = \frac{\partial \mathbb{F}}{\partial X}, v(X,t) = \frac{\partial \mathbb{F}}{\partial t}. \tag{1}$$

As is standard practice, we introduce the mass per unit length ρ, the axial force T, the internal energy per unit length u, the axial hear flux q and the lateral rate of heat transfer per unit length r, and, in the absence of body forces, we record the conservation laws of mass, momentum and energy (each measured per unit length in the reference state, in the forms

$$\rho = \rho_r/F, \tag{2}$$

$$\rho_r \dot{v} = T' \tag{3}$$

$$\rho_r \dot{u} = T\dot{F} - q' + r. \tag{4}$$

where the symbols $(\cdot)$ stands for the time derivative holding X fixed and $(\cdot)'$ stands for the derivative with respect to X. These can be derived by considering a small line element and applying the mass, momentum and energy balance to them and then by considering the limit as the length of the line element goes to zero.

Notice that, in principle, the above three equations provide evolution equations for the three unknowns—mass, velocity and the internal energy once the forms for the stress T, the heat flux q and the lateral heat supply r are known. The position of the particles can be obtained from these by simply integrating the velocity Of these, the latter two are usually very well known and are the fourier's Law of heat conduction (for q) and Newton's Law of cooling for r. But both of these require the introduction of yet another variable, the Temperature.

In this work, rather than proceed in an ad-hoc fashion by introducing the equations necessary, we proceed in a systematic manner by using the principles of thermodynamics and especially the second Law of thermodynamics to derive the forms of these equations.

THE EQUATION OF STATE AND SOME OF ITS CONSEQUENCES

The state variables

We will view the SMP as a dynamical system (i.e., one whose state changes are given by a suitable differential equation) and, borrowing terminology from the control systems literature, we will consider a state space approach to the problem. Thus, we are going to consider a class of models for the material whose general form is

$$\dot{\mathbb{S}} = \mathbb{G}(\mathbb{S}) \tag{5}$$

where $\mathbb{S}$ is a column vector of variables that model the current state of the system and $\mathbb{G}$ is a function of the state. The first step in this process is the determination of a list of state variables of the system.In order to do so, and to get an idea of how the material behaves we need to consider the response of the SMP in greater detail.

A preliminary list of state variables are given by the density ρ, the deformation gradient F, the internal axial force T, the internal axial heat flux q, and the temperature θ. Of these, we note that ρ is not an independent state variable since it is directly related to F through (2). Similarly, the velocity v and the internal energy u do not figure in the *independent state variable list* since, in principle, they can be computed from the balance laws.

A simple mechanical model for the SMP behavior

In order to provide a "physically intuitive" model of SMP behavior, me now proceed to develop a simplistic "cartoon-like" model of the behavior of a shape memory polymer. To begin with, we will focuss attention on SMPS that are actuated by glass transition. For these materials, let us assume that at high enough temperatures, the response is akin to that of an amorphous crosslinked polymer (seen as solid circles in Fig 1). To model this, we can create a "network" by joining each cross-linked location (node) with a certain number of its nearest neighbors (see for example Arruda and Boyce[1]) . As the wire is stretched, the network deforms due to the partial uncoiling of the polymeric chains between the junctions. These networks form the permanent backbone or skeleton of the polymer are ultimately responsible for its shape recovery.

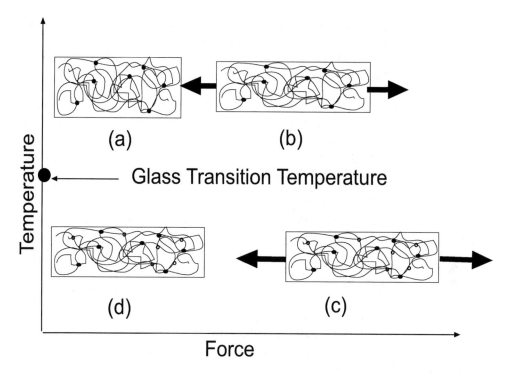

FIGURE 1. A schematic representation of the shape fixing effect of a rubbery polymer around the glass Transition. Figure (a) The stress-free rubbery polymer with the permanent crosslinks shown as dark circles. From (a) to (b) the polymer is stretched above T_g. From (b) to (c), the polymer is cooled below T_g by applying forces to keep the shape fixed. The force needed increases due to the prevention of thermal contraction. During this cooling process, new temporary physical links are form (show by open circles. These will "freeze" the rubbery network. (c) to (d): the forces are removed and the temporary shape is retained (approximately , there will be a slight "springback". (d)-(a): If the polymer is heated to above T_g the temporary links disappear and the material rebounds back to its original shape.

However, as the material is cooled, the mobility of the chains decreases and the rather small electrostatic attraction between the individual chains is sufficient to hold parts of the chains immobile for much longer periods of time, i.e., the chains exhibit "stickiness", so much so that below a certain temperature the chains lose their rubbery character and are "frozen". the material becomes a glassy polymer. We model this phenomenon, by introducing—apart from the permanent crosslinks described in the previous paragraph—certain temporary junctions (shown in as open circles in Figure 1) . The network formed by these temporary junctions forms a second temporary network that interpenetrates the first, permanent network. The Junctions are not fixed as in the previous case but are constantly breaking and reforming. Particularly they can be completely eliminated by heating and reformed at different locations by cooling. It is the temporary junctions that are responsible for the polymer to retain a temporary "frozen" shape upon cooling. In other words, if we raise the polymer to a high enough temperature to eliminate all the temporary junctions, deform it and cool it, new junctions are formed in the deformed state and cause the deformed state to be "locked in". The exact state of deformation that is "locked in" depends upon the relative effects of the permanent junctions which will try to return the polymer to the original state and the temporary junctions that try to keep it in the deformed state. Of course, in reality, due to thermal fluctuations, the temporary junctions will occasionally break and reform, giving rise to "creep" of the polymer with time.

Subsequent heating will "unlock" the polymer by eliminating the junctions. Then the permanent junctions will take over and return the polymer to its original state. Of course, This is an extraordinarily simplistic view of the response, but as we shall see in the sequel, this is adequate to create a reasonable macroscopic model which captures many of the features of the response of SMPs to various thermomechanical loads.

In view of this microscopic cartoon, we will begin by assuming that the two interpenetrating networks act like springs in parallel. Furthermore, the fact the temporary network can be broken and reformed and its "stickiness" is represented by the presence of a dashpot in series with one of the springs (see figure 2). At high temperatures, the frictional element has essentially zero friction (representing the materials ability to easily break the temporary bonds, so that for all practical purposes the second spring is "slack" at high temperatures. At low temperatures, the temporary bonds come into play and the frictional dashpot exhibits a high frictional threshold, causing the system to lock. It is clear that any model that we develop should have built in (a) two elastic modular: one which is apparent at low temperatures and another at high temperatures and (b) some kind of "yield" behavior which is temperature sensitive so that it allows strains to be "locked in" at low temperatures.

Figure (2a,b represent two possible models that have many of these features. In both models there are two springs : a rubbery spring and a glassy spring and a frictional element representing the "yield" behavior. They differ in their connectivity. In model (a) they are connected in series with an intervening frictional element while in model (b) they are in parallel. There are other models that can be obtained from just these three elements. It can be shown that *either of these models can be used to simulate the behavior of the SMP for some classes of responses* However, the authors have found that *model (a) appears to be simpler and more robust, from the point of view of material parameters* Indeed, with model (a) almost all the features of SMP response can be explained on the basis of just thermal expansion and temperature dependent yield stress, whereas in Model (b) needs to have a much more complex temperature response to be able to model the SMP behavior. Thus for this paper, we will focus only on Model (a).

While it is tempting to convert this mechanical model directly into a set of response functions, it is extremely dangerous to do so, since the properties are all very sensitive to temperature in the operational range and, because of the strong thermomechanical coupling, care must be taken in maintaining thermodynamical consistency (namely the impossibility of perpetual motion of various kinds).

We however notice that *in this model (model (b) in figure 2* the total force on the system is the sum of the forces due to the two networks whereas in Model (a) this is not the case.

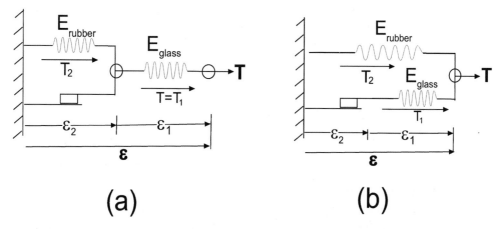

FIGURE 2. Two different spring dashpot models representing two possible candidates for SMP behavior. In Model (a), the springs are in series and in model (b) they are in parallel. We will use Model (a) since it seems to simulate the response in a simpler fashion. The stresses T_i and the strains ε_i corresponding to each spring in the model are also shown in the figure.

Thermodynamical Considerations

The Gibbs potential and the heat equation The foundation of the thermodynamical approach that is to be presented here is that, the non-dissipative properties of the material are derivable from a single potential, namely the Gibbs potential for the continuum, while the dissipative properties are postulated in the form of a kinetic equation. In the current case, we note that there are two networks each to be considered as a thermodynamical system in its own right. We shall thus assume that the Internal energy and the Gibbs potential is the sum of those for the two networks. Thus if u and $G_)$ is the internal energy and the Gibbs potential of the networks, we assume that[1]

$$G = G(T_i, \theta) \tag{6}$$

$$u = G - T_1\frac{\partial G}{\partial T_1} - T_2\frac{\partial G}{\partial T_2} - \theta\frac{\partial G}{\partial \theta} \tag{7}$$

where T_1 and T_2 are the stresses in the two springs (see figure 2). The definition of The Gibbs potential (see Callen ..) also implies that the entropy per unit mass is given by

$$s = -\frac{\partial G}{\partial \theta}. \tag{8}$$

---

[1] We will not discuss the reasons behind the assumptions here, referring the reader to e.g.,Callen [2]. for a detailed discussion of these issues.

. Now, by substituting equation (7) into the left hand side of (4) and using (8), we arrive at the following "heat equation"

$$\theta \dot{s} = T\dot{F} + T_1 \frac{d}{dt}\frac{\partial G}{\partial T_1} + T_2 \frac{d}{dt}\frac{\partial G}{\partial T_1} - q' + r. \tag{9}$$

The left hand side of the equation is the temperature times the rate of increase of entropy, which way be called the "rate of heating", while the right hand side has contributions from the axial heat flux q, the lateral heating r as well as the sum of three terms (involving the axial forces) that arise from mechanical working. This sum represents the "heating due to mechanical effects". It is this sum, that we are interested in since it is representative of the thermomechanical coupling.

The dissipation of mechanical power A close look at the first three terms in (9) reveal that the first term represents the mechanical power (or deformation power) supplied to the wire, while the second two terms represent the rate of decrease in recoverable mechanical work. We thus assume that the sum of these terms represent the net mechanical power dissipated (i.e., lost irretrievably) by the system. Note that the *energy* is not lost, but it is unavailable as work. We hence introduce the rate of dissipation ξ through

$$T\dot{F} + T_1 \frac{d}{dt}\frac{\partial G}{\partial T_1} + T_2 \frac{d}{dt}\frac{\partial G}{\partial T_2} = \xi \geq 0 \tag{10}$$

The above equation is referred to as the dissipation relation. The fact that ξ is non negative is an assumption that will allow us to satisfy the second law of thermodynamics. It is usually considered as the "strong" form of the second law , since it (together with fourier's law of heat conduction) is sufficient but not necessary for the satisfaction of the second law.

The form of the Gibbs potential shown above can be used to represent any simple mechanical model composed of two springs in any combination. For example,if we consider the simple mechanical model introduced for motivation and (where the two networks are represented by two springs in parallel) assume that $T = T_1 + T_2$, so that the equation(10) can be regrouped into

$$T_1 \frac{d}{dt}\left\{F + \frac{\partial G}{\partial T_1}\right\} + T_2 \frac{d}{dt}\left\{F + \frac{\partial G}{\partial T_2}\right\} = \xi \geq 0 \tag{11}$$

Thus the total rate of dissipation of mechanical power is also split into two parts: the dissipation due to the first or "backbone" network which is the first term on the left hand side of (11) while the second term on the left hand side of (11) represents the mechanical power dissipated by the temporary network.

On the other hand, if we were to consider the arrangement 2 (springs in series) and assume that $T = T_1$ Then we can write (10) as

$$T\frac{d}{dt}\left\{F + \frac{\partial G}{\partial T_1} + \frac{\partial G}{\partial T_2}\right\} + (T_2 - T)\frac{d}{dt}\left\{\frac{\partial G}{\partial T_2}\right\} = \xi \geq 0 \tag{12}$$

Here again the rate of dissipation is split into two parts, one corresponding to the first spring and the other corresponding to the second spring. Various other arrangements are possible. The point to note here is that *we have to choose a particular arrangement to reflect the elastic response of the system, motivated by our intuitive understanding of the system.* The authors have simulated the response corresponding to both these networks and have arrived at the conclusion that the network represented by (12) appears to simulate the response in a simpler manner with far fewer assumptions and with much simpler constitutive assumptions. For this reason, in the remainder of this article, we will assume that the response is to be simulated by using (12) and not (11).

In any case, separate assumptions have to made regarding the dissipative response using the notion of a kinetic equation. In any case, the fundamental entities are always the stresses and the total deformation. Ideas such as plastic or locked in strains etc. will appear only as convenient definitions.

A special simple form for the Gibbs potential

Rather than continue to develop the theory in general, we will illustrate the procedure for the development of the theory by choosing a special form for the Gibbs potential which will make the approach quite transparent and simple. To this end, we note that typical shape memory polymers are capable of sustaining only very low forces although the deformations can be of the order of 100% or higher. Further more they exhibit very strong temperature dependence. In view of this, we will assume that the Gibbs Potential is at most quadratic in the forces, i.e., we assume that G is of the form

$$G = C_0(\theta) + C_{11}(\theta)T_1 + C_{12}(\theta)T_2 + C_{21}(\theta)T_1^2 + C_{22}(\theta)T_2^2 \qquad (13)$$

Note that, in keeping with the requirement that the Gibbs potential be separable into two, we have included no cross terms of the form $T_1 T_2$ in the equation. substituting this into the equation (12), we get, for the series model (where $T = T_1$),

$$T\frac{d}{dt}\{F + C_{11}(\theta) + C_{21}T - \varepsilon_2\} + (T - T_2)\dot{\varepsilon}_2 = \zeta \text{ where } \varepsilon_2 = -\{C_{12}(\theta) + C_{22}T_2\} \quad (14)$$

In order to understand the terms that appear in the above equation, we note that if $\zeta = 0$ i.e., *if there is no damping* then the mechanical model suggests that $T = T_1 = T_2$ and so

$$\frac{d}{dt}\{F + C_{11}(\theta) + C_{21}T_1 - \varepsilon_2\} = 0 \qquad (15)$$

similarly, if $\zeta = \infty$, i.e., that the damper were essentially "locked" then the mechanical model suggests that $\varepsilon_2 = 0$, so that T_2 is indeterminate and irrelevant to the problem. Thus we can identify

1. $-C_{11}$ and $-C_{12}$ represent the thermal strains in the two networks, and for simplicity, we will assume them to be linear in the temperature i.e., $C_{1i} = -\alpha_i(\theta - \theta_0)$ where α_i are the coefficients of linear expansion of each of the two networks, and θ_0 is some reference temperature.

2. $-C_{21}$ and $-C_{22}$ are the compliances of the two networks . In this simplistic model, we will make a rather assumption that they are independent of temperature. Thus the model will be built on the idea that it will have essentially *two elastic responses*: one for low temperature (below Tg) and another for the high temperature.

65

Furthermore, we will assume that the glassy response is non-dissipative and that the rubbery response is the one that is the one that is dissipative and hysteretic, Accordingly, we will assume that the stress T_2 is due to the rubbery response, whereas T_1 is associated with the glassy response, The dashpot being a frictional dashpot that is temperature sensitive, i.e., , below the glass transition, the frictional dashpot *locks*, and prevents the rubbery network from deforming further, leading to a "locked in" strain. On the other hand, at high temperature, the dashpot ceases to operate (i.e., its resistance vanishes) and the material behaves like an elastic rubber.

Thus, we expect that the terms involving T_1 (the glassy network) will not appear in the rate of dissipation equation (14). We accomplish this in a way that also represents the elastic response of this network, i.e., we simply set the terms in the first bracket in equation (14) equal to zero and obtain

$$e = -C_{11} - C_{21}T_1 + \varepsilon_2. \tag{16}$$

The terms are easy to identify: the first is the thermal strain of the whole network, the second is the elastic strain of the glassy network and the second is the elastic strain of the rubbery network.

For the second spring, we note that since the temporary networks continually break and reform in new configurations, there is considerable dissipation. We therefore assume that

$$(T - T_2)\dot{\varepsilon}_2 = \zeta \geq 0 \tag{17}$$

We are thus led to the development of a *kinetic equation* for $\dot{\varepsilon}_2$ which will satisfy (17). It is to this aspect that we turn next.

THE KINETIC EQUATION

In the state variable approach presented here, the dissipative response of the temporary network is described by means of a kinetic equation that shows how the stress in the secondary network gives rise to a gradual change in the temporary nodes leading to a change in the amount by which the temporary network is stretched. As can be seen from the "cartoon" (fig 2), the change in amount of stretch of the temporary spring, is related to the rate of change of ε_2 and this in turn is determined by the nature of the frictional dashpot. Based on the need for providing both a threshold dry frictional behavior as well as rate dependent subsequent behavior beyond the threshold, we propose the following viscoplastic kinetic equation for the dashpot.

$$\dot{\varepsilon}_2 = \begin{cases} 0, & \|T - T_2\| < \alpha; \\ \eta(\theta)(T - T_2 - \alpha), & T - T_2 > \alpha; \\ \eta(\theta)(T - T_2 + \alpha), & T - T_2 < -\alpha; . \end{cases} \tag{18}$$

The above set of cases can be written in a compact form as

$$\dot{\varepsilon}_2 = \eta(\theta)\left\{ \langle T - T_2 - \alpha(\theta) \rangle - \langle -\alpha - T + T_2 \rangle \right\} \tag{19}$$

where $<x> = \frac{1}{2}(x + \|x\|)$. A schematic plot of $\dot{\varepsilon}_2$ versus T_2 is shown in Figure (3

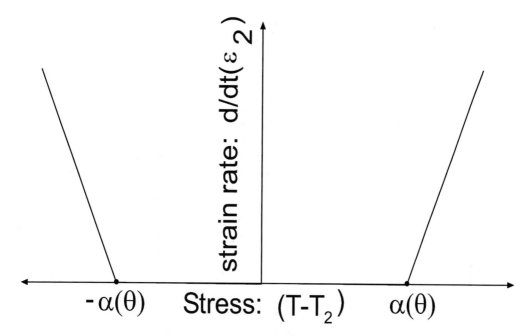

FIGURE 3. A schematic figure representing the response of the frictional Dashpot in (2a. Note that the "stress" on the dashpot is $T - T_2$ (plotted along the x-axis) while the strain rate associated with it is $\dot{\varepsilon}_2$ (plotted along the y-axis). Note that until the driving force exceeds $\alpha(\theta)$, $\dot{\varepsilon}_2 = 0$—the dashpot locks. Beyond that, the dashpot is assumed to be linear with a slope η. Of course, it is very easy to assume non-linear viscous behavior that reflect the response more accurately, but this simple structure is enough to express many of the properties of the SMP.

Qualitatively, we note that the dashpot "locks" if the magnitude of the driving force $T - T_2$ is less than α. *This is the origin of shape fixity in this model.* The state of the material is thus represented by the variables $\mathbb{S} = \{\theta, \varepsilon, T, \varepsilon_2, T_2\}$. Equations (16, and (19) together with (14)b represent three constitutive equations for the response of the material. Where do we get the other two to close the system? These are the equations of equilibrium (3) and the heat equation in the form (9) with G given by (13). Of course in the latter equation, we will specify the axial conduction and the lateral heating through

$$q = -k\theta', r = -h(\theta - \theta_\infty) \tag{20}$$

where k and h are the thermal conductivity and lateral heat transfer coefficients and θ_∞ is the temperature of the ambient atmosphere. For a straight wire in equilibrium, these latter two equations reduce to

$$T' = 0 \Rightarrow T = const. \tag{21}$$

$$-\frac{d}{dt}\frac{\partial G}{\partial \theta} = (T - T_2)\dot{\varepsilon}^p + k\theta'' - h(\theta - \theta_\infty). \tag{22}$$

The equations simplify further if we assume that the temperature is uniform so that there is no axial heat flux; in this case, (22) also simplifies into an ODE of the form

$$\theta' = 0 \Rightarrow -\frac{d}{dt}\frac{\partial G}{\partial \theta} = T_2\dot{\varepsilon}_2 - h(\theta - \theta_\infty). \tag{23}$$

This last assumption regarding the uniformity of the temperature is problematic and should considered carefully since these polymers are generally poor heat conductors and so local high temperatures can persist for substantial periods of time. However, in the interests of a "simplistic" model leading to a collection of ordinary differential equations, we will ignore this technologically significant issue and simply assume that the heating cooling and deformations are done so slowly that there is no axial temperature gradient.

The problem with the implementation of these equations is that some of them are in algebraic form whereas others are in rate form. Given the ease with which computer programs such as MATLAB are able to deal with differential equations, and in view of their versatility, we will write *all* the equations in rate form as follows:

Temperature specification: $\dot{\theta} = f(t)$

Force or strain control: $A(t)\dot{T} + B(t)\dot{\varepsilon} = g(t)$

State equation for $T = T_1$: $\dot{e} + \dfrac{d}{dt}\dfrac{\partial G}{\partial T_1} + \dot{\varepsilon}_2 = 0.$ (24)

Definition of $\dot{\varepsilon}_2$: $\dot{\varepsilon}_2 + \dfrac{d}{dt}\dfrac{\partial G}{\partial T_2} = 0;$

Kinetic equation for $\dot{\varepsilon}_2$: $\dot{\varepsilon}_2 = \eta(\theta)\left\{\langle T - T_2 - \alpha(\theta)\rangle - \langle -\alpha - T + T_2\rangle\right\}.$

The above set represents, in concise form, five differential equations for the five state variables $sS = \theta, \varepsilon, T, \varepsilon_2, T_2,$. A few comments are in order with regard to this set of differential equations. The first is that there is a single differential equation with time dependent coefficients for force or strain control. This is a very concise way to deal with complicated specifications, we simple set A or B to zero in the appropriate time interval when either strain or stress is specified. Second, in the Gibbs potential approach presented here, ε_2 is a defined variable and NOT a primitive, it can be completely eliminated by simply substituting (24)f into (24)e and (24)c and removing ε_2 from the list of state variables. It is however convenient to keep this variable in the list since it considerably simplifies the understanding of the shape memory effect and *represents a lower bound to the strain that can be retained at any instant*. In other words, the value of ε_2 will tell us what is the minimum strain that can be retained at any given state, so that, by monitoring its value during a process, we will be able to estimate whether the target shape retention has been reached or exceeded.

Our next remark pertains to the terms $d/dt\{\partial G/\partial T_i\}$ that appear on the left hand side of (24)d,e. Upon substituting (13) into the left hand side of (24)d, we obtain

$$\dot{e} + C_{21}\dot{T}_1 + \frac{dC_{11}}{d\theta}\dot{\theta} + T_1\frac{dC_{21}}{d\theta}\dot{\theta} - \dot{\varepsilon}_2 = 0 \tag{25}$$

we can identify the terms that appear in the equation in a straight forward manner: the second term is the elastic strain rate (elastic compliance times the stress rate) and the second term is the thermal strain rate (coefficient of thermal expansion times the temperature rate) However, the last term is due to changes in compliance with temperature and does not have a specific name as such. From a thermodynamical point of view, if the moduli depend upon the temperature, this term cannot be ignored (while retaining the temperature dependence of he compliance)since otherwise the results will be inconsistent leading to the possibility of a perpetual motion machine.

Finally, if one desires to model a process where the temperature of the wire is not known, but only the ambient temperatures is known, then one needs to replace the (24a with (23).

We now turn our attention to the simulation of the shape memory response of the polymer using the model whose differential equations are given by (24).

SIMULATION OF SMP RESPONSE

The key to the response of the material is the temperature dependence of α. Specifically, we suppose that α is such that it is practically zero above the glass transition temperature T_g and rises rapidly through the glass transition temperature to a much higher value when the temperature becomes below T_g. Thus, in qualitative terms, the operation of the model is explained as follows

1. Let us assume that the material is at a temperature that is higher than T_g in an unstretched state (point 1 in Figure 4a .

2. We apply forces and stretch it to a strain ε_h (point 2 in Figure 4a) . Note that at this temperature $\alpha \approx 0$ reflecting the absence of a temporary network[2]. The material responds like a rubbery elastic solid and upon removal of the load, returns to the original state.

3. We now keep the material fixed at ε_h and lower the temperature to below T_g(point 3 in Figure 4a). Now, persistent physical bonds are formed, the rubbery network is now "frozen" and the value of α rises. We now remove the load (point 4 in Figure 4a) . The rubbery network is unable to relax to its reference shape due to the formation of the new temporary bonds. This, together with the fact that the threshold for the physical bonds to break is now quite high, prevents the material from recovering its shape completely. The material recovers slightly and reaches a state of strain $\varepsilon_l < \varepsilon_h$. At this state, the rubbery network is "frozen in" by the bonds,since the magnitude of the stress in the rubbery network is below the threshold value α. Thus the material is in its temporary state and a strain is "locked in"

4. We can recover the permanent shape now by simply heating the material above T_g. When this occurs, the threshold value α decreases and the temporary bonds disappear, the frozen rubbery network regains its elasticity and relaxes, relieving

---

[2] In spite of this, there will be some damping due to the fleeting interactions between the chains, represented by the viscous dashpot, and this will be evident at higher strain rates, but no *persistent* physical bonds are formed.

69

the internal stress and strain and the material recovers its original shape.

In order to simulate this, we will use the set of differential equations (24). We need to choose a number of parameters in this model. We will assume very simple forms for these parameters—essentially assuming that the temperature dependence of the yield stress α is a sigmoidal shape asymptoting one to a low value α_r above the glass transition temperature and a different value high value α_g below it. In order to illustrate the response of the material, we first introduce a function of the temperature θ that is akin to a smoothed step function and represents a smooth transition from zero to one as the temperature ranges from below T_g to above T_g, this function is of the form

$$H(x,x_t) = \begin{cases} 0, & \text{if } x < -x_t; \\ \frac{1}{2}(1+\sin(\pi x/2x_t)), & -x_t < x < x_t; \\ 1, & x > x_t. \end{cases} \tag{26}$$

We will use this function to introduce the temperature dependence of α in this problem. Of course, this assumption is not validated by any statistical or structure-property relationship from polymer science. However,this is a very simple assumption that requires only one parameter—the transition value x_t. in order to represent the change in properties. We hasten to add that the model presented here is only a "cartoon" of the actual behavior of the of the SMP. It might be more suitable to assume, (based on the theory of activated processes,) functions of the form $e^{-\frac{\Delta H}{k\theta}}$ such as those suggested recently by Richeton et al [4]. but we wish to emphasize that the gross features of the response are independent of the form of the dependence, but is primarily due the abrupt change in yield strength α in this model.

In order to systematically non-dimensionalize all the variables, we note that (1) the temperatures are in and around the Glass transition temperature T_g (2) the strains in most of the experimental papers are around 4% to about 10% and the stresses are of the order of the hight temperature rubbery modulus times the strain. Thus, we introduce the following non-dimensionalizing variables: (1) the glass transition temperature T_g, (2) the high temperature or rubbery modulus E_∞, the typical strain $e_0 = 4\%$, and the typical stress $T_0 = E_\infty * e_0$. We will consider the non-dimensionalization of the time separately since this is connected with the kinetic response. Thus, we systematically nondimensionalizing stative variables as $\mathbb{S} = \{\theta^*\theta_0, \varepsilon^*e_0, T^*T_0, e_2^*e_0, T_2^*T_0\}$ and using the non-dimensional time $t^* = t/t_0$ where the starred variables are non-dimensional. We note that the term T_0e_0 has the dimensions of energy and thus, we will define a non-dimensional gibbs potential as $G^* = G/(T_0e_0)$. Furthermore, as previously indicated we will assume the following forces for the material parameters (Some of them have been already stated but in the interests of completeness and ease of comparison, we will restate them in tabular form here:

Property Name	Functional Form
Linear Coefficient of expansion (glass)	$C_{11} = -d(\theta - \theta_0)$
Linear Coefficient of expansion (rubber)	$C_{21} = 0$
Compliance (Glass)	$-2C_{12} = Const$
Compliance (Rubber)	$-2C_{22} = Const$

We now substitute these variables into the expression for (13) and the kinetic equation 19, and use them to define the following non-dimensional material parameters:

$$d^* = \frac{\theta_0}{e_0}d, \ S^* = -2E_\infty * C_{21},$$

$$2C_{22} = -1/E_\infty, \ \alpha^* = \alpha/T_0, \eta^* = \frac{\eta t_0}{E_\infty}.$$

(27)

Substituting these into the rate equations 24, we obtain the following governing ODES:

Temperature specification: $\dot{\theta}^* = f(t^*)$

Force or strain control: $A(t^*)\dot{T}^* + B(t)\dot{\varepsilon}^* = g(t^*)$

State equation for $T^* = T_1^*$: $e^* - d^*\theta^* - S^*T^* - e_2^* = 0.$ (28)

Definition of $\dot{e}_2^*$: $\dot{e}_2^* - \dot{T}_2^* = 0;$

Kinetic equation for $\dot{e}_2^*$: $\dot{e}_2^* = \eta^*(\theta^*) \{ \langle T^* - T_2^* - \alpha^*(\theta^*) \rangle - \langle -\alpha^* - T^* + T_2^* \rangle \}.$

where the superposed dot now represents differentiation with respect to the non-dimensionalized time t^*. We will assume further that η^* is a constant so that *the only function of temperature in the entire model is* α^*. For α^*, we choose a function of the form

$$\alpha^* = \alpha^0(1 - H(\theta^* - a, DT))$$

(29)

i.e, *alpha** vanishes above a temperature $\theta^* = a + DT$ and becomes a high positive value α_0 below a temperature $DT = a - DT$. This is used to represent the freezing of the rubbery network below T_g and the free rubbery network above T_g. Thus, the non-dimensional response of the material is characterized by the following constants, which were obtained from the data presented by Tobushi et al [5] for polyurethane:

Parameter	Definition	Value
Glassy Modulus	$-1/E_\infty C_2 2$	32
Coefficient of Expansion	$-C_{11}\theta_0/e_0$	0.3
Low temperature yield strength	α_0/T_0	2
non-dimensional time	$t_0\eta T_0/e_0$	50
Transition Temperatures	a	0.95
Transition Range	DT	0.05

The first three of these constants were determined by the experimental data for polyurethane, for which the glass transition temperature is $340K$. The constants a and DT refer to the interval of temperature where the yield strength of the material switches from α_0 to *zero*

We wish to emphasize that the model presented here is extremely idealized and has very few parameters. Indeed, we have not accounted for the rather strong dependence of the moduli or the temperature and neither have we accounted for a proper variation of the yield strength with temperature. It is possible to develop much more physically realistic models for the yielding behavior but our aim is to build the *simplest possible* model for the shape fixing phenomenon. Nevertheless it captures some essential features of the response quite well although the details of the response are not correct.

SIMULATION OF DIFFERENT THERMOMECHANICAL LOADING AND THE RESULTING RESPONSE

The standard shape fixing cycle

We first consider a simple case of fixing and recovering a shape. The process is as follows:

1. Initial conditions: The material is assumed to be in a stress free state at a temperature above the glass transition temperature $T_h = T_g + 20$. All strains are to be measured from this state.

2. Process A: High temperature stretching: we assume that the temperature is held fixed at T_h and the strain is increased steadily at a constant prescribed rate g_1 for a time $0 < t < t_1$. Thus, for this time interval. At the end of this time interval, the strain is $g_1 \times t_1$. Thus $f(t) = 0$, $A(t) = 1$, $B(t) = 0$ and $g(t) = g_1$ for this time interval.

3. Process B:Fixing the temporary shape: The strain is fixed at g_1 and the temperature is lowered to $t_l = T_g - 20$ at a predetermined rate a. During this process, due to fact that thermal contraction processes are NOT allowed to take place, the stress rises to T_{max}. Thus for $t_1 < t < t_2$, $(t_2 = 40/a)$ we have $f(t) = a$, $A(t) = 1$, $B(t) = 0$ and $g(t) = 0$; the last condition guarantees that the strain is fixed during this time interval.

4. Process C: Relaxing the stress: Now the temperature is fixed at T_l and the stress is gradually relaxed to zero at a predetermined rate $b < 0$. During this process, a permanent strain sets in. Thus for $t_2 < t < t_3,(t_3 = t_2 - T_{max}/b)$, we have we have $f(t) = 0$, $A(t) = 0$, $B(t) = 1$ and $g(t) = b$; the last condition guarantees that the strain is fixed during this time interval. The material has now attained its temporary shape.

5. Process: D Recovering the original shape: Now the body is heated at a rate a in a stress-free state back to the original temperature to recover the original shape. Thus for $t_3 < t < t_4$ we have $f(t) = -a$, $A(t) = 0$, $B(t) = 1$ and $g(t) = 0$. The strain slowly relaxes.

The entire cycle is shown in figure 4a,b,c,d where the processes are labeled.

The four graphs are draw together so that the reader may get a good idea as to how the processes occur. The results may be compared with the experimental data obtained by Tobushi et al [5] Specifically, comparisons can be made between figures 4 figures 5, 6, and 7 in Tobushi et al. [5] In each case, there are variations in the quantitative shape of the curves but the nondimensional values indicate quite good agreement, even with the extremely simplistic model developed here. As a comparison of complexity, note

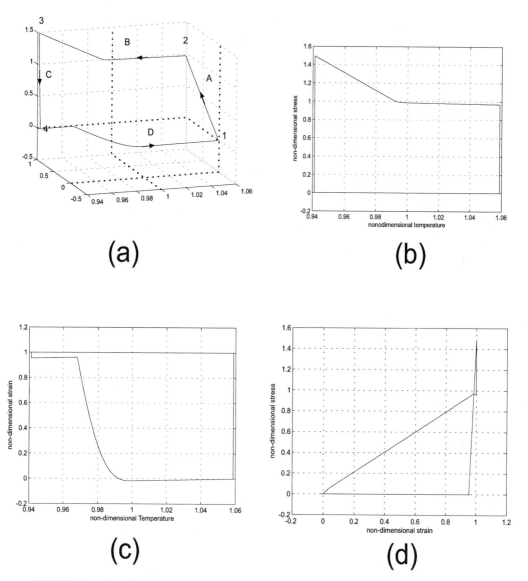

FIGURE 4. Simulation of the response of the SMP to a full thermomechanical cycle obtained by solving the ODES in equation(28) using ODE23 in Matlab. (a) The full 3-D view of the response. The letters A, B, C, D, refer to the processes listed in the text. (b) The stress versus temperature response obtained as a projection of the figure (a) along the strain axis (roughly perpendicular to the sheet of paper on which this figure is shown. (c) The strain temperature response obtained as a projection along the stress axis, roughly along the length of the paper and (d) the stress-strain response which is projection along the temperature axis (roughly along the width of the paper. Note the spike in the stress response in figure (d) caused by the prevention of thermal expansion. In figure (c) note that there is a "delayed" recovery caused by the temperature needed for the yield stress to decrease.

that the model proposed by Tobushi et al.[5] has a large number of constants and many functions of temperature. This is in contrast to the current model where the yield stress is the only temperature dependent parameter. A detailed study is underway to get more specific comparisons with experimental data.

CONCLUSION

We have shown how a relatively simple lumped mass model can simulate the gross features of shape memory polymers with relatively few constants. Of course, some of the features do not match up with experimental data. However the model shown above is the MINIMAL model that is required to show this behavior, if we eliminate any variable or simplify anything further, there will be substantial errors even in the qualitative response. It is hoped that this work can be built upon to develop more detailed realistic models for SMPs.

REFERENCES

1. M. C. Boyce, and E. M. Arruda, *Rubber Chemistry and Technology* **73** (3) , 504-523 (2000).
2. H. B. Callen, *Thermodynamics and an Introduction to Thermostatics*, 2nd edition John Wiley and Sons, (1985).
3. A. Lendlein, and S. Kelch, *Agnew. Chem. Int. Ed.* **41**, 2034-2057 (2002).
4. J. Richeton, S. Ahzi, L. Daridon, and Y. Ret'mond, *Polymer*, **46**, 6035–6043 (2005).
5. H. Tobushi, T, Hashimoto, S. Hayashi and E. Yamada, *Journal of Intelligent Material Systems and Structures*, **8**, 711–718 (1998).

SESSION A2

SHAPE MEMORY MODELING IN APPLICATIONS

Impact-induced Wave Patterns in a Slender Cylinder Composed of a Non-convex Elastic Material

Hui-Hui Dai* and Zhenli Xu†

*Department of Mathematics and Liu Bie Ju Centre for Mathematical Sciences, City University of
Hong Kong, 83 Tat Chee Avenue, Hong Kong, P.R. China
Email: mahhdai@cityu.edu.hk
†Department of Mathematics, University of Science and Technology of China, Hefei, Anhui
230026, P.R. China

Abstract. This paper studies nonlinear wave propagation in a slender elastic cylinder due to impacts. One purpose is to examine what kinds of wave patterns can arise when both the material nonlinearity caused by a non-convex energy function and the dispersion caused by the lateral movement are present. A high-order weighted essentially non-oscillatory (WENO) scheme is used to obtain the numerical solutions of a model equation which takes into account the dispersion and the above-mentioned nonlinearity. It turns out that a variety of wave patterns can appear, including the co-existence of a transformation front and solitons. We also find that pre-strains and the geometrical size can have some important influences. The interactions of transformation fronts and rarefaction waves (which have not been considered in literature before) are investigated and some interesting phenomena are found.

Keywords: nonlinear waves, solid-solid phase transition, WENO scheme, elastic cylinder
PACS: 45.20.dc, 64.70.km, 46.40.Cd, 64.70.kd

INTRODUCTION

It is known that in the continuum scale solid-solid phase transitions may be modelled by strain energy functions with multiple wells or in general by non-convex strain energy functions. For stress-induced dynamics of phase transitions in a bar, many authors have carried out studies within the framework of one-dimensional stress model. In the Lagrangian coordinates, one dimensional dynamics of phase transitions is governed by a system of conservation laws of mixed type as follows,

$$\gamma_t - u_x = 0,$$
$$u_t - \rho^{-1}\sigma(\gamma)_x = 0, \tag{1}$$

where $\gamma = w_x, u = w_t, \sigma, \rho$ are the strain, velocity, stress and density, respectively. Here w represents the axial displacement, and for a non-convex energy function the function $\sigma(\gamma)$ is non-monotonic such that the eigenvalues $\pm\sqrt{\sigma'(\gamma)}$ of the system are either real or pure imaginary; that is, the system is hyperbolic when $\sigma'(\gamma) > 0$ and becomes elliptic when $\sigma'(\gamma) < 0$. As is well-known, the system is ill-posed when the solution is in the elliptic region. This causes the non-uniqueness of the solution, and numerical simulations are difficult due to the blow-up phenomenon. For the Riemann problem,

CP1029, *Smart Devices: Modeling of Material Systems, An International Workshop*
edited by S. M. Sivakumar, V. Buravalla, and A. R. Srinivasa
©2008 American Institute of Physics 978-0-7354-0553-0/08/$23.00

77

the non-uniqueness was shown by Shearer [31]. To conquer this difficulty, Abeyaratne and Knowles [1, 2, 3] introduced the concept of the driving force $g(t)$ and a kinetic relation $g(t) = \phi(\dot{s})$ which is an extra condition to ensure the uniqueness of the solution (see also the review paper of Abeyaratne et. al. [4]). Based on this idea, impact-induced phase transitions in rubber-like materials were considered in Knowles [23].

A structure (e.g., a bar), no matter how thin it is, is always three-dimensional. Thus, the one-dimensional stress model is only an approximation. Sometimes, the effects from other dimensions may be very important and have to be taken into account. In some quasi-static experiments (Shaw and Kiyiakides [28, 29, 30] and Li and Sun [24]) on stress-induced phase transitions in thin structures (wires, strips and tubes) the necking was always observed when phase transformations took place, which implies that the lateral movement could play an an important role. An important geometrical size effect was reported in Sun et. al. [34] that smaller the radius-length ratio is sharper the stress drop is. Also, it was pointed out in Chang et. al. [8] that the axial extent of the transformation front in a NiTi wire is of the order of radius. These experimental results once again indicate the importance of the radial deformation for phase transitions in a wire (i.e., a slender cylinder). Further analytical explanations on why one should take into account the radial deformation as well as the traction-free condition on the lateral surface were also given in Dai [13], Dai and Cai [14], and Cai and Dai [7].

From the point view of nonlinear wave propagation, the radial movement implies that waves are dispersive. Within this spirit, in Dai [13] a model equation which further takes into account the dispersive terms were derived (see (3)). In this paper, we carry out numerical studies on some boundary and initial value problems of this model equation. More specifically, we consider the waves in a semi-infinite cylinder when impacts are applied at its end. The purpose is to find out whether some interesting wave patterns can arise due to the interaction of the dispersion and the material nonlinearity (caused by a non-convex energy function). We also explore how the wave patterns change when pre-strains are present in the cylinder. The tool of our investigation is a powerful and efficient WENO scheme.

This paper is arranged as follows. In section 2, some discussions on the governing equation and the jump conditions across a phase boundary are given. In sub-section 3.1, a simple linear stability analysis of a single mode is carried out. The numerical scheme is described in sub-section 3.2. To test its reliability, in sub-section 3.3 comparisons are made between the numerical solution and the exact two-phase solution and good agreements are found. In section 4, we investigate wave patterns in a semi-infinite cylinder due to impacts at its end by using the WENO scheme introduced in section 3. A few cases are considered, including different pre-strains, different impacts, and different values of the radius. A variety of wave patterns are found. Also, the interactions between transformation fronts and rarefaction waves are studied and the numerical solutions reveal some interesting phenomena. Finally, some conclusions are drawn.

THE GOVERNING EQUATION

For a moving strain discontinuity of (1), the Rankine-Hugoniot relation implies two jump conditions

$$[[\gamma]]\dot{s} + [[u]] = 0, \quad [[\sigma]] + \dot{s}\rho[[u]] = 0, \tag{2}$$

where $[[h]] = h^+ - h^-$, and $\dot{s}$ is the propagating speed of the discontinuity. Across a phase boundary, these two relations are not sufficient to determine all the state variables. Therefore, the solution of an initial-value problem is not unique.

For a slender circular cylinder, Dai [13] established a proper model equation to model solid-solid phase transitions, which can be written as

$$w_{\tau\tau} - \rho^{-1}\sigma_x + \frac{3}{4}a^2 C_T^2 w_{xxxx} - \frac{3}{4}a^2 w_{xx\tau\tau} + \frac{1}{8}a^2 C_T^{-2} w_{\tau\tau\tau\tau} = 0, \tag{3}$$

where $C_T = \sqrt{\mu/\rho}$ is the shear-wave speed, μ is the shear modulus and ρ is the density, w is the axial displacement and a is the radius of the slender cylinder. This equation models phase transitions in a slender circular cylinder with the small radius a composed of an incompressible phase-transforming material. The last three terms (representing the dispersive effects) arise due to the radial deformation and traction-free condition on the lateral surface. The presence of these dispersive terms can smoothen out a strain discontinuity, and correspondingly a phase boundary can have a small thickness. Thus, here we also call such a phase boundary as a transformation front. For a two-phase solution (a traveling-wave solution with a transformation front and crossing which the strain values γ^+ and γ^- are in two different phases), it is possible to deduce the relations among the state variables at two phases; see Dai [13]. Besides the same two conditions as (2), the third condition reads

$$[[W]] = [[\gamma\sigma]] - \frac{\rho}{2}\dot{s}^2[[\gamma^2]], \tag{4}$$

where $W(\gamma) = \int_{\gamma_0}^{\gamma} \sigma(\eta)d\eta$. These three conditions can be used to determine the speed of the transformation front.

With a scaling transformation $C_T\tau \to t$, equation (3) becomes

$$w_{tt} - C_T^{-2}\rho^{-1}\sigma_x + \frac{3}{4}a^2 w_{xxxx} - \frac{3}{4}a^2 w_{xxtt} + \frac{1}{8}a^2 w_{tttt} = 0. \tag{5}$$

Let $u = w_t$ be the "velocity". An equivalent system to (5) reads that

$$\begin{cases} \gamma_t - u_x = 0, \\ u_t - C_T^{-2}\rho^{-1}\sigma_x + \frac{3}{4}a^2\gamma_{xxx} - \frac{3}{4}a^2 u_{xxt} + \frac{1}{8}a^2 u_{ttt} = 0. \end{cases} \tag{6}$$

This just adds linear dispersive terms to system (1) by taking into account the influences of the radial deformation and the traction-free condition on the lateral surface.

Denote $U = (\gamma, u)^T$ and $F(U) = (-u, -C_T^{-2}\rho^{-1}\sigma)^T$. The Jacobian matrix of $F(U)$ reads

$$A(U) \equiv F_U(U) = \begin{pmatrix} 0 & -1 \\ -C_T^{-2}\rho^{-1}\sigma_\gamma & 0 \end{pmatrix},$$

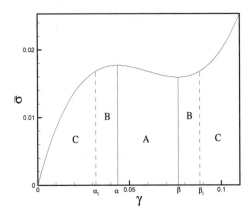

FIGURE 1. $\bar{\sigma} - \gamma$ (stress-strain) relation ($\bar{\sigma} = \sigma/3\mu$).

which has two eigenvalues $\lambda = \pm c$, where $c = \sqrt{C_T^{-2}\rho^{-1}\sigma_\gamma}$ is the "sound speed" when it is real, and σ_γ is the first-order derivative of σ with respect to γ. Let $R(U)$ denote the eigen-matrix consisting of two linearly independent right eigenvectors of the Jacobian A, then A can be diagonalized with a complete set of Riemann invariants. This diagonalization will be necessary for the local characteristic decomposition of the numerical scheme in the hyperbolic region.

For a non-convex strain energy function, the stress function $\sigma(\gamma)$ is non-monotonic. As a result, c may be a pure imaginary number. Then, the system becomes a hyperbolic-elliptic one. A simple form is a cubic polynomial

$$\sigma(\gamma) = 3\mu(\gamma + D_1\gamma^2 + D_2\gamma^3), \tag{7}$$

with $D_1 < 0, D_2 > 0$ and $3D_2 < D_1^2 < 4D_2$. This is sketched in Fig. 1 for

$$D_1 = -18, D_2 = 100. \tag{8}$$

We say the solution is in phase 1, 2 or 3 if γ lies in the interval $(0, \alpha)$, (α, β) or (β, ∞) with $\alpha = \frac{9-\sqrt{6}}{150} \approx 0.04367$ and $\beta = \frac{9+\sqrt{6}}{150} \approx 0.07633$. Here phase 1 and 3 correspond to hyperbolic regions and phase 2 an elliptic region. And the hyperbolic regions are also separated into metastable (B) and stable (C) regions by the two lines $\alpha_1 = 0.03172$ and $\beta_1 = 0.08828$, respectively. The Maxwell line is the horizontal line connecting the two points on the stress-strain curve with strain values α_1 and β_1.

Since the perturbation coefficient a is small, the temporal derivatives of the dispersive terms can be replaced with the spatial ones by using the governing equations (6):

$$u_t = C_T^{-2}\rho^{-1}\sigma_x + O(a^2),$$

$$u_{tt} = C_T^{-2}\rho^{-1}(\sigma_\gamma\gamma_t)_x + O(a^2) = C_T^{-2}\rho^{-1}(\sigma_\gamma u_x)_x + O(a^2),$$

$$u_{ttt} = C_T^{-2}\rho^{-1}(\sigma_{\gamma\gamma}u_x^2 + C_T^{-2}\rho^{-1}\sigma_\gamma\sigma_{xx})_x + O(a^2).$$

Then we obtain the approximation to (6) by eliminating the fourth-order small terms as follows.

$$
\begin{cases}
\gamma_t - u_x = 0, \\
u_t - C_T^{-2}\rho^{-1}\sigma_x + \frac{3}{4}a^2(\gamma - C_T^{-2}\rho^{-1}\sigma)_{xxx} + \frac{1}{8}a^2(C_T^{-2}\rho^{-1}\sigma_{\gamma\gamma}u_x^2 \\
\qquad\qquad + C_T^{-4}\rho^{-2}\sigma_\gamma\sigma_{xx})_x = 0.
\end{cases}
\tag{9}
$$

This system is more solvable than the original one (6), since its temporal derivatives is only of the first-order.

STABILITY AND NUMERICAL APPROXIMATION

Linear stability analysis

To do a stability analysis of the linearized version of system (9), we consider constant reference values $(\gamma, u) = (\gamma_0, u_0)$ in the elliptic region. Then we have $\alpha < \gamma_0 < \beta$ and $\sigma'(\gamma_0) < 0$. System (9) can be linearized into

$$
\begin{cases}
\gamma_t - u_x = 0, \\
u_t - p_0\gamma_x + \frac{3}{4}a^2(1-p_0)\gamma_{xxx} + \frac{1}{8}a^2 p_0^2 u_{xxx} = 0,
\end{cases}
\tag{10}
$$

where $p_0 = C_T^{-2}\rho^{-1}\sigma'(\gamma_0) < 0$. Basing on the linear superposition principle of linear equations, we need only consider the solution of a single Fourier mode

$$
(\gamma, u) = (\bar{\gamma}(t), \bar{u}(t))e^{ikx},
$$

where k is an arbitrary real number. Substituting the solution into (10), we obtain

$$
\begin{pmatrix} \bar{\gamma} \\ \bar{u} \end{pmatrix}_t = \begin{pmatrix} 0 & ik \\ ikp_0 + i\frac{3}{4}a^2(1-p_0)k^3 & i\frac{1}{8}a^2 p_0^2 k^3 \end{pmatrix} \begin{pmatrix} \bar{\gamma} \\ \bar{u} \end{pmatrix}.
$$

The eigenvalues of the coefficient matrix are

$$
\lambda_{1,2} = i\frac{1}{8}a^2 p_0^2 k^3 \pm i\sqrt{(\frac{1}{8}a^2 p_0^2 k^3)^2 + 4p_0 k^2 + 3a^2(1-p_0)k^4}.
$$

The solution is stable when $Re(\lambda) \leq 0$. This implies that

$$
k^2 \geq \frac{16}{a^2 p_0^4}[\sqrt{36(1-p_0)^2 - p_0^5} - 6(1-p_0)].
\tag{11}
$$

We can see that, if without the perturbation terms in (9), the solution must be discontinuous in the elliptic region, and then it is very difficult for directly numerical computations when the solution contains strain values in phase 2. And when $a > 0$, values in phase 2 are allowed and the solutions are continuous, but they must be short waves locally in this region according to (11). This illustrates that for the two-phase solution, the smooth connection of the two phases must have a large gradient.

High-order numerical approximation

For numerical simulations of phase transitions for a conservation law of mixed-type, many methods are based on viscosity-capillarity regularization (see [5, 9, 12, 19, 21, 33]), and references therein). Cockburn and Gau [12] have implemented a convergence study for different viscosity and capillarity coefficients. Jin [21] obtained robust solutions by utilizing the relaxation scheme [22], which automatically converge to the viscosity-capillarity solutions. It is not necessary to perform the local characteristic decomposition for the relaxation scheme. But, when the relaxation time is zero, the so-called relaxed scheme is similar to a component-by-component flux-splitting method. In this case, the local characteristic decomposition is an important approach to reduce oscillations for higher-order schemes [27]. In this paper, we use the fifth-order finite difference WENO scheme proposed by Jiang and Shu [20] to approximate the convective term when the equations are in the hyperbolic region, and higher-order central scheme to approximate the other terms. The scheme achieves higher accuracy and higher resolution in capturing shocks and phase boundaries. The method we use is an extension of Shu [32], in which the ENO scheme was implemented for mixed-type equations.

The WENO was first investigated by Liu, Osher and Chan [26] for a third-order finite volume case, and developed by Jiang and Shu [20], and Balsara and Shu [6], and others, for various finite difference or finite volume versions. Here we do not detail the WENO reconstruction, and interested readers can refer to the above-mentioned papers and the lecture notes [32] for more details.

One difficulty is the handling of the elliptic region. Basically, the central scheme should be used for there is no the wind direction. To couple this with the spatial discretization in the hyperbolic region, a simple way is, when splitting the flux $f(U)$ with $U = (\gamma, u)^T$ and $f(U) = (-u, C_T^{-2}\rho^{-1}\sigma)$ by the local Lax-Friedrichs splitting,

$$f^{\pm}(U) = \frac{1}{2}(f(U) \pm \lambda U),$$

where λ is the maximum eigenvalue of the Jacobian $\partial f(U)/\partial U$ in the local point stencil in the hyperbolic region, that we directly set λ zero if the stencil contains the elliptic region. And the local characteristic decomposition is used in the hyperbolic region to reduce non-physical oscillations.

The higher order perturbation terms with respect to x in (9) are discritized by second-order central differences. Since the approximation to the first order spatial term is a fifth-order WENO scheme, the order of the truncate error to (9) in space is $O(\Delta x^5 + \Delta x^2 a^2)$ when the solution is smooth, and thus the order to (6) is $O(\Delta x^5 + \Delta x^2 a^2 + a^4)$. The third-order TVD-Runge-Kutta method is applied to discretize the temporal integration, which is not discussed further here, and interested readers can refer to [33] and [17] for details.

Numerical test of the two-phase solution

It is possible to obtain all bounded traveling-wave solutions of (3) by a similar analysis in Dai and Cai [14]. Here we consider the two-phase solution. Denote γ^+ and γ^-

respectively the right and left values of the strain at infinity, which satisfy the condition

$$\gamma^+ + \gamma^- = -\frac{2D_1}{3D_2}. \tag{12}$$

The two-phase solution of (6) (or (3)) takes the form

$$\gamma = \frac{\gamma^+ + \gamma^-}{2} + \frac{\gamma^+ - \gamma^-}{2}\tanh\left(-\frac{\gamma^+ - \gamma^-}{a}\sqrt{\frac{3D_2}{6 - 6\dot{s}^2 + \dot{s}^4}}(x - \dot{s}t)\right),$$
$$u = -\dot{s}\gamma + \dot{s}\gamma^- - V, \tag{13}$$

where the traveling velocity

$$\dot{s} = \sqrt{3D_2\gamma^{+2} + 2D_1\gamma^+ + \frac{9D_2 - 2D_1^2}{3D_2}},$$

and V is related to the Rankine-Hugoniot jump condition

$$(\sigma^+ - \sigma^-) + C_T^{-2}\rho^{-1}\dot{s}(u^+ + V) = 0,$$

where we take $u^+ = 0$. (Note: One can check by Mathematica that the solution (13) satisfies (6) exactly.)

In order to analyze the stability of the two-phase solution as well as to test the correctness of our numerical scheme, we compute the numerical evolution for an initial two-phase wave under random noises in $[-0.5, 1.5]$. The parameters are chosen as $a = 0.001$, $V = 0.05$; thereby $r^+ = 0.0291987$, $r^- = 0.0908013$ and $\dot{s} = 0.211225$. We take the initial strain with random noises as

$$\gamma(x, 0) = \gamma_0(x)\left(1 + \frac{\eta(x)}{10}e^{-200x^2}\right), \tag{14}$$

where $\gamma_0(x)$ is the initial strain profile (i.e., letting $t = 0$ in (13)), and $\eta(x)$ is a stochastic function in $[0,1]$. The results are shown in Figs. 2(a) and 2(b) for different grid points N for $t = 0$ and $t = 0.4$, respectively. From them one can see that the initial profile goes to the exact solution profile quickly. It illustrates that the two-phase wave is a very stable structure for a phase boundary and also shows that the numerical solution agrees with the exact two-phase solution well.

IMPACT-INDUCED WAVES IN A SEMI-INFINITE CYLINDER

In this section, we consider the wave patterns in a semi-infinite cylinder when phase transitions are induced by impacts at one end by using the numerical scheme discussed in section 3.

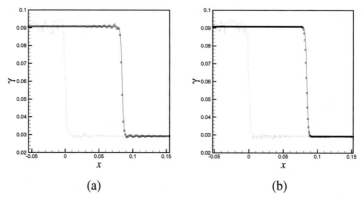

FIGURE 2. (a) $N = 128$; (b) $N = 256$. Dashed line: initial values; square: numerical solutions; solid lines: exact solution without noises

Zero pre-strain and the influence of pre-strains on the speed of the transformation front

We consider a semi-infinite slender elastic cylinder at $[0, \infty]$ which is initially at rest in its undeformed state so that

$$\gamma(x,0) = 0, \quad u(x,0) = 0. \tag{15}$$

At the left end $x = 0$, a sudden impact with constant velocity $u(0,t) = -V$ is loaded. For a rubberlike material with a concave-convex stress-strain relation, Knowles [23] showed that there are three regimes of response, depending on the intensity of the loading. For weak impacts, the response is a pure rarefaction wave, as in a material with a concave stress-strain relation; for the strong impacts, the response is a pure shock wave, as in a material with a convex stress-strain relation; and for impacts with intermediate strength, the response exhibits a two-wave structure consisting of a rarefaction wave followed by a phase transformation front. Dai and Kong [15, 16] discussed the propagation of impact-induced tensile waves for all levels in detail.

In the numerical test, we take $V = 0.08$ in $[0, L]$ with $L = 2$ for simulated time from $t = 0$ to $t = T = 1$. Here $L = 2$ is large enough such that there is no reflective wave from the right boundary. To satisfy the well-posedness of the equations, we also impose the natural boundary condition $\gamma_x = 0$ at the left end. Then we also have $u_{xx} = \gamma_{xt} = 0$ at $x = 0$. So we can achieve the boundary condition from this symmetry for the WENO scheme. Our numerical results are shown in Fig. 3 in different gird sizes for two perturbation coefficients $a = 0.0006$ and 0.001. The wave pattern consists of a right-going rarefaction wave followed by a tranformation front. We can see the numerical solution converges to the "analytical" solution (the solution so-constructed according to conditions (2) and (4) for a transformation front) when $\Delta x (= L/N)$ is close to a, and thus the solution with a larger a converges to the "analytical" solution more quickly when the grid size is refined.

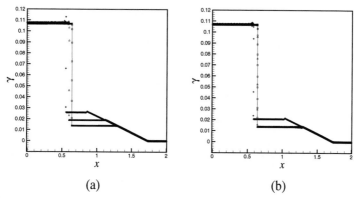

FIGURE 3. Numerical solutions for $a = 0.0006$ (a) and $a = 0.001$ (b) with three different grid points: $N = 1024$ for the cross shape, $N = 2048$ for the delta shape, and $N = 4096$ for the square shape. The solid line is the "analytical" solution

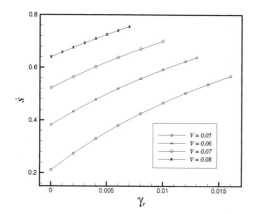

FIGURE 4. $\gamma_r - \dot{s}$ relationships for different impacts

If we first stretch the semi-infinite cylinder to a strain value γ_r and then apply an impact velocity at the left end, we find that the propagating speed of the transformation front depends on not only the impact strength, but also the pre-strain. The $\gamma_r - \dot{s}$ relationships for different impacts are shown in Fig. 4. It can be seen that the speed $\dot{s}$ is an increasing function of both the impact velocity V and pre-strain γ_r.

Same pre-strain with different impacts

When the initial strain is not zero, such as $\gamma = 0.03$ in (15), then the wave pattern becomes one with a phase boundary and a shock. These cases are shown in Fig. 5 for

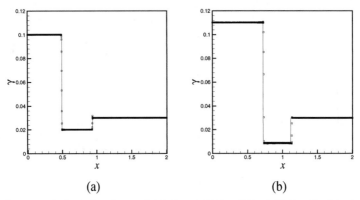

FIGURE 5. Numerical solutions for $a = 0.001$, $V = 0.03$(a) and $V = 0.05$(b) with grid points $N = 2024$. The solid line is the "analytical" solution

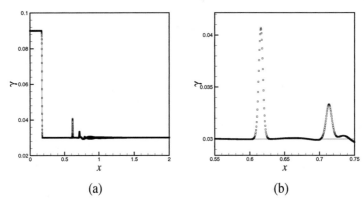

FIGURE 6. Numerical solution for $a = 0.001$ and $V = 0.0103923$ with grid points $N = 4048$. The solid line is the "exact" solution. The enlarged part for the soliton-like pattern is shown in (b)

$V = 0.03$ and 0.05. We see the jump $[[\gamma]]$ of the shock is reduced as V decreases. When $V = 0.0103923$, the shock disappears (the jump $[[\gamma]]$ is zero), but some soliton-like waves appear, as shown in Fig. 6. The solitons induced by boundary conditions were studied by many publications (see [10, 11]), but here the solitary waves and a transformation front coexist due to the impact on the boundary, and it appears that this phenomenon has not been reported elsewhere.

Different pre-strains with the same impact

We consider the impact problem with different pre-strains. Here we maintain the impact speed $V = 0.02$. In this case, by varying the pre-strain $\gamma(x,0) = \gamma_0$, there are many situations. When γ_0 is relatively large, the "analytical" solution has the structure of a

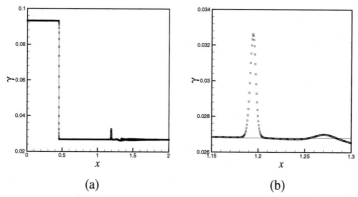

FIGURE 7. Numerical solution of for $a = 0.001$ and $V = 0.0267979$ with grid points $N = 4048$. The solid line is the "exact" solution. The enlarged part for the soliton-like pattern is shown in (b)

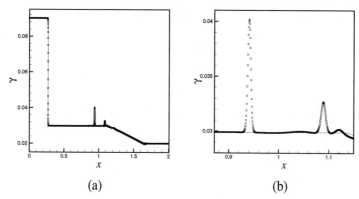

FIGURE 8. Numerical solution for $a = 0.001$ and $V = 0.02$ with grid points $N = 4048$. The solid line is the "analytical" solution. The enlarged part for the soliton-like pattern is shown in (b)

phase boundary plus a shock. When γ_0 is relatively small, the solution structure consists of a transformation front plus a rarefaction wave. And, when γ_0 is very small, there is only a rarefaction wave. So, there should be a critical pre-strain value γ^* (between a relatively large pre-strain and a relatively small pre-strain), which we find to be $\gamma^* = 0.0267979$, such that when $\gamma_0 = \gamma_*$ there is only a transformation front with neither a shock nor a rarefaction wave. Here, we consider the numerical solutions when $\gamma_0 = \gamma^*$ and $\gamma_0 = 0.02$, since in these two cases there are some interesting wave patterns. The numerical results for the two cases with $a = 0.001$ and $N = 4048$ in $[0, 2]$ at $t = 1.5$ are shown in Figs. 7 and 8. In both cases, we observe that soliton-like humps are generated ahead of the transformation front.

Interactions of rarefaction waves and transformation fronts

In this sub-section, we will simulate the interactions of rarefaction waves and transformation fronts. We consider a semi-infinite elastic cylinder, which is first stretched to a strain value $\gamma_0 = 0.03$. Then, as before, a sudden impact with speed -0.04 is loaded at the left end. But, the impact is only maintained during the time interval $t \in [0, 0.5]$ and then becomes zero afterwards. We take $a = 0.001$, $T = 3.6$, and $N = 4096$ in the computational region $x \in [0, 4]$. The numerical results are shown in Figure 9 for $t = T/8$, $2T/8$, $\cdots$, T. For $t \leq 0$, a wave pattern with a right-going transformation front (TF1) and a right-going shock is generated. Once the impact velocity becomes zero for $t > 0.5$ at the left end, a new transformation front (TF2) is formed together with a right-going rarefaction wave (RRW1) (Fig. 9(a)). The rarefaction wave propagates to the right faster than both transformation fronts such that it interacts with the first transformation front (TF1) with a back strain value γ^- about 0.1054. After the interaction, the TF1 changes to a new one with γ^- about 0.099. At the same time, a smaller rarefaction wave (LRW1) propagating to the left is created, which will interact with TF2 (Fig. 9(b)). Their interaction, in turn, generates a right-going rarefaction wave (RRW2; see Fig. 9(b)), which is to interact with the TF1. During the interaction processes rarefaction waves are created one by one until the TF1 and TF2 converge to the states with the same γ^- and γ^+ values (Figs. 9(c-d)). This also implies that the velocities of the TF1 and TF2 become more and more close to each other (see Fig. 10(a) in which the results at $t = 1.5T$ and $2T$ are computed in the interval $x \in [0, 8]$ using the same mesh size). It should be noted that the "analytical" solution describing the interaction processes is impossible to be obtained. Fig. 10(b) shows the evolution of the total energy. We see the total energy decays when the entropy fluxes are zero at two boundaries. This agrees with the property of the energy dissipation for a numerical scheme.

CONCLUSIONS

Numerical solutions for boundary and initial value problems of a model equation by using an efficient WENO scheme are carried out. This equation contains both the material nonlinearity due to the non-convexity of the strain energy function and dispersion terms due to the finite geometrical size of the cylinder. For such a nonlinearity, one would expect that a phase boundary (transformation front) can arise. On the other hand, it is well-known that the interaction of nonlinearity and dispersion can cause the appearance of solitary waves. Thus, it would be interesting to find that when both effects are present whether there is a co-existence of transformation fronts and solitary waves. Our numerical results show that under a proper pre-strain and impact such a wave pattern can indeed arise. Also, it is found that the pre-stains can change wave patterns. Under different pre-strains and impacts, a variety of wave patterns can appear, such as a transformation front and a rarefaction wave, a transformation front and a shock wave, a transformation front and solitary waves, a combination of a transformation front, solitary waves and a rarefaction wave. It is also found that as the pre-strain increases the speed of the transformation front also increases. The interactions of rarefaction waves and transformation fronts are studied. It is found that the head-on collision of a rarefaction wave and a transforma-

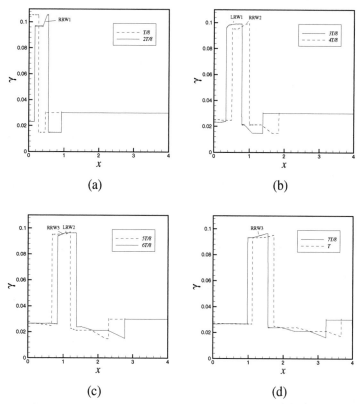

(a)

(b)

(c)

(d)

FIGURE 9. Temporal evolution for $a = 0.001$ with grid points $N = 4096$ at $t = T/8, 2T/8, \cdots, T$

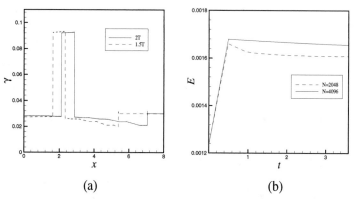

(a)

(b)

FIGURE 10. (a) The results at $t = 1.5T$ and $2T$; (b) Total Energy evolution

tion front results in the increasing of the jump amplitude of the latter. On the hand the catching-on collision results in the decreasing of the jump amplitude of the transformation front. This work once again demonstrates that the coupling of nonlinearity and dispersion can cause some interesting nonlinear wave phenomena.

ACKNOWLEDGMENTS

The work described in this paper is fully supported by a grant from the Research Grants Council of Hong Kong SAR (Project No.: CityU 100804) and a strategic grant from City University of Hong Kong (Project No.: 7001645)

REFERENCES

1. Abeyaratne, R. & Knowles, J.K. 1990 On the driving traction acting on a surface of strain discontinuity in a continuum, *J. Mech. Phy. Solids* **38**, 345-360.
2. Abeyaratne, R. & Knowles, J.K. 1991 Kinetic relations and the propagation of phase boundary in solids, *Arch. Rational Mech. Anal.* **114**, 119-154.
3. Abeyaratne, R. & Knowles, J.K. 2000 On a shock-induced martensite phase transition, *J. Appl. Phys.* **87**, 1123-1134.
4. Abeyaratne, R., Bhattacharya, K. & Knowles, J.K. 2001 Strain-energy functions with multiple local minima: modeling phase transformations using finite thermoelasticity. In *Nonlinear Elasticity: Theory and Applications* (ed. Y. Fu & R. Ogden), pp.433-490. Cambridge: Cambridge University Press.
5. Affouf, M. & Calfisch, R.E. 1991 A numerical study of Riemann problem solutions and stability for a system of viscous conservation laws of mixed type, *SIAM J. Appl. Math.*, **51**, 605-634.
6. Balsara, D. & Shu, C.-W. 2000 Monotonicity preserving weighted essentially non-oscillatory schemes with increasingly high order accuracy, *J. Compt. Phys.*, **160**, 405-452.
7. Cai, Z.X. & Dai, H.-H. 2006 Phase transitions in a slender cylinder composed of an incompressible elastic material II. Analytical solutions for two boundary-value problems, *Proc. R. Soc. London A* **462**, 419-438.
8. Chang, B.C., Shaw, J.A., & Iadicola, M.A. 2006 Thermodynamics of shape memory alloys wire: modeling, experiments, and application, *Continuum Mechanics and Thermodynamics* **18**, 83-118.
9. Chalons, C. and LeFloch, P.G. 2001 A fully discrete scheme for diffusive-dispersive conservation laws, *Numer. Math.*, **89**, 493-509.
10. Chou, R.L. & Chu, C.K. 1990 Solitons induced by bounday conditions from Boussinesq equation, *Phys. Fluid A*, **2**, 1574-1584.
11. Chu, C.K. and Xiang, L.W. & Baranskv 1983 Solitary waves induced by boundary motion, *Commun. Pure Appl. Math.*, **36**, 495-504.
12. Cockburn, B. and Gau, H. 1996 A model numerical scheme for the propagation of phase transitions in solids, *SIAM J. Sci. Comput.*, **17**, 1092-1121.
13. Dai, H.-H. 2004 Non-existence of one-dimensional stress problems in solid-solid phase transitions and uniqueness conditions for incompressible phase-transforming materials, *C. R. Acad. Sci. Paris, Ser. I* **338**, 981-984.
14. Dai, H.-H. & Cai, Z.X. 2006 Phase transitions in a slender cylinder composed of an incompressible elastic material I. Asymptotic model equation, *Proc. R. Soc. London A* **462**, 75-95.
15. Dai, H.-H. and Kong, D.-X. 2005 Global structure stability of impact-induced tensile waves in rubber-like material, *IMA J. Appl. Math.*, **71**, 14-33.
16. Dai, H.-H. and Kong, D.-X. 2006 The propagation of impact-induced tensile waves in a kind of phase-transforming material, *J. Compt. Appl. Math.*, **190**, 57-73.
17. Gottlieb, Shu, C.-W. & Tradmor, E. 2001 Strong stability preserving high order time discretization method, *SIAM Review*, **43**, 89-112.

18. Hayes, B.T. & LeFloch, P.G. 1998 Nonclassical shocks and kinetic relations: finite difference schemes, *SIAM J. Numer. Anal.*, **35**, 2169-2194.
19. Hsieh, D.Y. & Wang X.P. 1997 Phase transition in van der Waals fluid, *SIAM J. Appl. Math.*, **57**, 871-892.
20. Jiang, G.S. and Shu, C.-W. 1996 Efficient implementation o f weighted ENO schemes, *J. Compt. Phys.*, **126**, 202-228.
21. Jin, S. 1995 Numerical integration of systems of conservation laws of mixed type, *SIAM J. Appl. Math.*, **55**, 1536-1552.
22. Jin, S. & Xin, Z.P. The relaxation schemes for systems of hyperbolic conservation laws, *Commun. Pure Appl. Math.*, **48**, 235-278.
23. Knowles, J.K. 2002 Impact-induced tensile waves in a rubberlike material, *SIAM Appl. Math.*, **62**, 1153-1157.
24. Li, Z.Q. & Sun, Q.P. 2002 The initiation and growth of macroscopic martensite band in nano-grained NiTi microtube under tension, *Inter J. Plasticity*, **18**, 1481-1498.
25. Lin, P. and Shu, C.-W. 2002 Numerical solution of a virtual internal bond model for material fracture, *Phys. D*, **167**, 101-121.
26. Liu, X.-D., Osher, S. & Chan, T. 2000 Weighted essentially non-oscillatory schemes, *J. Compt. Phys.*, **115**, 200-212.
27. Qiu, J.X. & Shu C.-W. 2002 On the construction, comparison, and local characteristic decomposition for high-order central WENO schemes, *J. Comput. Phys.*, **183**, 187-209.
28. Shaw, J.A. & Kyriakides, S. 1995 Thermomechanical aspects of NiTi, *J. Mech. Phys. Solids* **47**, 1243-1281.
29. Shaw, J.A. & Kyriakides, S. 1997 On the nucleation and propagation of phase transformation fronts in a NiTi alloy, *Acta mater.* **45**, 683-700.
30. Shaw, J.A. & Kyriakides, S. 1998 Initiation and propagation of localized deformation in elasto-plastic strips under uniaxial tension, *Inter. J. Plasticity*, **13**, 837-871.
31. Shear, M. 1986 Non-uniqueness of admissible solutions of conservation laws of mixed type, *Arch. Rational Mech. Anal.*, **114**, 119-154.
32. Shu, C.-W. 1997, Essentially non-oscillatory and weighted essentially non-oscillatory schemes for hyperbolic conservation laws, *Technical Report ICASE*, Report No.: 1997-65.
33. Shu, C.-W. 1988, Total-Variation-Diminishing time discretization, *SIAM J. Sci. Comput.*, **9**, 1073-1084.
34. Sun, Q.P., Li, Z.Q. & Tse, K.K. 2000 On superelastic deformation of NiTi shape memory alloy micro-tubes and wires–band nucleation and propagation, in: *Proc. of IUTAM Symposium on Smart Structures and structure systems* (U. Gabbert and H.S. Tzou (eds.)), pp. 1-8, Magdeburg, Germany; Kluwer Academic Publishers.

Modeling the Behaviour of an Advanced Material Based Smart Landing Gear System for Aerospace Vehicles

Byji Varughese, G.N.Dayananda, and M.Subba Rao

National Aerospace Laboratories, P.B. No: 1779, Airport Road, Bangalore-560017, India

Abstract: The last two decades have seen a substantial rise in the use of advanced materials such as polymer composites for aerospace structural applications. In more recent years there has been a concerted effort to integrate materials, which mimic biological functions (referred to as smart materials) with polymeric composites. Prominent among smart materials are shape memory alloys, which possess both actuating and sensory functions that can be realized simultaneously. The proper characterization and modeling of advanced and smart materials holds the key to the design and development of efficient smart devices/systems. This paper focuses on the material characterization; modeling and validation of the model in relation to the development of a Shape Memory Alloy (SMA) based smart landing gear (with high energy dissipation features) for a semi rigid radio controlled airship (RC-blimp). The Super Elastic (SE) SMA element is configured in such a way that it is forced into a tensile mode of high elastic deformation. The smart landing gear comprises of a landing beam, an arch and a super elastic Nickel-Titanium (Ni-Ti) SMA element. The landing gear is primarily made of polymer carbon composites, which possess high specific stiffness and high specific strength compared to conventional materials, and are therefore ideally suited for the design and development of an efficient skid landing gear system with good energy dissipation characteristics. The development of the smart landing gear in relation to a conventional metal landing gear design is also dealt with.

Keywords: Carbon composites, Shape Memory Alloys, Modeling, Landing gear.

Nomenclature:

A_s = Austenite start temperature	δ_h = Horizontal deflection
A_f = Austenite finish temperature	δ_v = Vertical deflection
E = Young's modulus	σ = Stress
M_s = Martensite start temperature	ε = Strain
M_f = Martensite finish temperature	σ_{ult} = Ultimate stress

INTRODUCTION

This paper explains the course followed in the design and analysis of a smart landing gear starting from a conventional design and then taking recourse to the polymer composite design and analysis. The landing gear serves the purpose of dissipating impact energy during landing. The Super elastic (SE) SMA based smart landing gear development at NAL has already been reported [1]. The RC blimp is a non rigid airship. It is used for weather monitoring, pollution monitoring, disaster

CP1029, *Smart Devices: Modeling of Material Systems, An International Workshop*
edited by S. M. Sivakumar, V. Buravalla, and A. R. Srinivasa
©2008 American Institute of Physics 978-0-7354-0553-0/08/$23.00

management, traffic management, and other purposes. Helium gas is used to generate the lift for the blimp while a fixed engine provides the thrust. The RC blimp shown in (Fig.1.) has a 320m³ volume, measures 18m long, and has a maximum diameter of 6m.

The landing gear has to have a minimum height because of the ground clearance required to accommodate the engine and other accessories. A typical conventional landing gear construction using aluminium metallic tubes (for example) consisting of two horizontal landing beams, vertical members and cross members (gusset like) could be constructed as shown in (Fig.1.).

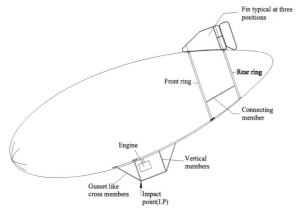

FIGURE 1. The RC-Blimp with Aluminium metallic tubes.

The conventional metallic design shown in Fig.1 results in a highly rigid configuration and it cannot undergo large elastic deformations at the Impact point (I.P) and dissipate energy. It is therefore required to build a mechanism in the landing gear using advanced materials having minimum weight and performing energy dissipation functions. It is imperative that a high energy dissipating landing gear must undergo large elastic deformations. The SE SMA element and carbon composites have been effectively integrated to realize large elastic deformations. The NiTi SE SMA is configured in such a way that it is forced into a tensile mode of large elastic deformations. The dissipation of energy is because of the hysteresis between the loading and the unloading path in this material. The detailed process is illustrated in section II.

The landing gear is a part of the structural framework shown in (Fig.2.) and comprises of two rings, inter-connecting members and is built primarily using carbon composites. The structural framework also serves as a base to mount the fin and rudder units in addition to holding the envelope. The impact energy during landing can cause severe shock and therefore render these gadgets ineffective and even damage them. Therefore, the landing gear (comprising landing beam, arch, and superelastic SMA element) shown in (Fig.2.) has been designed in such a way that it absorbs/dissipates the maximum impact energy during landing. The design and analysis of landing gear in aerospace vehicles is an actively pursued area of research.

Philips et al. [2] deals with the design of a crashworthy landing gear for helicopters which would lessen the magnitude of crash forces. In this design the stiffness of the skid was idealized as a bilinear curve. The first part of the curve represents elastic deformation and the second part, plastic deformation of the skid.

93

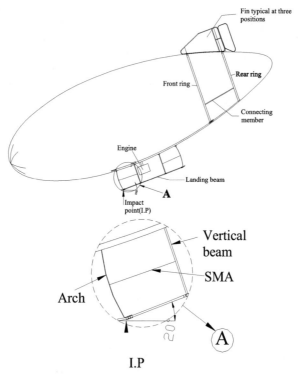

FIGURE 2. RC-Blimp with Carbon composites.

Airoldi et al. [3] presents a numerical approach to the optimization of the skid landing gears. The optimization technique is applied to investigate the tradeoff between landing performance and gear strength.

Ashish et al. [4] discuses a nonlinear finite element based method of analyzing the structural behavior of helicopter skid gears during a high-energy landing.

In this paper a simple but innovative approach towards analyzing the polymeric carbon composite landing gear with super elastic SMA and curved arch (comprising carbon composites and cane) is presented. The analysis is backed by some limited experimental results.

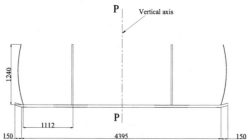

FIGURE 3. Details of landing gear (in mm).

The landing gear (Fig.3.) incorporates advanced materials as already mentioned with a wide range of mechanical properties. This includes the high stiffness and low strain polymeric materials, such as carbon and glass composites and the large strain materials such as superelastic SMAs and naturally occurring cane (plant material). The material non-linearities and the geometric non-linearities are both taken into account. The material non-linearities are with regard to the SMA and cane, whereas the geometric non-linearity is due to the curved configuration of the arch and the varying cross-section of the beam (Fig.3.).

Advanced polymer carbon composites have now gained increased acceptance in the aerospace industry. This is because of their high strength and stiffness characteristics compared with other traditional structural materials such as aluminum and steel. These composites have high performance reinforcements, which are of a very thin diameter (typically in the range of 0.005-0.010mm) in a matrix material such as epoxy. Examples of advanced polymer composites are graphite/epoxy, carbon/epoxy & Kevlar/epoxy. These materials have now found applications in commercial industries as well. A representation of the strength and stiffness of many composite materials including carbon composites is on the basis of effectiveness per unit weight [5]. The strength of the carbon fiber when placed in unidirectional orientation is more than that of most other materials. Even in the biaxial isotropic configurations, the stiffness of carbon/epoxy is comparable to that of steel. The comparison of the strength properties of carbon and glass fiber reinforced plastic (CFRP & GFRP) composites in different orientations with SE SMA and naturally occurring cane which have been used here is shown in (Fig.4.).

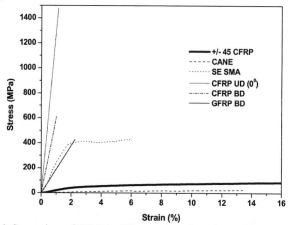

FIGURE 4. Comparison of CFRP in different orientations along with SE SMA & Cane.

Like other shape memory materials, Ni-Ti SMAs have the unique ability to recover their original shape after undergoing large deformations at a given temperature either through heating (referred to as the shape memory effect) or by the removal of the external load (referred to as the super-elastic effect). These properties are due to a reversible martensitic phase transformation occurring in the system between a crystallographically high-symmetry (cubic crystal structure) austenite phase to low symmetry (monoclinic crystal structure) martensite phase. The mechanical behavior as a function of temperature, strain and stress is summarized in (Fig.5.). The SMA exhibits the shape memory effect when deformed below the martensite finish temperature, M_f. These deformations (below M_f) are recovered by heating the material above the austenite finish temperature, A_f. The SMA is in its parent austenite phase above A_f. Stress-induced martensite is formed when the austenite is stressed to a certain level (above A_f). On removal of load, the stress induced martensite reverts to austenite at a lower stress, thereby resulting in the super elastic behavior. The resulting non-linear stress-strain relationship results in a hysteresis.

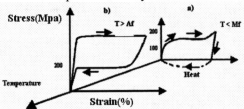

FIGURE 5. Schematic representation of stress-strain temperature showing:
a) shape memory, b) Superelasticity.

The dissipation of energy in the super elastic SMA is due to the stress hysteresis and is typically of the order of 150 to 300 MPa. The Ni-Ti SMA has very high specific energy storage capacity in the elastic region compared to both helical steel spring and aluminum flat specimen [6]. This is shown in Table 1.

TABLE 1. Specific energy stored in elastic region

SPECIMEN	Length (mm)	Mass (N)	Specific energy stored in elastic region (J/N)
NiTi SMA	50	0.00091	934
Steel helical spring	50	0.2	5
Aluminum strip	50	0.015	133

Fig.6. shows the comparison of the Ni-Ti super elastic SMA behavior with that of a helical tensile spring (wire dia=2.5 mm, coil dia=12 mm and length=100mm) whose overall stiffness is comparable with that of a Ni -Ti super elastic SMA wire.

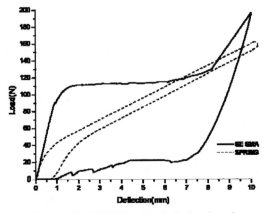

FIGURE 6. Comparison of Ni-Ti SE SMA with a helical spring of comparable stiffness

It is clear that the NiTi SMA has a far higher energy dissipation characteristic up to about 6% strain. Extending the comparison up to say, 10% strain (i.e., deformation up to 3 mm for 100 mm length of spring) the Ni-Ti SMA stiffness (deflection for a unit load) increases with increasing deformation beyond 6%. This stiffness increase enhances the stiffness of the device in which it is used similar to the steel springs. The variation of stiffness is a function of the strain rate in NiTi SMA and is discussed in detail in references [6].

DESIGN CONSIDERATION

The inputs considered in the design of landing gear are as below [7]:
1) Taking into account the maximum gas lift for a 320 m^3 airship; the reaction load is calculated to be about 1600 N. Applying a factor of safety of 1.5 on this load, an ultimate load of 2400 N is considered.
2) During landing, the skid is assumed to make an angle of 20 deg with the ground (Fig.2.).
3) Because there are two identical sub segments of the skid, each segment will take half of the load, that is, 1200 N.

The landing gear incorporating advanced materials is designed to have two load paths, namely, a primary load path and a secondary load path. The primary load path comprises the landing beam and vertical beam as shown in (Fig.2.). The secondary load path is the arch made of carbon composites and cane and incorporates the SMA element. The primary load path is designed to resist a larger share of the impact load and have a longer life. The secondary load path members which are the arch and the SMA are designed to resist relatively lesser amount of load while undergoing large elastic deformations and in the process dissipate the impact energy in the form of heat. The elements in the secondary load path are to be replaced after a fixed number of cycles, which is typically a couple of hundred cycles.

In order to realize the smart landing gear with energy dissipation features (comprising of arch, horizontal beam & vertical beam), the analysis approach that was adopted is the following:

1. First the arch containing natural cane and CFRP was analysed.

2. Next the arch was integrated with the beam of constant cross-section as shown in Table 2.

3. The beam was configured as a stepped member as shown in Table 2.

4. The SMA was integrated with the stepped beam and the arch.

For the arch, the section considered was rectangular as shown in Table 2 from bending considerations. This was analysed for different moduli; 60 GPa corresponds to CFRP, 20 GPa corresponds to GFRP & 14 GPa corresponds to value obtained from actual cut segments of the arch. For the same maximum stress values, the weight of GFRP section was comparably higher. Therefore, it was decided to use CFRP. In order to address the problem of achieving large elastic deformations, the curved shape of the arch was considered to enhance the elastic deformations. The results from the analysis of the arch are not presented here.

97

The arch was first integrated with the beam of constant cross-section. The sections considered were as shown in Table 2.

It is clear that given the high stiffness of carbon composites, the elastic deformations are not appreciable and there is little energy dissipation. In view of this, it was decided to have a stepped CFRP beam as shown in Table 2 which can give larger tip deflections.

With the stepped beam (where 'h' represents the depth of the rectangular beam cross section and 't' represents the CFRP layup thickness), a revised analysis was carried out. There was an improvement in the elastic deformation at the impact point but still it was far less than that required from the point of view of efficient energy dissipation (as there is no hysteresis involved between the loading and unloading of the CFRP material).

From the above studies it was clear that a smart material such as SMA with large recoverable elastic deformations and high energy dissipation capabilities had to be used to substantially enhance the elastic deformations and the energy dissipation features. The results of analysis for the configuration integrating SMA, arch and stepped beam is shown in Table 2. It is clear that the SMA substantially increases the load carrying capacity of the landing gear sub unit while retaining the elastic deformations.

The detailed process of energy dissipation is given in reference [6].

It is assumed that the impact takes place for a period of 3 s. The process of energy dissipation is illustrated in Table 1a.

MATERIAL AND TESTING DETAILS

In order to validate the analysis some limited tests were carried out.

The material used for the study was 0.6 mm superelastic NiTi SMA Wire. The chemical composition of the wire was Ni 54.3% and Ti 45%. The transformation temperatures in deg ^{0}C obtained from the DSC tests were as follows:

Mf	Ms	As	Af
6.8	12.5	11.9	17.5

The length of the wire used for the testing was 1000 mm. The specimens were straight and of uniform cross section. The modulus of the arch and beam obtained from test specimens cut from the actual component using a Zwick Universal Testing Machine (UTM) were used in the analysis. The experiments were conducted for two configurations. For configuration P, the subunit comprises a segment of the landing beam, CFRP arch with cane, and SMA. Configuration O is the same as configuration P but without SMA. In both the configurations the load was applied at the point as shown in Table 2. The length of the landing beam sub segment considered for tests was 1112 mm, and the height of the arch was 1240 mm. Load cell 1 monitors the total vertical load. Load cell 2 monitors the load taken by the arch. Load cell 3 monitors the SMA load for configuration P. The δv was measured using linear scales as shown in (Fig.7.)

TABLE 1a: The sequence of the energy dissipation process in the landing gear subunit.

Time 't' in s	σ in Mpa	ϵ in SMA	Stress-strain plot	Shape of Arch, beam, SMA sub unit
0	≈100	≈0.5%		
1	>400	≈1%		
2	≈500	>5%		
3	≈100	≈0.5%		

+

TABLE 2: Summary of load sharing (derived from testing and analysis results) between Arch, Beam and SMA.								
Figure	Total load, N	Young's mod. (GPa)	Analysis			Experiment		
			Load sharing, N		Defl. mm	Load sharing, N		Defl. mm
			Arch	Beam	δ_v	Arch	Beam	δ_v
t=4.625mm h=40.75mm Arch δv **Uniform c/s of Beam**	1000	Arch = 14 Beam = 30	401	604	28	-	-	-
h = 47.5 mm t = 1.25 mm / h1 = 46.75 mm t1 = 1.625 mm / h2 = 45.25 mm t2 = 2.375 mm / h3 = 43.75 mm t3 = 3.125 mm / h4 = 42.25 mm t4 = 3.875 mm / h5 = 40.75 mm t5 = 4.625 mm Arch b5 b4 b3 b2 b1 Beam δv h5 h4 h3 h2 h1 **Stepped Beam c/s**	1000	Arch = 14 Beam = 30	482	526	42	503	497	42
Arch SMA t5 b4 b3 b2 b1 Beam δv h5 h4 h3 h2 h1	1200	Arch = 14 Beam = 30	721	497	38	670	530	45

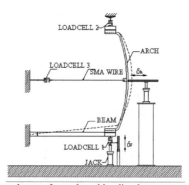

FIGURE 7. Experimental setup for arch and landing beam segment testing using SMA.

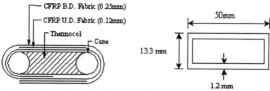

FIGURE 8. a) Cross-section view of arch; b) equivalent section of arch used in FE model.

ANALYSIS

During the preliminary design stage the sizing of the arch and beam was done based on the classical strength of materials approach. To simplify analysis during this stage, the arch and beam were considered separately. The sizing of the two components was based on the assumed load-sharing requirement and their individual deflections were computed.

The landing gear system comprising of beam, arch and SMA is designed to cater for the requirement of high energy absorption during landing as already mentioned. As the geometry clearly shows, the arch by virtue of its shape and material can undergo large deformation. In this way the system has geometrically non-linear behavior. The energy absorbing capability of the system is substantially enhanced by introducing the superelastic SMA wire in the geometrically non linear system. The SMA wire will, during its deformation stiffen the arch and increase the load carrying capacity of the arch considerably. The SMA being a non-linear elastic material, the entire system is non-linear geometrically and materially. Therefore, the behavior of the system is complex.

Modeling of such a system with classical method is not feasible. This is because, in the classical method, the assumption is that the structure is in linear elastic regime. It is assumed that the structure undergoes only small deflections. The classical solutions for large deflection behavior are quite complex. Therefore, in the present work, the analysis of the landing gear system is not addressed through classical methods. In order to study the behavior of the landing gear structure, non-linear analysis using finite element method (FEM) has been carried out.

ANALYSIS & TESTING: COMPARISON

Static tests conducted on the arch and the arch-SMA combination showed a nearly nonlinear load-deflection pattern. In the entire setup of the landing gear, the nonlinear behavior is primarily due to the geometric nonlinearity associated with the arch and the varying cross section beam. To compare the predictions of the FE model with the test results a nonlinear FE analysis is carried out using MSC/NASTRAN. The analysis is done for various configurations as shown in the Table 2. The arch and beam are modeled with the beam elements (CBEAM) and the SMA wire is modeled with the CELAS element in NASTRAN. The beam consists of 15 elements, the arch consists of 20 elements and the SMA wire is modeled with 1 element. Preliminary convergence studies were done before finalizing the FE model. The beam cross section is modeled in the finite element method (FEM) as a hollow section of a stepped beam of 50 mm*50 mm. The beam is fabricated by a wet layup process using bidirectional CFRP material with [0/90] orientation. For the beam, an equivalent modulus of 30 GPa obtained from tests conducted on specimens cut from actual test component is used. It may be noted that the modulus obtained here is lower than that normally found in published literature for CFRP and GFRP material. The reduction in modulus can be attributed to factors such as fabrication process, deviation in fiber orientation, environmental effects and so on. The arch is modeled here as an equivalent rectangular cross sections of 50 mm*13.3 mm*1.2 mm. The area of the equivalent arch section is adopted through the rule of mixtures considering the areas and moduli of both cane and CFRP layers wound over the cane section as shown in Fig.8. A modulus of 14 GPa is used for the arch section in the analysis. The SMA behavior is modeled in a simplified bilinear way. The comparison between analysis and experiments is provided in Table3.

CONCLUSIONS

1. The modeling approach has provided the inputs for building an efficient energy dissipating device combining materials with a wide range of properties within the volume and weight constraints.
2. The material and geometric nonlinearities have been effectively captured by the analysis.
3. The SMA behavior is well captured for the loading path.
4. The modeling of the beam, arch and SMA has been validated by the experiment.

ACKNOWLEDGMENTS

The author wish to acknowledge Dr A.R. Upadhya, Director, NAL for his constant encouragement.

Thanks are also due to Dr M.R. Madhava, Head, Advanced Composites Division for his guidance and advice. The useful discussions with Dr S. Selvarajan of C-CADD, NAL is also acknowledged. Thanks are also due to Satisha and G.T.Harsha of Advanced Composites Division.

REFERENCES

1. G. N. Dayananda, B. Varughese, and M. Subba Rao, " Shape Memory Alloy Based Smart Landing Gear for an Airship," AIAA JOURNAL OF AIRCRAFT Vol. 44, No. 5, September–October 2007
2. Philips, N.S., Carr, R.W., and Scranton, R.S., "Crashworthy Landing Gear Study," U.S. Army Air Mobility Research and Development Laboratory Tech. Rept. 72-61, April 1973.
3. Airoldi, A., and Lanzi, L., "Multi-Objective Genetic Optimization for Helicopter Skid Landing Gears," *Proceedings of the 46th AIAA/ASME/ASCE/AHS/ASC Structures, Structural Dynamics, and Materials Conference, American Institute of Aeronautics and Astronautics, Inc.,* Washington, D.C., 2005.
4. Sareen, A.K., Smith, M.R., and Howard, J.V., "Helicopter Skid Gear Dynamic Drop Analysis and Test Correlation," *Proceedings of the American Helicopter Society 54th Annual Forum,* American Helicopter Society International, Inc., Alexandria, VA, 1998.

5. Jones, R.M., "Mechanics of Composite materials, 2nd ed., Taylor and Francis. Inc., Philadelphia (1999).
6. G. N. Dayananda, M.S. Rao, Effect of strain rate on properties of Superelastic NiTi thin wires, Material Science and Engineering A,doi:10.1016/j.msea.2007.09.006
7. Khoury, G. A., and Gillet, J. D., "Airship Technology," Cambridge Univ. Press, Cambridge, England, U.K., 1999.

Phenomenological Modeling of Shape Memory Alloys

Vidyashankar Buravalla and Ashish Khandelwal

India Science Lab, General Motors R&D, Creator Bldg., ITPB, Bangalore 560 066, INDIA.

Abstract. Shape memory alloys exhibit two characteristic effects, viz., shape memory and superelasticity or pseudoelasticity, due to a reversible solid-solid transformation brought about by either temperature or stress or both. The two important aspects involved in modeling the macroscopic SMA behavior are the constitutive equation describing the stress-strain-temperature relationship and the evolution kinetics describing the phase transformation as a function of the driving forces. Phenomenological models for macroscopic behavior of SMAs are frequently used wherein the aforementioned aspects of SMA behavior are treated independently. Using empirical data, a phase diagram is constructed to describe evolution of martensitic phase fraction (ξ) as a function of stress and temperature. A constitutive equation is derived using the appropriate form of free energy. In this paper, salient aspects in phenomenological models are discussed and a robust model for SMA behavior is presented. Using a distance based memory parameter, rate based kinetics is provided along with a differential form of constitutive equation. Also, several critical issues in phenomenological modeling like prescribing consistent kinetics and catering to arbitrary thermomechanical loading are highlighted. Through numerical studies, it is shown that the proposed model provides consistent kinetics and caters to arbitrary thermomechanical loading.

Keywords: **Shape memory, constitutive models, evolution kinetics**.
PACS: Replace this text with PACS numbers; choose from this list:
http://www.aip.org/pacs/index.html Text should remain 10-pt.

1 INTRODUCTION

Shape memory alloys (SMAs) are a class of smart materials, which show a tendency to remember its shape corresponding to its stable/parent phase. They exhibit two interesting characteristic behaviors, viz., Shape Memory Effect (SME) and SuperElasticity (SE) or PseudoElasticity (PE) which enable them for a variety of engineering and medical applications. In general, medical applications exploit superelasticity, whereas, in actuators, SME is used to do useful work by generating recovery stress and/or recovery strain. The reason behind the presence of such fascinating properties is the first order diffusionless solid to solid phase transformation between its stable (parent) austenitic phase and the product martensitic phase. The Austenite (A) is a high symmetry cubic phase, in contrast to the low-symmetry monoclinic Martensite (M). This phase transformation results in a lattice distortion caused predominantly by the deviatoric stress and could lead to macroscopic shape

CP1029, *Smart Devices: Modeling of Material Systems, An International Workshop*
edited by S. M. Sivakumar, V. Buravalla, and A. R. Srinivasa
©2008 American Institute of Physics 978-0-7354-0553-0/08/$23.00

change. This is also accompanied by exchange of latent heat and hysteretic thermomechanical response. Concomitant to this phase change are significant changes in mechanical, electrical and thermal properties that render them as prime candidates for the development of smart structures and devices. A few common examples of this type of SMAs are Nitinol (NiTi), NiTiCu, CuAlZn and CuAlNi. Some ferrous alloys exhibit similar properties when activated by a magnetic field and hence are called as magnetic shape memory materials (eg NiMnGa). In this work, focus is on thermomechanically activated SMAs. Extensive literature exists that describes the SMA behavior (Duerig et al. [1990], Otsuka and Wayman [1998], Birman [1997], Shaw and Kyriakides [1995], Bernardini and Pence [2002], Devonshire [1954], Muller and Seelecke [2001], Friend [2001], Graesser and Cozzarelli [1991], Zhang and McCormick [2000a,b]).

As noted earlier, both thermal and elastic energies play a vital role in phase transformation in SMAs. Further, there is a strong interaction or coupling between these two energies resulting in a highly non-linear coupled stress-strain-temperature behavior. Hence, material models that are applicable to conventional metallic materials are not, in general, appropriate for these materials. These features pose significant challenges to the development of mathematical models for shape memory material behavior. Several review papers exist that provide a broad review of models for SMA behavior (Prahlad and Chopra [2001], Brinson and Huang [1996], Matsuzaki and Naito [2004], Elahinia and Ahmadian [2004] and Buravalla and Khandelwal [2007]). In the following section an overview of some of the frequently used phenomenological models is provided.

2 A BRIEF OVERVIEW OF THE PHENOMENOLOGICAL MODELS FOR SMA

A brief background on the modeling macroscopic SMA behavior and a short overview of the some of the phenomenological models are given below. Most of macroscopic modeling approaches for constitutive behavior can be classified into two broad categories, viz., phenomenological approach and thermodynamically consistent energy based approach. In order to capture the stress-temperature-strain behavior in terms of phase transformation, usually an internal variable is introduced. In general, this internal variable is usually the martensitic phase fraction, ξ. Some models distinguish the temperature induced or self accommodating martensite (ξ_T) and stress induced or detwinned martensite (ξ_S). Another frequently used distinction among martensitic variants are the martensitic twins, M^+ and M^-. It may be noted that ξ_T is will be equivalent to $M^\pm$ and ξ_S corresponds to M^+. Fundamental free energy based models use a suitable free-energy form like Gibb's or Helmholtz potential and using thermodynamic principles, derive the constitutive equation and appropriate phase evolution kinetics. These models, so far, are fairly complex and computationally intensive and hence do not easily lend to the development of simple design tools. Hence more empirical approaches are developed based on the phenomenology. Phenomenological approach is simple and amenable to represent the macroscopic phenomena, but rely heavily on experimental data to derive necessary material parameters, thereby restricting their applicability to specific materials and conditions. However, these models are quick and accurate within the requirements of engineering accuracy.

Such phenomenological models are the focus of the present study. They separate out the following two features in context of modeling:

a) Constitutive relationship between the stresses and strains due to elastic forces, the thermal expansion and the strain due to phase transformation

b) The driving forces and the evolution (kinetics) of the phase transformation

A commonly used phenomenological approach to investigate the SMA behavior is the phase diagram based model using the stress-temperature phase diagram. Typically the phase diagram is constructed by determining the phase boundaries using experimental data, wherein tests are done under simple loading conditions, like constant stress thermal cycle or a constant temperature mechanical (stress) cycle. A linear fit to the test data is used to obtain idealized phase boundaries. This is used to identify active transformation zones wherein the stress induced martensite (ξ_S), temperature induced martensite (ξ_T) and austenite form. Subsequently, in each of the active evolution regions, local kinetics for the phase fraction is determined by expressing ξ as a suitable function of stress and temperature. This is similar to the concept of 'flow-rule' in rate independent plasticity. Figure 1 illustrates a typical phase diagram with different active transformation zones. The loading directions that induce phase transformation are indicated in each of the active zones. An arbitrary load path is also indicated on the phase diagram that represents arbitrary thermomechanical loading of SMA material. The model is developed using several experimentally obtained material parameters (ref. Fig. 1) like, two critical stresses (σ^{cr}_s and σ^{cr}_f) governing formation of stress induced martensite, temperatures governing the formation of martensite (M_s, M_f), austenite (A_s and A_f) and the constants C_M and C_A. The subscripts 's' and 'f', respectively, denote start and finish transformation.

Tanaka [1986] was among the first to use this approach for SMA to study superelasticity with martensitic fraction (ξ_S) as an internal variable. Analogous to the exponential Koistinen-Marburger relation, an evolution law is deduced in terms of stress and temperature. Further, using strain and temperature as the control variables, a constitutive relation was derived with constant material functions. This model was further adapted by Liang and Rogers [1992] and Brinson [1993]. To obtain a better fit to the experimental data, Liang and Rogers [1992] modified the phase kinetics to a cosine based function. Both Tanaka [1986] and, Liang and Rogers [1992] were successful in capturing the superelastic behavior. Based on the broad framework of Tanaka [1986] and Liang and Rogers [1992], Brinson [1993] proposed a modified model to also account for the shape memory effect. The essential distinguishing feature in Brinson [1993] model is splitting the martensitic phase fraction into two parts, viz., the temperature induced twinned fraction (ξ_T) and the stress induced (detwinned) fraction (ξ_S). This differentiation of the phase fractions is necessary to capture recovery stress/strain. To describe the constitutive behavior, Brinson [1993] derived a constitutive model with Voigt type non-constant material functions. Prahlad and Chopra [2001] provide a comparative evaluation of these models. It may be mentioned that the Brinson's model has been used extensively in the literature to model SMA material and devices (Brinson and Lammering [1993], Prahlad and Chopra [2001], Elahinia and Ahmadian [2004]). Wang et al [2006], have modified Brinson's kinetics to account for rubber-like effects below M_f for certain class of SMAs. This model is further used by Prahlad and Chopra [2001] as a basis to include the effects wherein control variables are not quasistatic. Wu and Pence [1998] and Chenchiah and Sivakumar [2004] have proposed a similar differentiation of phase fractions (M^+ and M^-) to describe shape memory behavior.

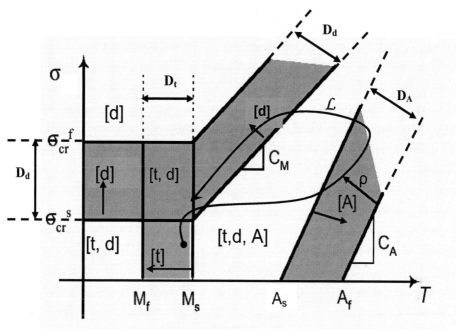

FIGURE 1. Idealized phase diagram for typical NiTi SMA material under uniaxial stress condition showing the dependence of transformation temperature on stress and regions with different phase mixtures. L is the load path, [A]- Austenite, [t] – twinned martensite; [d] – detwinned martensite. The shaded region represents the region where transformation is said to occur. ρ is the distance of a point on the load path from finish transformation boundary.

Based on the Brinson [1993] model, a consistent rate form Lumped Parameter Model (LPM) for macroscopic thermomechanical behavior of SMAs is provided in Sections 3 and 4. A set of evolution equations in terms of rate of martensite fraction ($\xi = \xi_S + \xi_T$) using new transformation conditions based on an additional memory parameter is given in Section 3. The 'rate' is in terms of driving forces, viz., stress (σ) and temperature (T). A constitutive equation in differential or rate form is derived in Section 4. Section 5 illustrates the capability of the proposed model to capture the essential SMA behavior, viz., superelastic and shape memory effects. Section 6 discusses several important issues in phenomenological models and shows the efficacy of the proposed model in addressing these issues.

3. EVOLUTION KINETICS

In this section, expressions for the evolution of martensitic volume fractions ξ_S and ξ_T as a function of the driving forces stress and temperature are provided. As noted earlier, common assumed kinetics are of linear, exponential and cosine forms. Most of plasticity based models incorporate exponential type kinetics (Boyd and Lagoudas [1996] and Aurrichio and Lubliner [1997]). From the available test data, cosine based kinetics seems to offer the best fit to the data. The kinetics can be given either in the algebraic form or in the rate form (rate in terms of σ and T) given in one form, the other can always be derived. In this rate form, the notion of time is introduced only as a parameter and these models describe rate independent (quasistatic) evolution. From an algorithmic consideration, rate form offers several advantages like easy implementation in situations where complicated load paths are traversed involving partial or inner transformation loops. In view of this, in this paper, a rate form of kinetics based on the cosine function is provided in the following section. It may be noted that the expressions per se, are based on the Brinson's kinetics in algebraic form. However, the present model differs significantly from the Brinson's kinetics, specifically in the conditions that determine, given a load increment in terms of stress and temperature, whether the transformation occurs or not. As will be explained in Section 6, additional transformation conditions are stipulated to provide adequate memory during arbitrary thermomechanical loading.

This global kinetic law governs the application of transformation functions for the prediction of SMA behavior to arbitrary stress-temperature loading (Bekker and Brinson [1998]). This is achieved by splitting the load-path into convenient segments wherein the phase fraction is a single valued function of stress and temperature (local kinetic laws). The assembly of the chain of these local kinetic laws provides the global law. The evolution expressions for the martensitic phase fraction and the constitutive equations are given below. This model has been found to correlate well with quasi-static and isothermal experimental data on SMA wires.

The evolution kinetics initially proposed by Brinson [1993] and later modified by Khandelwal and Buravalla [2008] for two-variant martensitic transformation is adapted here. To account for arbitrary thermomechanical loading Distance Based Algorithm (DBA) in the rate form is given below.

3.1 Distance based evolution kinetics

Figure 1 depicts a typical idealized phase diagram with associated parameters that are used to describe the kinetics. l is the arbitrary load path on the phase-diagram. $\mathcal{L}$ is the load path already traversed in a given transformation zone. In the proposed approach, for each point on the load path inside a given transformation zone, a notion of distance from the finish transformation boundary is associated. Inside each transformation zone the minimum value of distance of all the points over the load history is defined as the memory parameter. Using this memory parameter, an additional transformation condition is defined to obtain a robust description of the kinetics. The distance (ρ) of a point i(σ,T) on the load path ($\mathcal{L}$) from the finish transformation boundary is used to define the memory parameter ($\wp$). The definition of ρ and $\wp$ are given below:

$$\rho_\alpha^i = \frac{\vec{r}^i \bullet \vec{n}_\alpha^f}{D_\alpha}, \quad \wp_\alpha = \min(\rho_\alpha), \quad \forall \ i \in L \ \text{ and } \quad \alpha = \{A,d,t\} \tag{1}$$

where, $\vec{r}^i$ is the position vector of point i , $\vec{n}_\alpha^f$ is the normal to finish boundary and D_α is the width of transformation strip (Fig. 1). The memory parameter ($\wp_\alpha$) corresponds to the maximum amount of transformation that has occurred in the zone α. Due to monotonicity of the ξ inside that zone, $\wp_\alpha$ represents the distance from the finish boundary of the last point at which the transformation has occurred.

For multi-variant transformation, the conservation of mass implies,

$$\dot{\xi} = \dot{\xi}_T + \dot{\xi}_S;$$

$$\dot{\xi}_A + \dot{\xi}_T + \dot{\xi}_S = 0$$

where, ξ_A is volumetric fraction of austenite.

3.1.1 Formation of Martensite

The kinetics for formation of martensite is given below.

For $T \geq M_s$ and $\sigma_s^{cr} + C_M(T - M_s) \leq \sigma \leq \sigma_f^{cr} + C_M(T - M_s)$ and $\rho_d < \wp_d$,

$$\dot{\xi}_S = \chi_2^{AS}(1-\xi_S)\frac{\pi}{(\sigma_s^{cr} - \sigma_f^{cr})} \cot\left\{\frac{\pi}{2(\sigma_s^{cr} - \sigma_f^{cr})}\left[\sigma - \sigma_f^{cr} - C_M(T - M_s)\right]\right\} \left(\dot{\sigma} - C_M \dot{T}\right),$$

$$\dot{\xi}_T = \chi_2^{AS}\xi_T\frac{\pi}{(\sigma_s^{cr} - \sigma_f^{cr})} \cot\left\{\frac{\pi}{2(\sigma_s^{cr} - \sigma_f^{cr})}\left[\sigma - \sigma_f^{cr} - C_M(T - M_s)\right]\right\} \left(\dot{\sigma} - C_M \dot{T}\right), \tag{2}$$

where
$$\chi_2^{AS} = \begin{cases} 1 & \text{when} \begin{cases} (\dot{\sigma} - C_M \dot{T}) > 0 \\ \sigma_s^{cr} + C_M(T - M_s) \leq \sigma \leq \sigma_f^{cr} + C_M(T - M_s) \end{cases} \\ 0 & \text{otherwise} \end{cases}$$

109

For $T \leq M_s$ and $\sigma_s^{cr} \leq \sigma \leq \sigma_f^{cr}$ and $\rho_d < \wp_d$,

$$\dot{\xi}_S = \chi_1^{AS}\left(1-\xi_S\right)\frac{\pi}{\left(\sigma_s^{cr}-\sigma_f^{cr}\right)}\cot\left\{\frac{\pi}{2\left(\sigma_s^{cr}-\sigma_f^{cr}\right)}\left[\sigma-\sigma_F^{cr}\right]\right\}\dot{\sigma} \,,\qquad (3)$$

where $\chi_1^{AS} = \begin{cases} 1 & \text{when}\begin{cases}\sigma>0 \\ \sigma_s^{cr}\leq\sigma\leq\sigma_f^{cr}\end{cases} \\ 0 & \text{otherwise}\end{cases}.$

For $M_f \leq T \leq M_s$ and $\sigma \leq \sigma_f^{cr}$ and $\rho_t < \wp_t$,

$$\dot{\xi}_T = -\chi^{AM}\left(1-\xi_T\right)\frac{\pi}{M_s-M_f}\cot\left[\frac{\alpha_M}{2}\left(T-M_f\right)\right]\dot{T} \,,\qquad (4)$$

where $\chi^{AM} = \begin{cases} 1 & \text{when}\begin{cases}\dot{T}<0 \\ M_s\leq T\leq M_f\end{cases} \\ 0 & \text{otherwise}\end{cases}.$

3.1.2 Formation of Austenite

For $T \geq A_s$ and $C_A(T-A_f) \leq \sigma \leq C_A(T-A_s)$ and $\rho_A < \wp_A$,

$$\dot{\xi} = \chi^{SA}\xi\frac{\pi}{C_A\left(A_s-A_f\right)}\left\{\cot\left[\frac{\pi}{2\left(A_s-A_f\right)}\left(T-A_s-\frac{\sigma}{C_A}\right)\right]\right\}^{-1}\left(\dot{\sigma}-C_A\dot{T}\right),\qquad (5)$$

$$\dot{\xi}_S = \frac{\xi_S}{\xi}\dot{\xi} \qquad (6)$$

$$\dot{\xi}_T = \frac{\xi_T}{\xi}\dot{\xi} \,,\qquad (7)$$

where $\chi^{SA} = \begin{cases} 1 & \text{when}\begin{cases}(\dot{\sigma}-C_A\dot{T})<0 \\ C_A(T-A_f)\leq\sigma\leq C_A(T-A_s)\end{cases} \\ 0 & \text{otherwise}\end{cases}.$

Having estimated the volume fraction(s) for any σ and T, it is necessary to obtain strain to complete the thermomechanical description. To facilitate this, a constitutive law is needed to relate stress, strain and temperature in terms of the internal variable, ξ. In the following, one such law is derived.

4. CONSTITUTIVE LAW BASED ON THERMODYNAMICS

In the context of one-dimensional constitutive modeling, the principle of mechanical energy balance and Clausius-Duhem inequality (concerning entropy production) is simplified and in the current configuration, can be stated as:

110

$$\rho \dot{U} - \hat{\sigma} L + \frac{\partial \boldsymbol{q}_{sur}}{\partial x} - \rho q = 0; \qquad \rho \dot{S} - \rho \frac{q}{T} + \frac{\partial}{\partial x}\left(\frac{\boldsymbol{q}_{sur}}{T}\right) \ge 0 \qquad (8)$$

where, $\dot{U}$, $\hat{\sigma}$, L, q, and q_{sur}, respectively, represents the internal energy density, the Cauchy stress tensor, the rate of deformation tensor, the heat production term and the heat flux, and S, T, x and ρ represent the entropy density, temperature, the material coordinate and the density in current configuration, respectively. It is further assumed (Tanaka [1986]) that the thermomechanical behavior is completely described by the set of variables (ε,ξ,T) where ε is the Green strain and ξ is martensitic fraction, ranging from zero (complete austenite) to one (complete martensite). Using the definition of Helmholtz's free energy ϕ = U- TS, the Eq. (8), in the reference configuration, takes the following form

$$\left(\frac{\sigma}{\rho_0} - \frac{\partial \varphi}{\partial \varepsilon}\right)\dot{\varepsilon} - \left(S + \frac{\partial \phi}{\partial T}\right)\dot{T} - \frac{\partial \phi}{\partial \xi}\dot{\xi} - \frac{1}{T\rho_0}\boldsymbol{q}_{sur}(Cof\ F)^T \frac{\partial T}{\partial X} \ge 0 \qquad (9)$$

where σ is the second Piola-Kirchoff stress tensor, $Cof\ F$ is the cofactor of the deformation gradient tensor F and ρ_0 is the density and X is the material coordinate in the reference configuration.

A necessary condition for Eq. (8) to hold for any arbitrary $\dot{\varepsilon}$ and $\dot{T}$ is:

$$\sigma = \rho_0 \frac{\partial \varphi(\varepsilon,\xi,T)}{\partial \varepsilon}$$

$$S = \frac{\partial \varphi}{\partial T}$$

From first principles, the above takes the form,

$$d\sigma = \frac{\partial \sigma}{\partial \varepsilon}d\varepsilon + \frac{\partial \sigma}{\partial \xi}d\xi + \frac{\partial \sigma}{\partial T}dT$$

Without loss of generality, this equation can take the form:

$$d\sigma = D(\varepsilon,\xi,T)\,d\varepsilon + \Omega(\varepsilon,\xi,T)\,d\xi + \Theta(\varepsilon,\xi,T)\,dT \qquad (10)$$

where, the characteristic material functions are, in general, defined as follows.

$$D(\varepsilon,\xi,T) = \rho_0 \frac{\partial^2 \varphi}{\partial \varepsilon^2}, \quad \Omega(\varepsilon,\xi,T) = \rho_0 \frac{\partial^2 \varphi}{\partial \varepsilon \partial \xi}, \quad \Theta(\varepsilon,\xi,T) = \rho_0 \frac{\partial^2 \varphi}{\partial \varepsilon \partial T}$$

For simplicity, D can be assumed to be linear function of ξ and Θ can be assumed to be a constant. Brinson [1993] considers that only the stress induced martensite ξ_S contributes to the recovery stress and hence defines the transformation tensor Ω to obtain the recovery stress. The differential form of the constitutive equation is as follows.

$$d\sigma = D(\xi)d\varepsilon + \Omega(\xi,\varepsilon)d\xi_S + \Theta dT \tag{11}$$

Where, D and Ω are defined as follows:

$$D = D_a + \xi(D_m - D_a)$$

$$\Omega(\xi,\varepsilon) = -\varepsilon_l D(\xi) + (\varepsilon - \varepsilon_l \xi_s)(D_m - D_a)$$

where, ε_l is the maximum residual strain, a material property determined experimentally. One can obtain a simplified form of constitutive equation by integration. On integration of Eq.11 and imposition of appropriate initial conditions, the following constitutive equation as given by Brinson [1993] is obtained:

$$\sigma - \sigma_0 = D(\xi)\varepsilon - D(\xi_0)\varepsilon_0 + \Omega(\xi)\xi_S - \Omega(\xi_0)\xi_{S0} + \Theta(T - T_0) \tag{12}$$

It may be noted that though the integrated form of this equation is identical to that derived by Brinson [1993], the definition of transformation modulus Ω is different in the two models. Buravalla and Khandelwal [2007] provide a detailed analysis regarding the need for the consistent definition of Ω which is used here.

5. NUMERICAL CASE STUDIES

The efficacy of this model to capture the essential superelastic effect and the shape memory effect is illustrated by considering two typical examples. The SMA material properties used in the present study are listed in Table 1.

TABLE 1. SMA Material properties used in the analysis.

Property	Martensite	Austenite
Modulus	$E_m = 26.3$ GPa	$E_a = 67$ GPa
Theta (Θ)	0.55 MPa/ ^{0}C	0.55 MPa/ ^{0}C
Transformation Temperatures	$M_f = 9.0\ ^0$C	$A_s = 34.5\ ^0$C
	$M_s = 18.4\ ^0$C	$A_f = 49\ ^0$C
Critical Stresses for detwinning	$\sigma_s^{cr} = 100$ MPa	-
	$\sigma_f^{cr} = 170$ MPa	
Stress-Temperature slopes	$C_M = 8$ Pa/^{0}C	$C_A = 13.8$ Pa/^{0}C
Transformation strain	$\varepsilon_L = 0.067$	$\varepsilon_L = 0.067$

5.1 Superelastic behavior

The stress-strain behavior for several constant temperatures is captured. The SMA is subjected to mechanical loading and unloading at a constant temperature and the strain and the volume fraction are computed for each load increment. Figs. 2a-d illustrate the classical superelastic behavior. The load path on the phase diagram, stress-temperature-volume fraction and the stress-strain curves are plotted to illustrate the Superelastic behavior. Figures 2b,c illustrate the superelastic behavior wherein, during loading, the Stress Induced Martensite (SIM) fraction ξ_S begins to evolve at a certain critical stress and the entire austenite is transformed into SIM at a particular stress. This transformation is reversed during unloading. Note that the critical stresses for forward and reverse transformations are different.

5.2 Shape memory behavior

A typical case of stress-free shape or strain recovery cycle is captured. The SMA is subjected to mechanical loading and unloading at a constant temperature below M_F. This induces a remnant strain in the SMA which can only be recovered when heated above A_F. Cooling back to temperature below M_F will result in no macroscopic strain and thus completes one shape memory cycle. Since the heating and cooling is performed in stress-free condition, it is called stress free recovery. Figs. 3a-d illustrate the classical superelastic behavior. The load path on the phase diagram, stress-temperature load history, stress-temperature-volume fraction and the stress-strain curves are plotted to illustrate the shape memory behavior.

From these examples it is clear that the model is able to capture the essential shape memory alloy behavior. In the following section, some important issues pertaining to the phenomenological models in general, phase diagram based models in particular, are discussed. It is shown that the model presented here overcomes the drawbacks in some of the existing models.

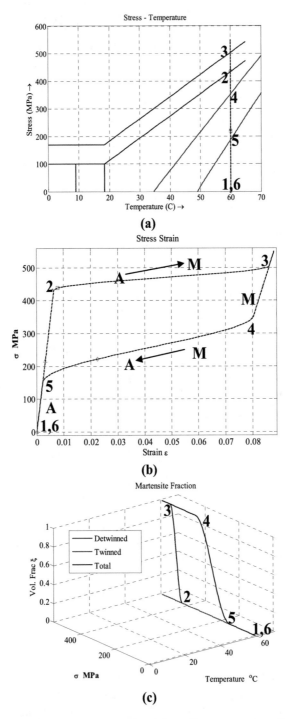

FIGURE 2. Superelasticity in Mat-A.at T-60 C. (a) Stress-Temperature phase diagram with load path (1-6); (b) Stress-strain plot. (c). Evolution of volume fraction ξ as a function of stress and temperature

114

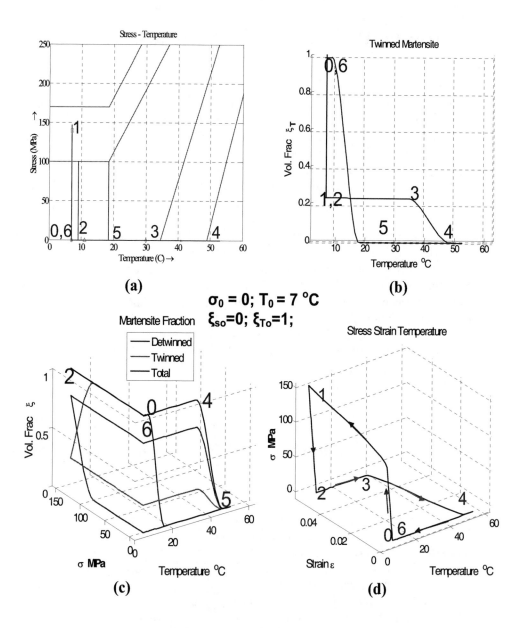

FIGURE 3. Free recovery cycle in Mat-A. (a) Stress-Temperature phase diagram with load path (1-6); (b) Twinned fraction vs Temperature showing detwinning and subsequent transformation to A and in turn to twinned martensite. . (c). Evolution of volume fraction ξ as a function of stress and temperature (d) Stress-strain temperature showing the memory cycle

6. SOME CRITICAL ISSUES IN PHASE DIAGRAM BASED MODELS

Researchers have faced several issues in developing appropriate phenomenological models for SMA behavior, especially under realistic arbitrary thermomechanical loading conditions. Due to empirical approach, several critical issues can occur in for SMAs. A few of the important aspects that need to be considered in arriving at a satisfactory model for macroscopic SMA behavior are:
1. Consistency in the definition of material functions
2. Identification of appropriate phase transformation conditions under arbitrary loading
3. Consistency in evolution kinetics defined in each local transformation zones, especially in zones where more than one variant is evolving
4. Capability to model incomplete transformations (partial and/or inner loops)
Buravalla and Khandelwal [2007] discuss in detail the first aspect in the context of the frequently used Brinson's model (Brinson [1993]). The second and third aspects are related to the evolution kinetics based on phase diagram based approach and will be discussed in the following. Modeling partial and inner hysteresis loops that could occur frequently in realistic applications of SMA has been addressed by several authors in the past (Ortin et al [2006], Visintin [2006], and Bekker and Brinson [1998]) and is not discussed in this paper.

6.1 Transformation conditions for arbitrary thermomechanical loading

An important aspect of transformation in SMAs is the path or history dependency. Thus, it is necessary to develop models which account for transformation under arbitrary loading. Due to the nature of the evolution kinetics and phenomenological description of phase diagram, existing phase diagram based models function adequately only under simplistic conditions of strictly monotonic loading. However, in reality, load perturbations could occur, leading to arbitrary non-monotonic thermomechanical loading. Under such conditions, even if there is variation in loading, transformation may not occur due to the existence of active and dead directions. In view of this, Bekker and Brinson [1998] have proposed models to predict evolution under arbitrary thermomechanical loading. They propose three different algorithms (Y, YF and Z algorithms) for arbitrary thermomechanical loading. These algorithms are based on discretizing the given load path into segments by defining various types of switching points. The kinetics is then applied to each segment (local kinetics) and then a rule to assemble the local kinetics into a global kinetics is presented. They investigated several cases of arbitrary thermomechanical loading and have highlighted several issues in predicting the evolutionary behavior. For instance, if there are load fluctuations inside the transformation zone, there is continuous creation of new switching points leading to ratcheting phenomenon. These could be spurious since it has not been experimentally verified and are highly unlikely in polycrystalline materials. This may be attributed to a lack of appropriate transformation conditions to determine the evolution under fluctuating loads (Govindjee and Kasper [1999]). These issues in the above model need to be addressed in developing a robust kinetics using a phase diagram. Two important aspects in prescribing kinetics under arbitrary loading are:

1. Appropriate memory to incorporate path dependency
2. Stability of the material behavior during repeated loading

It is essential to incorporate an appropriate notion of memory of the transformation within a zone while prescribing the kinetics. In general, due to history dependent material behavior, information about each point on the previously traversed load path is necessary to describe subsequent transformation. However, it is possible to capture the evolution adequately using simpler ways. The memory incorporated in the kinetics provided in the present paper involves a notion of distance of a point on the load path from the finish transformation boundary. This distance is used as a measure of transformation within a given transformation zone. Since ξ is monotonic in any given transformation zone, additional condition is specified for the occurrence of transformation in any active zone using the distance measure (refer Section 3.1). This will be clear if one compares the transformation conditions provided by Brinson [1993] and those prescribed in the present work. It is shown in the following that this additional criterion would eliminate the prediction of spurious evolution behavior.

6.1.1 Superelastic behavior with arbitrary stress variation

The prediction of the influence of load fluctuations on superelastic behavior is investigated. As shown in Figure 4a-c, after a complete load-unload cycle, several load fluctuations are introduced during the second load cycle. This clearly brings out the differences in the predictions of the three algorithms in the Bekker and Brinson [1998], (B-B) approach and the proposed DB kinetics. Due to repeated change of load directions, each time the load starts to increase, switching points are created in the B-B model. These switching points restart the transformation and hence all the three switching point based models (Y, Z and YF) show increasing strain due to increasing martensite formation. However, under stress controlled loading, the amount of transformation and hence the strain accumulation varies showing ratcheting behavior. This is also referred to as 'shakedown' in the literature. Of the three B-B models, Y algorithm shows least pile-up. However, due to the incorporation of appropriate memory, the distance based model shows no strain accumulation over all the load fluctuations (Fig. 4b,c). Table 2 lists the volume fraction and strain at the end of two different load fluctuation cycles, predicted by both B-B and proposed distance based model. It is clear that the DB model predicts elastic response (no transformation and hence no pile-up) under this type of load fluctuations. Further, transformation occurs only when the load exceeds the previous maximum.

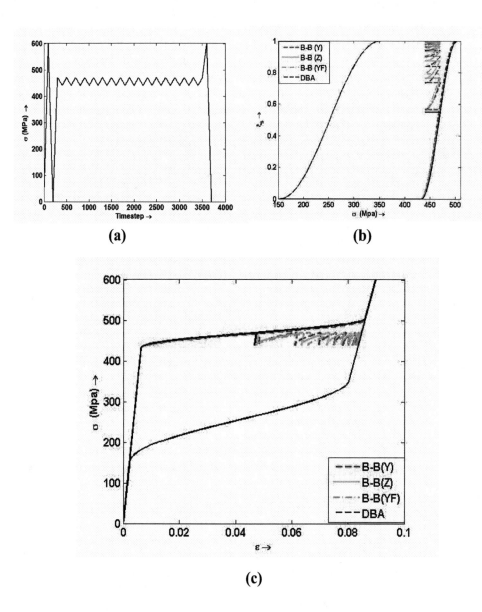

(a)

(b)

(c)

FIGURE 4. (a) Superelastic load history with second cycle having repeated load excursions in the second load cycle. The excursions are between the same maximum and minimum levels. (b) Detwinned martensite evolution as a function of stress. Pile up in Y, Z and YF algorithms. (c) Stress strain response for a superelastic load history with load excursions. Strain ratcheting is clearly seen the switching point based algorithms. DBA shows elastic load-unload behavior.

TABLE 2. Volume fraction ξ_S and strain, ε predicted at the end of different load fluctuation cycles

Load Excursion cycle Number	Bekker-Brinson (Y-Algorithm)	Bekker-Brinson (YF-Algorithm)	Bekker-Brinson (Z-Algorithm)	Distance Based Algorithm
2	ξ_S =0.8381	ξ_S =0.9048	ξ_S =0.8733	ξ_S =0.5314
	ε =0.0704	ε =0.0761	ε =0.0735	ε =0.0460
6	ξ_S =0.9774	ξ_S =0.9956	ξ_S =0.9957	ξ_S =0.5314
	ε =0.0828	ε =0.0845	ε =0.0839	ε =0.0460

6.2 Clarity in the definition of Phase boundary for combined martensite evolution zone

It may be noted that in phase diagram based models, the kinetics of evolution of each phase is defined independently for each active zone. Khandelwal and Buravalla [2007] and Chung et al [2007], have shown that it is essential to ensure that the kinetics prescribed for simultaneous evolution of ξ_S and ξ_T do not yield aphysical values (ξ >1). There is another important aspect to be considered here. If the load path traverses from an active zone wherein only either the twinned or the detwinned martensite evolves, to a zone wherein both of them evolve, it is essential to have consistent kinetics. For instance, typically two different kinetics are prescribed in zones [t] and [t,d] (Figure YY) for the evolution of twinned martensite ξ_T (Brinson [1993]). Let us consider the load path LP1. There will not be any transformation from point 'i' to 'a' and only ξ_S evolves from 'a' to 'b'. Subsequently, inside the zone [t,d], both ξ_S and ξ_T evolve. Due to the prescription of independent kinetics for ξ_S in [d] and [t,d], there will be a discontinuity in the value of ξ_S due to the definition of boundary value ξ_{S0} used in the kinetics. Hence, it is essential to identify and define appropriate boundaries while transitioning into combined evolution zones. In the algebraic or exact kinetics (prescribing ξ) this can be achieved by ignoring the boundary between [d] and [t,d] for ξ_S. Similar rationale applies to evolution of ξ_T in the case of load path like LP2. This can be achieved naturally in rate form of kinetics and is numerically illustrated below.

Figure 6a shows a linear load path P(80 MPa, 20 ^{0}C) to S(190 MPa, 11 ^{0}C). In this example, at P, the material is assumed to be completely austenite. There will not be any transformation till Q. From Q till R, only ξ_S will evolve. Beyond R, along with ξ_S, ξ_T also evolves. Due to independent kinetics for ξ_S in [t,d], as in Brinson [1993], there is a jump in ξ_S which is clearly seen (Fig. 6b,c) with respect to both stress and temperature at R. Due to the nature of transformation zone, initially ξ_T evolves from austenite and then gets detwinned to ξ_S. It is clear from these results that in Brinson [1993], due to independent kinetics for ξ_S and ξ_T in [t,d], the total martensite fraction (ξ_S + ξ_T) exceeds 1, which is physically inadmissible. Khandelwal and Buravalla[2008] corrected this anomaly and presented an algebraic kinetics which provides physically admissible values of ξ (Fig 6b,c). It may be mentioned that such problems do not occur in the rate form of kinetics (proposed DBA) which gives both a smoother and physical admissible evolution. It may be noted that the predictions from Algebraic and DBA models, though cosine based, are different for the evolution of ξ_T.

119

Thus, it is essential to have a kinetics which ensures smooth and consistent evolution. This issue will also be critical when modeling rubber-like elasticity as discussed in Wang et al [2006].

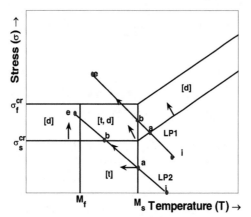

FIGURE 5. Phase diagram showing martensite evolution zones. . LP1 and LP2 are the load paths, [t] – twinned martensite; [d] – detwinned martensite. a and b are the points where the load path enters a given transformation zone. [t,d] is the combined evolution zone.

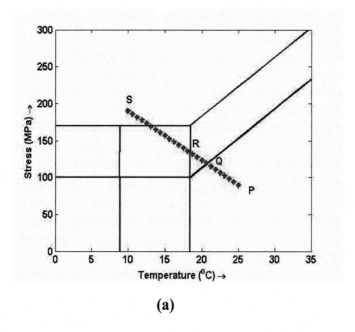

(a)

FIGURE 6A. Phase diagram showing load path PQRS.

120

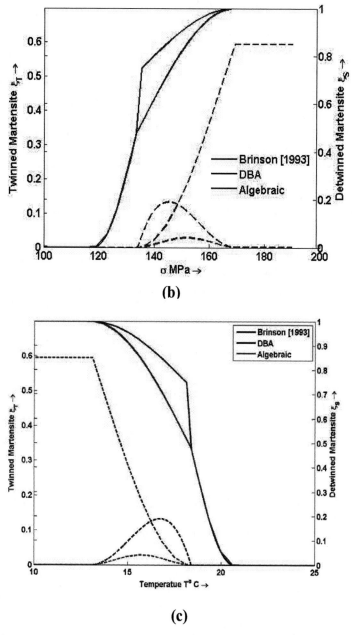

(b)

(c)

FIGURE 6. Evolution of martensite. (b) ξS and ξT (- - -) as a function of σ for load path PQRS. (c) ξ_S and ξ_T (- - -) as a function of T for load path PQRS

CONCLUSIONS

Phenomenological models are frequently used to predict the macroscopic behavior of SMAs. In this paper, based on the Brinson [1993], a new LPM for SMA is provided which incorporates a rate form of kinetics and constitutive equation. The cotangent form kinetics describes the evolution of martensitic fraction as a function of driving forces, viz., stress and temperature using a phase diagram. To account for the evolution under arbitrary thermomechanical loading, a suitable memory parameter is defined based on the distance of a point on the load path already traversed in a given transformation zone. This memory parameter is used to define additional transformation condition. Also, critical issues like consistency in defining local evolution kinetics, especially in zones where multiple types of martensite evolve are highlighted. Through several numerical studies, it is shown that the proposed model overcomes the drawbacks highlighted in some of the existing models. It is shown that the proposed model offers a simple and convenient model to investigate SMA behavior under realistic situations involving arbitrary thermomechanical loading. This LPM can be suitably implemented in a finite element like framework to investigate SMA based devices. Due to simplicity, this model can also be used as a design aid in developing SMA based devices. Further, this model can be easily extended to investigate incomplete transformations which manifest as inner or partial hysteresis loops.

ACKNOWLEDGMENTS

The authors gratefully acknowledge the valuable discussions with Dr. Prakash Mangalgiri and Prof. Sivakumar M S. and Mr. Amol Thakre.

REFERENCES

1. W. T. Duerig, N. K. Melton, D. Stockel and M. C. Wayman, "Engineering aspects of shape memory alloys," Butterworth-Heinemann (1990).
2. K. Otsuka and C. M. Wayman, "Shape Memory Materials," Cambridge University Press, (1998), Cambridge, 284pp.
3. V. Birman, *Appl. Mech. Reviews* **50**, 629-645 (1997).
4. J. A. Shaw and S. Kyriakides, *J. Mech. Phy. Solids* **43** (8), 1243–1281 (1995).
5. D. Bernardini and T.J.Pence, "Shape memory materials, modeling," in Encyclopedia of Smart Materials, John Wiley & Sons, 2002, pp. 964–980.
6. A. F. Devonshire, *Adv. Phys.* **3**, 86–130 (1954).
7. I. Muller and S. Seelecke, *Math. Comput. Model.* **34**, 1307–1355 (2001).
8. C. Friend "Shape memory alloys," in Encyclopedia of Materials, Science and Technology (2001) Elsevier Science, pp. 1–7.
9. E. Graesser and F. Cozzarelli, *J. Eng. Mater.* **117**, 2590–2608 (1991).
10. S. Zhang and G. P. McCormick, *Acta Mater.* **48**, 3081–3089 (2000a).
11. S. Zhang and G. P. McCormick, *Acta Mater.* **48**, 3091–3101 (2000b).
12. H. Prahlad and I. Chopra, *J. Intell. Mater. Syst. Struct.* **11**(4), 272–282 (2001).
13. L . C. Brinson and M. S. Huang, *J. Intell. Mater. Syst. Struct.* **7**, 97-107 (1996).
14. Y. Matsuzaki and H. Naito, *J. Intell. Matl. Syst. Struct.* **15**,141-155 (2004).

15. M. H. Elahinia and M. Ahmadian,"On the shortcomings of shape memory alloy phenomenological models" in Proc. IMECE04, ASME International Mechanical Engineering Congress and Exposition (2004) ASME, Anaheim, CA, USA, pp. 13–20.
16. V. R. Buravalla and A. Khandelwal, *Int. J. Sol. Struct.* **44**, 4369-4381 (2007).
17. K. Tanaka, Res. Mech. 18, 251-263 (1986).
18. C. Liang and C. A. Rogers, *J. Intell. Mater. Syst. Struct.* **2**, 207-234 (1990).
19. L. C. Brinson, *J. Intell. Mater. Syst. Struct.* **4** (2), 229-242 (1993).
20. L. C. Brinson and R. Lammering, *Int. J. Sol. Struct.* **30** (23), 3261-3280 (1993).
21. X. Wu and T. J. Pence, *J. Intell. Mater. Syst. Struct.* **9**, 335-354 (1998).
22. I. V. Chenchiah and M. S. Siva, *Mech. Res. Comm.* **26**, 301–307 (1999).
23. J. G. Boyd and D. Lagoudas, *Int. J. Plast.* **12**, 805-841 (1996).
24. F. Auricchio and J. Lubliner, *Int. J. Sol. Struct.* **34**(27), 3601-3618 (1997).
25. A. Bekker and L.C. Brinson, *Act. Mat.* **46**(10), 3649-3665 (1998).
26. A. Khandelwal and V. R. Buravalla, *J. of Intell. Mater. Syst. Struct.* **19**, 46-49 (2008).
27. J. Ortin, A. Planes and L. Delaey, " Hysteresis in shape-memory mateials" in The Science of hysteresis: Hysteresis in Materials. Vol. iii: edited by, Bertotti , G., Mayergoyz, I.D., Academic Press Elsevier Inc., 2006, pp. 467-553.
28. A. Visintin, "Mathematical models of Hysteresis" in The Science of hysteresis: Mathematical modeling and applications vol. I, edited by, Bertotti , G., Mayergoyz, I.D., Academic Press Elsevier Inc., 2006, pp. 1-90.
29. S. Govindjee and E. P. Kasper, *Comput. Meth. Appl. Mech. Engrg.* **171**, 309-326 (1999).
30. J. Chung, J. Heo and J. Lee, *Smart Mat. Struct.*,**16**, N1,N5, (2007)
31. R. Wang, C. Cho, C. Kim and Q. Pan, *Smart. Mat. Struct.* **15**, 393-400 (2006).

123

SESSION B

MAGNETIC MATERIALS

Modeling of MR Fluids and Devices

John C. Ulicny*, Daniel J. Klingenberg†, David Kittipoomwong†, Mark A.
Golden*, Anthony L. Smith*, Chandra S. Namuduri* and Zongxuan Sun**

*GM R&D and Planning, 30500 Mound Road, Warren, MI 48090-9055
†University of Wisconsin, 2006 Engineering Hall, Madison, WI 53706-1691
**University of Minnesota, 111 Church Street, S.E., Minneapolis, MN 55455

Abstract. Magnetorheological (MR) fluids and MR fluid devices have been under development at
General Motors R&D since the early 1990's. As part of this work, a number of modeling efforts
have been undertaken to better understand the fundamentals of MR fluids and MR device behavior.

In this paper, we include a discussion of four modeling examples: MR fluids with bimodal
particle size distributions; an analysis of the transient shear behavior of MR fluids; a model of
the behavior of MR particles in a gravitational field; and a lumped parameter model of a MR clutch.
For each example, we will discuss the motivation that led to the study, the methods used to try to
answer the question that led to the work and conclude with a brief discussion of other potential work
that could be done.

Keywords: magnetorheological fluid
PACS: 47.65.-d

INTRODUCTION

We will discuss the results of four MR fluid or MR device modeling efforts in this report.
For each, we will summarize the work using the following general format: motivation;
methods; results; and potential next steps. Only a relatively brief summary of each case
will be discussed here. Details can be found in published literature [1, 2, 3, 4].

MODELING OF BIDISPERSE MR FLUIDS

Motivation

The starting point for this investigation was the observation by Foister [5] that when
a bimodal distribution of iron particles is used in an MR fluid, a maximum in the yield
stress is achieved at a certain volume ratio of large-to-small particles (see Figure 1). The
question is: Why does this work?

Our initial guess was that a bimodal distribution with an appropriate particle size ratio
would provide for tighter packing of the particles when a magnetic field was applied
thus leading to a higher density and resulting higher yield stress. In order to check this
idea, the shearing of an MR suspension containing magnetizable particles of different
particles sizes was modelled using particle level simulation [1].

CP1029, *Smart Devices: Modeling of Material Systems, An International Workshop*
edited by S. M. Sivakumar, V. Buravalla, and A. R. Srinivasa
©2008 American Institute of Physics 978-0-7354-0553-0/08/$23.00

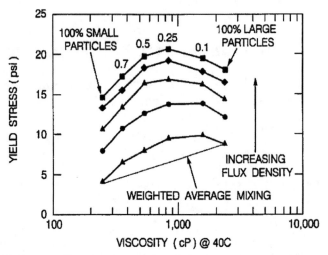

FIGURE 1. Yield stress versus viscosity at various bimodal particle size fractions and magnetic flux densities for a series of 55 volume percent MR fluids. The curves demonstrate the experimentally determined maximum in yield stress found using 25% volume fraction of small particles in the mixture [5].

Modeling Method: Particle Level Simulation

The simulation assumes that a collection of neutrally-buoyant, magnetizable spheres with a field-dependent permeability immersed in a non-magnetizable Newtonian liquid is subjected to a uniform magnetic field. The magnetic moment induced on sphere is proportional to the *local field*, that is, the applied field plus the disturbance fields created by all the other spheres. This field is equivalent to that of a point-dipole of appropriate magnitude located at the sphere center (commonly referred to as the "point-dipole approximation"). We further assume that the motion of each particle is governed by Newton's law of motion. The three forces acting on each particle are hydrodynamic (Stokes' drag), short-range repulsive (hard sphere) and magnetic (including interactions with the shearing surfaces). The final governing equations form a set of non-linear ordinary differential equations which are solved numerically (Euler's method). To determine the dynamic yield stress, one simulates shear flow at successively smaller shear rates, extrapolating the shear stress obtained to zero shear rate.

Results

We examined the effects of particle size ratio, composition, and field strength on the dynamic yield stress. The dynamic yield stress of bidisperse suspensions was found to be larger than that of monodisperse suspensions at the same particle volume fraction. The smaller particles induce the larger particles to form more chainlike aggregates than those formed in monodisperse suspensions (Figure 2). This is an unexpected result and certainly a much different mechanism than simply an increase in packing efficiency.

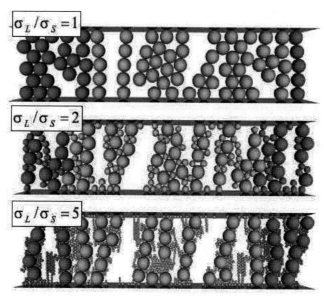

FIGURE 2. Results of particle level simulations at various particle size ratios. The smaller particles tend to force the large particles to form single columns resulting in higher yield stresses as compared to monosized particles [1].

Potential Next Steps

A next potential step in this investigation would be to verify the predictions of the simulations experimentally. If this could be done, it would provide us with the confidence to use these simulations to explore other combinations of particle sizes, for example, to determine if any further increases in yield stress could be gained in this way.

TRANSIENT SHEAR FLOW IN MR FLUIDS

Motivation

The motivation for this next modeling investigation stemmed from the observation that, under certain circumstances, the yield stress of an MR fluid in a device would increase over time when a field was applied. For example, if current of a certain level was applied to an otherwise uncontrolled MR clutch for a period of time, the torque of the clutch would increase with time to a lock-up condition. The simple question we wished to answer was: Why does this occur and under what conditions?

Method: Experiment and stability analysis

The phenomenon of transient yield stress had first been observed with electrorheological (ER) fluids. An analysis by von Pfeil et al. [6] showed that under a strong enough electric field, the ER particles would tend to separate into stripes in shear flow and the onset would occur at a critical value of the Mason number. The onset of stripe formation was later found [2] to correlate with the onset of the yield stress transients.

To determine if similar behavior occurred in MR fluids, current-step transient experiments were conducted in a concentric cylinder rheometer (Figure 3). A schematic of the experimental conditions used is shown in Figure 4.

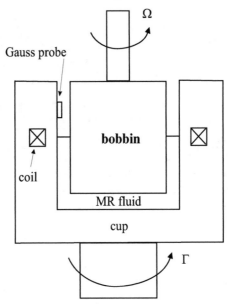

FIGURE 3. Schematic of concentric cylinder rheometer used in transient shear flow experiments [2].

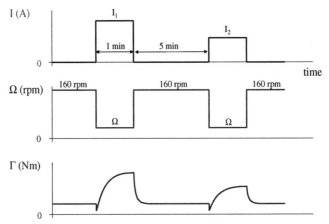

FIGURE 4. Schematic of the experimental conditions used in the concentric cylinder rheometer to measure transient shear stress at various current levels at constant rotational speed [2].

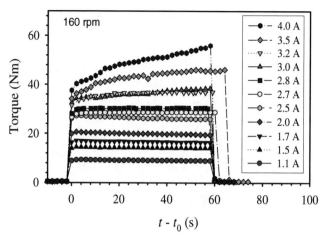

FIGURE 5. Example of the transient shear stress observed at various levels of current for a 45 volume percent MR fluid in the concentric cylinder rheometer [2].

Results: Transients and Stripes

For small applied field strengths, the shear stress increases rapidly to a steady value. Above a critical field strength, the rapid initial increase in shear stress is followed by a slow, transient increase in stress due to stripe formation (Figures 5, 6, 7). The critical Mason Number corresponding to the critical magnetic field strength at the onset of this transient depends on the particle volume fraction as well as the shear rate (Figures 7 and 8).

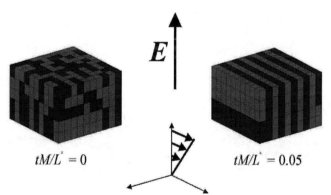

FIGURE 6. Schematic of stripe formation (after von Pfeil [6]).

Potential Next Steps: Look for visual experimental evidence of MR stripes

Experimental evidence of stripe formation might be obtained by shearing an MR fluid that could be frozen in a magnetic field of sufficient strength so that stripes are expected

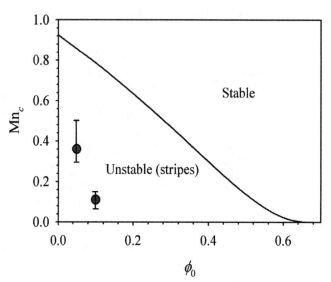

FIGURE 7. Critical Mason Number versus particle volume fraction indicating the boundary between stripe forming and non-stripe forming regions [2].

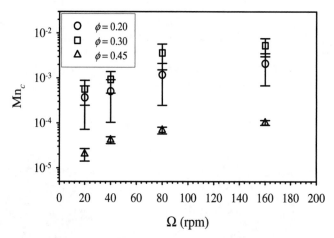

FIGURE 8. Critical Mason Number versus rheometer rotation rate indicating that Mnc varies with shear rate contrary to the predictions of the continuum model [2].

to form. The "freezing" can be accomplished by using a carrier fluid that solidifies such as a wax or a curable adhesive to fix the particles in place.

GRAVITATIONAL EFFECTS IN ROTATING MR DEVICES

Our motivation for this investigation began with some unexplained behavior observed in MR clutches at high speeds (i.e., high g forces). We found that at these high gravitation acceleration, the clutches would tend to lose torque capacity even when a field was applied. We suspected that the gravitational forces were responsible but we wanted some confirmation of this and experiments to confirm our suspicions were expensive to conduct.

Method 1: Dimensional analysis

The relative magnitudes of the gravitational and magnetic forces present in a MR fluid were compared in the form of a dimensionless group ("G") which is the ratio of the body forces and the magnetostatic forces (Figure 9). The value of G was estimated for a range of values typical of MR fluids used in practice and it was found that, in general, $G \ll 1$ especially at values of magnetic flux density near saturation. These estimates suggest that, except perhaps at very low values of flux density combined with very high values of gravitational force, the magnetic forces should always be able to overcome the gravitational forces.

$$G = \frac{\text{Body Forces}}{\text{Electro/Magnetostatic Forces}}$$

$$= \frac{\pi \sigma^3 \Delta \rho a}{6 F_0}$$

$$G = \begin{cases} \frac{8 \sigma \Delta \rho a}{9 \epsilon_0 \epsilon_c \beta^2 E_0^2} & \text{(linear polarization),} \\ \frac{8 \sigma \Delta \rho a}{9 \mu_0 \mu_c \beta_M^2 H_0^2} & \text{(linear magnetization,)} \\ \frac{8 \sigma \Delta \rho a}{\mu_0 \mu_c M_s^2} & \text{(saturated magnetization).} \end{cases}$$

ER Fluids		MR Fluids	
Parameter	Value(s)	Parameter	Value(s)
E_0	0–10 kV/mm	H_0	0–800 kA/m
σ	0.1–100 μm	σ	0.1–100 μm
$\Delta\rho$	1000 kg/m^3	$\Delta\rho$	7000 kg/m^3
ϵ_c	2	μ_c	1
β	1	β_M	1
r	0.05–0.2 m	r	0.05–0.2 m
ω	0–1000 rad/s	ω	0–1000 rad/s

FIGURE 9. Definition of the gravity number G for a MR fluid [3].

133

$$\sigma = 10^{-5} \text{ m}, \; r = 0.1 \text{ m}$$

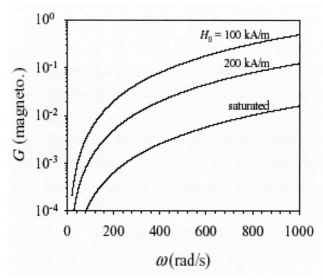

FIGURE 10.   Typical values of the gravity number G for a MR fluid [3].

Method 2: Particle level simulation

Particle level simulations were also performed for comparison against the dimensional analysis. The assumptions and approximations used in this case are similar to those used for the modeling of bidisperse MR fluids but with the addition of a body force to simulate the gravitation acceleration in a device (See Figures 11 and 12).

Results: Gravitation effective on MR chains

Body forces in electro- and magnetorheological fluids are relatively small in magnitude compared to the field-induced electric and magnetic forces as indicated by the dimensional analysis. However, particle-level simulations indicate that these relatively small forces can have large effects on the rheological response of these fluids in shear flow (Figures 13, 14). The reason for these unexpectedly large effects is that when the gravitational force is oriented parallel to a chain of particles, each particle experiences not only the body force on itself but also the body force on all of the particles above that particle. This means that the gravitational force can reduce the effective yield stress at much lower values of G than expected.

This effect on the particle chains can be seen in both the simulation results (Figure 13) and in experiments even at 1 g (Figure 14). As shown in the simulations, the yield stress is reduced because the top of the particle chain eventually loses contact with the shearing surface as either the body force is increased or the magnetic flux density is decreased.

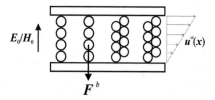

FIGURE 11. Schematic diagram of an ER/MR fluid in which the particles experience a body force in the direction opposite to that of the applied field [3].

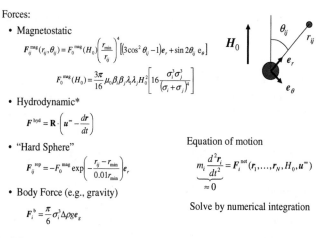

Forces:
- Magnetostatic

$$F_{ij}^{mag}(r_{ij},\theta_{ij}) = F_0^{mag}(H_0)\left(\frac{r_{min}}{r_{ij}}\right)^4 \left[(3\cos^2\theta_{ij}-1)e_r + \sin 2\theta_{ij}\, e_\theta\right]$$

$$F_0^{mag}(H_0) = \frac{3\pi}{16}\mu_0\beta_i\beta_j\lambda_i\lambda_j H_0^2 \left[16\frac{\sigma_i^3\sigma_j^3}{(\sigma_i+\sigma_j)^4}\right]$$

- Hydrodynamic*

$$F^{hyd} = \mathbf{R}\cdot\left(u^\infty - \frac{dr}{dt}\right)$$

- "Hard Sphere"

$$F_{ij}^{rep} = -F_0^{mag}\exp\left(-\frac{r_{ij}-r_{min}}{0.01r_{min}}\right)e_r$$

- Body Force (e.g., gravity)

$$F_i^b = \frac{\pi}{6}\sigma_i^3\Delta\rho g e_g$$

Equation of motion

$$\underbrace{m_i\frac{d^2r_i}{dt^2}}_{\approx 0} = F_i^{net}(r_1,\ldots,r_N,H_0,u^\infty)$$

Solve by numerical integration

*or Stokesian dynamics

FIGURE 12. Equations for the particle level simulation of gravity on a column of MR particles in a magnetic field [3].

simulation

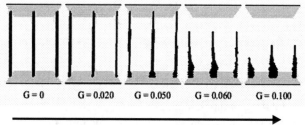

| G = 0 | G = 0.020 | G = 0.050 | G = 0.060 | G = 0.100 |

increasing g-force or decreasing field

FIGURE 13. Simulation results showing the effect of gravity on a column of MR particles in a magnetic field.

Potential Next Steps: MR clutch experiments

A further way to probe the effect of gravitational forces in a practical situation would be to set up a MR clutch with independent input/output speed control. In this way the

2 wt.% BaTiO$_3$ particles in mineral oil
(53 – 63 μm diameter)

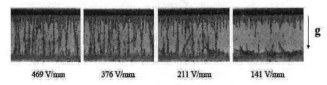

469 V/mm   376 V/mm   211 V/mm   141 V/mm

experiment

FIGURE 14. Experimental results showing the effect of gravity on columns of ER particles at various electric field strengths.

gravitational forces (due to the rotational speed of the clutch housing) and the shear rate on the MR fluid (due to the relative speeds of the input rotor and the clutch housing) could be varied independently in addition to varying the magnetic flux density.

LUMPED PARAMETER MODEL OF AN MR FAN CLUTCH

This final modeling exercise was motivated by our interest in identifying the factors which determine our ability to control an MR automotive fan clutch. Because of the computational intensity involved in a detailed model of such a device, a lumped-parameter model was chosen for this purpose. With this type of model, only the average properties of the fluid and other materials and average temperatures are used rather than describing the temperature of the fluid, for example, as a function of position and time. The resulting models are much simpler to solve at the cost of a more restricted level of description, as well as the introduction of parameters that must often be treated as adjustable.

Modeling Method

Averaging the energy balance over the fluid and rotor volume yields

$$m_f \hat{C}_{p,f} \frac{dT_f}{dt} = (UA)_{cf}(T_c - T_f) + SV, \tag{1}$$

where m_f is the mass of the fluid plus rotor, $\hat{C}_{p,f}$ is the specific heat capacity of the fluid plus rotor, T_f is the average temperature of the fluid plus rotor, $(UA)_{cf}$ is the overall heat transfer coefficient times the area for heat transfer from the fluid to the clutch, T_c is the average clutch temperature, S is the average rate of thermal energy production per unit volume by viscous dissipation within the MR fluid, and V is the fluid volume. Note that we have not included the ohmic heating of the magnetic coil since the contribution from this source is about two orders of magnitude smaller than that from viscous heating.

Averaging the energy balance over the clutch volume (fan and coil housings) yields

$$m_c \hat{C}_{p,c} \frac{dT_c}{dt} = -(UA)_{cf}(T_c - T_f) - (UA)_\infty(T_c - T_\infty), \tag{2}$$

where m_c is the mass of the clutch, $\hat{C}_{p,c}$ is the specific heat capacity of the clutch, T_c is the average clutch temperature, $(UA)_\infty$ is the overall heat transfer coefficient times the area for heat transfer from the clutch to the surrounding fluid, and T_∞ is the temperature of the surroundings.

Finally, a macroscopic angular momentum balance on the fan yields

$$I\frac{d\Omega_o}{dt} = \Gamma - k\Omega_o^2, \tag{3}$$

where I is the moment of inertia of the fan clutch assembly, Γ is the torque exerted by the MR fluid on the fan, k is the fan torque constant, and $k\Omega_o^2$ is the resistance torque on the fan.

In order to obtain the optimal heat transfer coefficients, a series of simulations was performed, varying each coefficient over a range. The difference between the predictions and the experimental data for each simulation was calculated for a range of fan speeds and clutch temperatures in the form of a least-squares type objective function:

$$PI = 0.2\sum_{t_1}^{t_2}(\Delta w_f)^2 + 1.0\sum_{t_3}^{t_4}(\Delta T_c)^2$$

where PI is the performance index, Δw_f is the difference between the predicted and experimental fan speed, ΔT_c is the difference between the predicted and experimental clutch temperatures, $t_1 = 100\ s$, $t_2 = 180\ s$, $t_3 = 20\ s$ and $t_4 = 218\ s$.

Results: Comparison of simulations with data

A plot of the simulation results for a fit of the speed and temperature data is shown in Figure 15. We note that a reasonably good correlation between the simulation results and the data can be obtained for each parameter (fan speed and temperature) but that the values of the adjustable parameters (heat transfer coefficients) in each case is different.

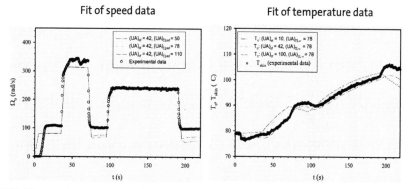

FIGURE 15. Results of fit of the simulation results for step experiments for the fan speed and temperature independently. Note that the values of the adjustable parameters used (heat transfer coefficients) is different in each case [4].

A similar situation was found in our attempts to simulate hysteresis experiments. In these cases, the fan clutch was operated with a constant input speed while the coil current was ramped at a set rate (either "fast" or "slow"). In the experiments, the fan speed increased to a point and then jumped to the input speed when the MR fluid suddenly locked-up. The lock-up current was lower for the slow ramp than for the fast ramp and the clutch unlocked at a slightly higher current for the fast ramp. Comparisons of the simulation results with data are shown in Figure 16.

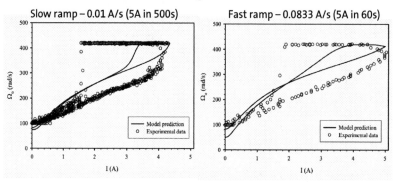

FIGURE 16. Results of fit of the simulation results for the hysteresis experiments at slow and fast current ramp speeds [4].

It is apparent that the correlation between the simulation and the data for the hysteresis experiments is less satisfactory than for the step experiments. It seems clear that there is some feature or features missing from the simulations.

We speculate that one or more known phenomena may be responsible for the lack of agreement such as particle migration, sedimentation and resuspension, shear-induced migration or stripe formation. Another possibility is due to the wide range of fluid temperatures we would expect during the experiments; a larger range than was used to measure the fluid properties (viscosity and on-state yield stress in particular). It will also be the case that, in the experiments, the fluid and device temperatures will take a long time to reach steady state which may not be in close agreement with the assumptions inherent in the simulations.

Potential next steps

Physical property data for the MR fluids at temperatures in the typical operating range of a MR clutch (in the range of 100-300°C for example) is needed to improve the simulations. Also, it would be valuable to compare the simulations with more experimental cases to better identify the features that are missing from the simulations.

CONCLUSIONS

These and other modeling exercises we have undertaken have demonstrated that this work provides us with valuable information and insight into our work with MR fluids and

MR devices. As we have seen, scaling methods and simulations help decipher MR fluid and device behavior and particle level simulations can reveal unexpected mechanisms in even the simplest MR fluid flow situations. The results have also revealed that in order to improve our predictive capabilities, a wider range of MR fluid property data will be required. It also seems likely that we may need to incorporate other particle behavior and properties such as sedimentation, migration, magnetophoresis, interparticle friction, and colloidal effects.

ACKNOWLEDGMENTS

The authors wish to acknowledge Mr. Keith Snavely and Mr. Kenneth Shoemaker for their contributions to the laboratory work included in this paper.

REFERENCES

1. D. Kittipoomwong, D. Klingenberg, and J. Ulicny, *Journal of Rheology* **49**, 1521–1538 (2005).
2. J. Ulicny, M. Golden, C. Namuduri, and D. Klingenberg, *Journal of Rheology* **49**, 87–104 (2005).
3. D. Klingenberg, J. Ulicny, and A. Smith, *Applied Physics Letters* **86**, 1–3 (2005).
4. J. Ulicny, D. Klingenberg, A. Smith, and Z. Sun, "Lumped Parameter Model of a Magnetorheological Fluid Clutch," in *Proceedings of the 2006 International MecProceedings of the 2006 International Mechanical Engineering Congress and Exposition, Chicago, IL*, IMECE2006 13225, 2006.
5. R. Foister, Magnetorheological fluids, U.S. Patent 5,667,715 (1997).
6. K. von Pfeil, M. Graham, D. Klingenberg, and J. Morris, *Journal of Applied Physics* **93**, 5769–5779 (2003).

MODELING OF MAGNETOSTRICTIVE MATERIALS AND STRUCTURES

S. GOPALAKRISHNAN

Department of Aerospace Engineering
Indian Institute of Science
Bangalore 560 012, India

Abstract: The constitutive model for a magnetostrictive material and its effect on the structural response is presented in this article. The example of magnetostrictive material considered is the TERFENOL-D. As like the piezoelectric material, this material has two constitutive laws, one of which is the sensing law and the other is the actuation law, both of which are highly coupled and non-linear. For the purpose of analysis, the constitutive laws can be characterized as coupled or uncoupled and linear or non linear. Coupled model is studied without assuming any explicit direct relationship with magnetic field. In the linear coupled model, which is assumed to preserve the magnetic flux line continuity, the elastic modulus, the permeability and magneto-elastic constant are assumed as constant. In the nonlinear-coupled model, the nonlinearity is decoupled and solved separately for the magnetic domain and the mechanical domain using two nonlinear curves, namely the stress vs. strain curve and the magnetic flux density vs. magnetic field curve. This is performed by two different methods. In the first, the magnetic flux density is computed iteratively, while in the second, the artificial neural network is used, where in the trained network will give the necessary strain and magnetic flux density for a given magnetic field and stress level. The effect of nonlinearity is demonstrated on a simple magnetostrictive rod.

Keywords: Magnetostriction, Constitutive laws Magnetostrictive Materials, TERFENOL-D, Magnetic flux density Enter Keywords here.

INTRODUCTION

Some magnetic materials show elongation and contraction in the magnetization direction due to an induced magnetic field. This is called the *magnetostriction*, which is due to the switching of a large amount of magnetic domains caused by spontaneous magnetization below the Curie point of temperature. Thus magnetostrictive materials have the ability to convert magnetic energy into mechanical energy and vice versa. This coupling between magnetic and mechanical energies represents the transudation capability and this allows a magnetostrictive material to be used in both actuation and sensing devices. Due to magnetostriction and its inverse effect (also called Villery effect)[Villery (1865)], magnetostrictive materials can be used both as an actuator and as well as a sensor. The theoretical and experimental study of magnetostrictive materials has been the focus of considerable research for many years. However, only with the recent development of giant magnetostrictive materials (e.g. TERFENOL-D), it is now possible to produce sufficiently large strains and forces to facilitate the use of these materials in actuators and sensors. This has led to the application of

CP1029, *Smart Devices: Modeling of Material Systems, An International Workshop*
edited by S. M. Sivakumar, V. Buravalla, and A. R. Srinivasa
©2008 American Institute of Physics 978-0-7354-0553-0/08/$23.00

magnetostrictive materials to such devices as micro positioners, vibration controller, sonar projectors and insulators, etc. Magnetostrictive material has found its way in many structural applications such as vibration control, noise control and structural health monitoring.

The Constitutive relationship of magnetostrictive materials consists of a sensing and an actuation equation. In sensing equation, magnetic flux density is function of applied magnetic field and stress where as, in actuation equation, strain is a function of applied magnetic field and stress. Both sensing and actuation equations are coupled through applied magnetic field and mechanical stress level. Analysis of smart structures with magnetostrictive material is generally performed using uncoupled models. Uncoupled models are based on the assumption that the magnetic field within the magnetostrictive material is proportional to the electric coil current times the number of coil turn per unit length [Ghosh and Gopalakrishnan (2004)]. Due to this assumption, the actuation and the sensing equations get uncoupled. For the actuator, the strain due to magnetic field (which is proportional to coil current) is incorporated as the equivalent nodal load in the finite element model for calculating the block force. Thus, with this procedure, analysis can be carried out without taking the smart degrees of freedom in the finite element model. Similarly for sensor, where generally coil current is assumed zero, the magnetic flux density is proportional to mechanical stress, which can be calculated from the finite element results through post-processing. This assumption on magnetic field, leads to the violation of flux line continuity, which is one of the four Maxwell's equations in electromagnetism. On the other hand, in coupled model, it is considered that magnetic flux density and/or strain of the material are functions of stress and magnetic field, without any additional assumption on magnetic field, like uncoupled model. Benbouzid et al. modeled the static [Benbouzid, Reyne, and Meunier (1993)] and dynamic [Benbouzid, Kvarnsjo, and Engdahl (1995)] behavior of the nonlinear magneto-elastic medium for magneto-static case using finite element method. Magneto-mechanical coupling was incorporated considering both permeability and elastic modulus as functions of stress and magnetic field. However, all these work do not provide a convenient way for analysis of magnetostrictive smart structure considering coupled magneto-mechanical features. This article deals with the numerical characterization of the constitutive relationship considering the coupled features of magnetostrictive materials, which can be directly used in any mathematical/numerical formulation such as finite element method for structures with magnetostrictive material considering both magnetic and mechanical degrees as unknown degrees of freedom. In addition, it will be shown that the magnetic field is not proportional of applied coil current (which is the assumption of uncoupled model), and it depends on the mechanical stress on the magnetostrictive material. This study here will also demonstrate as to how the coupled model preserves the flux line continuity, which is one of the drawbacks of uncoupled model.

The constitutive relations of magnetostrictive materials are essentially nonlinear [Butler (1988)]. The prediction of behavior of magnetostrictive material, in general, is extremely complicated due to its hysteretic non-linear character. In structural application, due to these nonlinear material properties, modeling of the system will become nonlinear, for which exact non-linear constitutive relationships are essential. Toupin (1956) and Maugin (1985) had done extensive work related to electrostrictive and piezoelectric phenomena which have similarities in form with the magnetostriction phenomena. Earlier study to model uncoupled nonlinear actuation of magnetostrictive material was done by Krishnamurty et. al (1999) by considering a fourth order polynomial of magnetic field for each stress level. In this approach, stress level for which curve is not available, the coefficients of the curve will have to be interpolated from the coefficients of nearest upper and lower stress level curves.

In this article, magnetic flux density and strains are computed from magnetic field and stress level through an iterative procedure due to these nonlinear curves. To avoid this nonlinear iteration, one three layer artificial neural network (ANN) is trained to get this nonlinear mapping directly. ANN is a universal approximater, which can give a nonlinear parameterize mapping from a given input data to an output data. In this study, ANN is used to get the direct mapping for constitutive relationship of magnetostrictive materials, where inputs in the network are magnetic field and applied stress level and outputs in the network are the strain and the magnetic flux density. Hence, nonlinearity in elastic modulus and permeability is replaced by this trained network.

The article is organized as follows. In the next section, the complete architecture of the ANN developed for this work is explained, which includes the process of training and validation of network. This is followed by a detailed section on the numerical characterization of both Linear and non-linear coupled and uncoupled constitutive model. Finally the developed model is validated with each other and also with available results.

ARTIFICIAL NEURAL NETWORK (ANN)

Artificial Neural Networks can provide non-linear parameterized mapping between a set of inputs and a set of outputs with unknown function relationship. A three-layer network (Figure 1) with the sigmoid activation functions can approximate any smooth mapping. A typical supervised feed-forward multi layer neural network is called as a back propagation (BP) neural network. The structure of a BP neural network shown in Figure1 mainly include an input layer for receiving the input data; some hidden layer for processing data; and an output layer to indicate the identified results. In this study, the tusk of identifying nonlinear magnetostriction through ANN is performed by training the neural network using the known samples.

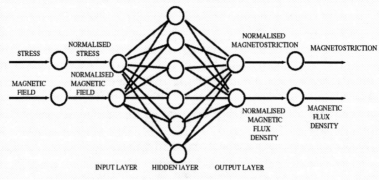

FIGURE 1. Artificial Neural Network Architecture

142

TRAINING OF THE NETWORK

The training of a BP neural network is a two-step procedure [Rumelhart, Hinton, and Williams (1986)]. In the first step, the network propagates input through each layer until an output is generated. The error between the output and the target output is then computed. In the second step, the calculated error is transmitted backwards from the output layer and the weights are adjusted to minimize the error. The training process is terminated when the error is sufficiently small for all training samples. In practical applications of the back-propagation algorithm, learning is the result from many presentations of these training examples to the multi-layer perceptron. One complete presentation of the entire training set during the learning process is called an *epoch*. The learning process is maintained on an epoch-by-epoch basic until the synaptic weights and bias levels of the network stabilizes and the averaged squared error over the entire training set converges to some minimum value. For a given training set, back-propagation learning can be done in sequential or batch mode.

VALIDATION OF TRAINED NETWORK

To validate the trained network, the data set is separated into two parts, one for training and the other for testing the network performance. The network will be trained using training sample and the trained network will be validated with the test sample. A network is said to generalize well when the input-output mapping computed by the network, is corrected with the test data that was never used in creating or training the network. Although the network performs useful interpolation, because of multilayered perceptrons with continuous activation functions, it leads to output functions that are also continuous.

ANHYESTERIC COUPLED CONSTITUTIVE MODEL

Here experimental data is taken from Etrema manual [Butler (1988)] for TERFENOL-D, a giant magnetostrictive material to verify the proposed model. Experimental data of the magnetostriction vs. magnetic field for different stress level given in the manual is reproduced in Figure 2 and the stress vs. strain curves for different magnetic field level is reproduced in Figure 3.

Application of magnetic field causes strain in the magnetostrictive material (TERFENOL-D) and hence the stress, which changes magnetization of the material. As described by Butler (1988), Moffett, Clark, Wun-Fogle, Linberg, Teter, and McLaughlin (1989), and Hall and Flatau (1994), the three-dimensional coupled constitutive relationship between magnetic and mechanical quantities for magnetostrictive material are given by

$$\{\varepsilon\} = \left[S^{(H)}\right]\{\sigma\} + \left[d\right]^T \{H\} \qquad (1)$$

$$\{B\} = \left[\mu^\sigma\right]\{H\} + \left[d\right]\{\sigma\} \qquad (2)$$

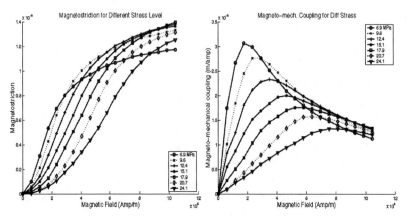

FIGURE 2. Magnetostriction and Magneto-mechanical coupling as a function of magnetic field-Data plotted from ETREMA data

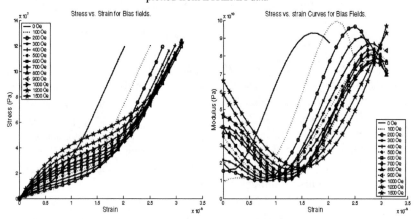

FIGURE 3. Stress-Strain relationship for different magnetic field-Data Plotted from ETREMA Data

where $\{\varepsilon\}$ and $\{\sigma\}$ are strain and stress respectively. $[S^{(H)}]$ represents elastic compliance measured at constant $\{H\}$ and $[\mu^{(\sigma)}]$ represents the permeability measured at constant stress $\{\sigma\}$. Here $[d]$ is the magneto-mechanical coupling coefficient, which provides a measure of the coupling between the mechanical strain and magnetic field. In general, $[S]$, $[d]$ and $[\mu]$ are nonlinear as they depend upon $\{\sigma\}$ and $\{H\}$. Equation (1) is often referred to as the direct effect and Equation (2) is known as the converse effect. These equations are traditionally used for actuation and sensing purpose, respectively. It should be noted that the elastic constants used, correspond to the fixed magnetic field values and the permeability correspond the fixed stress values.

COUPLED CONSTITUTIVE MODEL

Analysis of smart structures using magnetostrictive materials as either sensors or actuators has traditionally been performed using uncoupled models. Uncoupled models make the assumption that the magnetic field within the magnetostrictive material is constant and proportional to the electric coil current times the number of coil turn per unit length [Ghosh and Gopalakrishnan (2002)]. Hence actuation and sensing problems are solved by two uncoupled equations, which are given by the last part of Equations (1) and (2), respectively. This makes the analysis relatively simple, however this method has its limitations. It is quite well known that $[S]$, $[d]$ and $[\mu]$ all depend on stress level and magnetic field. In the presence of mechanical loads, the stress changes and so is the magnetic field. Estimating the constitutive properties using uncoupled model in such cases will give inaccurate predictions. Hence, the constitutive model should be represented by a pair of coupled equations given by Equations (1) & (2) to predict the mechanical and magnetic response. It is therefore necessary to simultaneously solve for both the magnetic response as well as the mechanical response regardless of whether the magnetostrictive material is being used as a sensor or actuator. Due to in-built non-linearity, the uncoupled model may not be capable of handling certain applications such as (1) modeling passive damping circuits in vibration control and (2) development of self-sensing actuators in structural health monitoring. In these applications, the coupled equations require to be solved simultaneously. The solution of coupled equations simultaneously is a necessity for general-purpose analysis of adaptive structures built with magnetostrictive materials.

In general, the errors that result from using uncoupled models, as opposed to coupled ones, are problem dependent. There are some cases where very large differences exist in situation, where an uncoupled model is used over a coupled model [Ghosh and Gopalakrishnan (2002)]. In this work, the coupled case is analyzed with both linear and non-linear model. In linear-coupled model, magneto-mechanical coefficient, elasticity matrix and permeability matrix are assumed as constant. In nonlinear-coupled model, mechanical and magnetic nonlinearity are decoupled in their respective domains. The nonlinear stress-strain relationship is generally represented by modulus of elasticity and the nonlinear magnetic flux-magnetic field relationship represented by permeability of the material. Magneto-mechanical coupling coefficient will be assumed as constant in this case.

Linear Model

From Equation (1) and Equation (2), the 3D constitutive model for the magnetostrictive material can be written as

$$\{\sigma\} = [Q]\{\varepsilon\} - [e]^T \{H\} \tag{3}$$

$$\{B\} = [e]\{\varepsilon\} + \left[\mu^\varepsilon\right]\{H\} \tag{4}$$

Where $[Q]$ is Elasticity matrix, which is the inverse of compliance matrix $[S]$, $[\mu^\varepsilon]$ is the permeability at constant strain. $[\mu^\varepsilon]$ and $[e]$ are related to $[Q]$ through

$$[e] = [d][Q] \tag{5}$$

$$\left[\mu^\varepsilon\right] = \left[\mu^\sigma\right] - [d][Q][d]^T \tag{6}$$

For ordinary magnetic materials, where magnetostrictive coupling coefficients are zero, $[\mu^\varepsilon] = [\mu^\sigma]$, the permeability.

Consider a magnetostrictive rod element of length L, area A, with Young modulus Q. If a tensile force F is applied the rod develops a strain ε, and hence stress σ. Total strain energy in the rod will be

$$V_e = \frac{1}{2} \int \varepsilon \sigma \, dv = \frac{1}{2} \int \varepsilon \{Q\varepsilon - eH\} dv$$

$$= \frac{1}{2} ALQ\varepsilon^2 - \frac{1}{2} AL e\varepsilon H \tag{7}$$

Magnetic potential energy in magnetostrictive rod is

$$V_M = \frac{1}{2} \int BH dv = \frac{1}{2} \int \{e\varepsilon + \mu H\} H dv$$

$$= \frac{1}{2} ALHe \, \varepsilon + \frac{1}{2} AL \, \mu^\varepsilon H^2 \tag{8}$$

The Magnetic and mechanical external work done for N number of coil turn with coil current I is

$$W_m = IN\mu^\sigma HA, \qquad W_e = F\varepsilon L \tag{9}$$

Total potential energy of the system comes $T_p = -(V_e - W_e) + (V_m - W_m)$.

$$T_p = -\frac{1}{2} ALQ \, \varepsilon^2 + \frac{1}{2} ALe \, \varepsilon H$$

$$+ \frac{1}{2} ALHe \, \varepsilon + \frac{1}{2} ALH \, \mu^\varepsilon H^2 - IN \mu^\sigma HA + F\varepsilon L \tag{10}$$

Using Hamilton's Principle, two equations in terms of H and ε can be written as

$$- ALQ \, \varepsilon + ALeH + FL = 0 \tag{11}$$

$$ALe \, \varepsilon + ALeH + FL = 0 \tag{12}$$

Dividing both equations by AL, the equations will become

$$- Q \, \varepsilon + eH = -\frac{F}{A} \tag{13}$$

$$e\varepsilon + H \mu^\varepsilon = -\frac{F}{A} \tag{14}$$

As right hand side of Equation (14) is not function of ε and left hand side is magnetic flux density (Equation (4), the magnetic flux density in this model is not function of ε. Hence, it is preserving the flux line continuity. Eliminating H from Equation (13) and substituting this in Equation (14), stress - strain relationship of the magnetostrictive material can be written as follows.

$$H = (Q\varepsilon - F/A)/e \tag{15}$$

$$\varepsilon = \frac{IN\mu^\sigma Ae + L\mu^\varepsilon F}{ALe^2 + AL\mu^\varepsilon Q} = \frac{IN\mu^\sigma eA + F\mu^\varepsilon L}{AL\mu^\sigma Q} \tag{16}$$

From Equation (16), total strain for applied coil current I and tensile stress F/A can be written as

$$\varepsilon = \lambda + \varepsilon\sigma \tag{17}$$

where λ is the strain due to coil current, which is called the magnetostriction and ε_σ is the strain due to tensile stress (elastic strain).

$$\lambda = \frac{IN\mu^\sigma Ae}{AL\mu^\sigma A} = INd/L \tag{18}$$

146

$$\varepsilon_\sigma = \frac{L\mu^\varepsilon F}{AL\mu^\sigma Q} = \frac{F}{AQ^*} \tag{19}$$

Let Q^* be the modified elastic modulus and substituting the value of e and μ^ε from Equation (5) and Equation (6), Q^* will be

$$Q^* = \frac{Q\mu^\sigma}{\mu^\varepsilon} = \frac{Q\mu^\sigma}{\mu^\sigma - d^2 Q} = Q + \frac{e^2}{\mu^\varepsilon} \tag{20}$$

If the value of μ^σ is much greater than d^2Q, μ^ε can be assumed equal to μ^σ and $Q*$ can be assumed as equal to Q. If the value of μ^σ is much greater than d^2Q the total strain of the rod will be same as for the uncoupled model. The first term in the above expression is the strain due to magnetic field, and the second term is the strain due to the applied mechanical loading. However, for TERFENOL-D [Butler (1988)], the value of d^2Q is comparable with μ^σ. Substituting the value of strain from Equation (16) in the Equation (15), the value of magnetic field will be

$$H = \frac{F}{Ae}(1 - \frac{\mu^\varepsilon}{\mu^\sigma}) + \frac{IN}{L} \tag{21}$$

Note that although the magnetostriction value (INd/L) in Equation (18) is same for coupled and uncoupled case, the value of magnetic field is different. Let r be the ratio of two permeabilities or two elastic modulii. From Equation (20), r can be written as.

$$r = \frac{\mu^\sigma}{\mu^\varepsilon} = \frac{Q^*}{Q} \tag{22}$$

If the value of r is one, the results of coupled analysis are similar with uncoupled analysis. In Figure 4, the value of r is shown in contour plot for different values of constant strain permeability and modulus of elasticity for coupling coefficient of 15 x 10^{-9} m/Amp. In the left figure, value of r is shown for different values of permeability and elastic modulus. In the right figure, value of r is shown for different values of permeability and modified elasticity. In these plots it is clear that for a particular value of elasticity if the value of permeability increases, the value of r will decrease. But for a particular value of permeability, if the value of elastic modulus increases, the value of r will increase. In Figure 5, the value of r is shown in contour plot for different value of permeabilities and coupling coefficient considering modulus of elasticity as15GPa. In the left figure, value of r is given for different values of constant strain permeability and coupling coefficient. In the right figure, value of r is shown for different values of constant stress permeability and coupling coefficients. In these plots it is clear that for a particular value of permeability, if the value of coupling coefficient increases the value of r will increase. But for a particular value of coupling coefficient if the value of permeability increases, the value of r will decrease.

In Figure 6, the value of r is shown in contour plot for different value of elastic modulus and coupling coefficient considering constant strain permeability as 7x10−6 henry/m. In the left figure, value of r is given for different values of modulii of elasticity and coupling coefficient. In the right figure, value of r is shown for different values of modulii of elasticity and coupling coefficients. In these plots it is clear that for a particular value of elastic modulus, if the value of coupling coefficient increases, the value of r will increase. Similarly, for a particular value of coupling coefficient, if the value of elastic modulus increases, the value of r will increase.

From experimental data given in Etrema manual [Butler (1988)], the best value of Q, μ^σ and d is calculated by minimizing the difference between the experimental data and the data according to Equation (16) by least square approach. In the first set of experimental data, only magnetostriction values was reported, which is given in Equation (18). The value of coupling coefficient is calculated by minimizing the total square error, λ_{Error} as given by

$$\lambda Error = \sum \left(\lambda \exp - \lambda \right)^2 \tag{23}$$

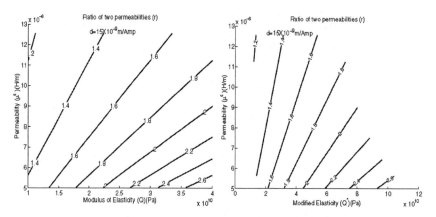

FIGURE 4. Ratio of two permeabilities (r): Permeability vs. modulus of elasticity (left) Permeability vs. Modified elasticity (right), for $d=15X10^{-9}$ m/Amp

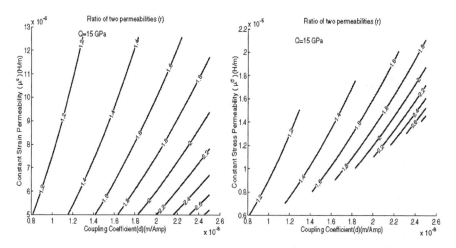

FIGURE 5. Ratio of two permeabilities (r): Permeability vs. Coupling Coefficients (left) Constant Stress vs. Coupling Coefficients (right), for Q=15Gpa

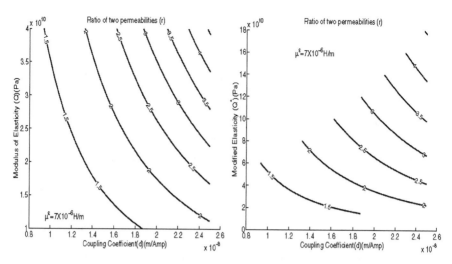

FIGURE 6. Ratio of two permeabilities (r): Modulus Q vs Coupling Coefficent (left) Modulus Q* vs. Coupling Coefficient (right), for $\mu^\Box = 7 \times 10^{-6}$ henry/m

Similarly in the second set of experimental data, the strain due to compressive stress (ε_σ) was reported. The expression for the value of elastic strain, ε_σ is given in Equation (19). In this expression, the value of $Q*$ is calculated by minimizing the total square error $\varepsilon_\sigma^{Error}$

$$\varepsilon_\sigma^{Error} = \sum \left(\varepsilon_\sigma^{\exp} - \varepsilon_\sigma\right)^2 \tag{24}$$

From Equation (23), using first set of experimental data (plotted in Figure 2), the value of d was calculated as $14.8X10-9$ (m/amp). From Equation (24), using the second set of experimental data (plotted in Figure 3), the value of $Q*$ is 33.4 GPa. Assuming constant strain permeability (μ^ε) of the material is $7x10-6$ henry/m, the value of r is 1.6, constant stress permeability (μ^σ) is $11.2x10-6$ henry/m and Q is 20.8 GPa. From these studies, it is clear that for a giant magnetostrictive material, like TERFENOL-D, the coupled analysis will give better result than the uncoupled analysis. However, for a magnetostrictive material with low coupling coefficient, the uncoupled analysis will give a similar result as the coupled analysis.

The coupled linear model cannot model the high nonlinearity of magnetostriction λ, which is required for design of actuators. Even considering nonlinear magnetic (magnetic field-magnetic flux) and mechanical (stress-strain) relationships with linear coupling coefficient, nonlinear relationships of magnetostriction cannot be modeled as it is a function of coil current, coil turn per unit length of actuator and magneto-mechanical coefficient (Equation (18)). In the next section, we introduce a nonlinear model with a constant coupling coefficient, which can model the non-linear constitutive model exactly for constant magnetic coupling.

Nonlinear Coupled Model

The model developed in this section is based on a coupled magneto-mechanical formulation, which allows accurate prediction of both the mechanical and the magnetic response of a magnetostrictive device with nonlinear magnetic and mechanical properties. Non-linearity in this model is introduced using two nonlinear curves, one for stress-strain relation and the second for magnetic field-magnetic flux relation, which enables to decouple the non-linearity in mechanical and magnetic domains. Magneto-mechanical coefficient is considered as a real parameter scalar value. Two-way coupled magneto-mechanical theory is used to model Magnetostrictive material. The formulation starts with the constitutive relations.

In earlier coupled-linear model, stress (σ) and magnetic flux density (B) was expressed as a function of the components of strain (ε) and magnetic field (H) as per Equation (3) and Equation (4), respectively. Main draw back with such an approach is that the non-linearity between magnetic domain ($\mu\Sigma$) and mechanical domain (Q) are not uncoupled. Hence, it is difficult to model non-linearity in earlier representation. To address these issues, a different approach is used in which Equation (3) and Equation (4) are rearranged in terms of the mechanical strain (ε) and the magnetic flux density (B). In doing so, the mechanical non-linearity is limited to the stress-strain relationship and the magnetic non-linearity is limited to magnetic field and magnetic flux relationship. One-dimensional nonlinear modeling is again studied using one-dimensional experimental data from Etrema manual. The constitutive equation can now be rewritten in terms of magnetic flux density (B) and strain (ε), as

$$\sigma = E\varepsilon - f^T B \quad (25)$$

$$H = -f\varepsilon + gB \quad (26)$$

where

$$g = \left(\mu^\varepsilon\right)^{-1}$$

$$f = gdQ = e/\mu^\varepsilon \quad (27)$$

$$E = Q + Qdf = Q*$$

Like linear case, considering a Magnetostrictive rod element of length L, area A, applied tensile force F, strain ε, stress σ, elastic modulus E. Total strain energy in the rod will be

$$V_e = \frac{1}{2} AL\varepsilon\sigma = \frac{1}{2}\varepsilon(E\varepsilon - fB)$$

$$= \frac{1}{2} ALE\varepsilon^2 - \frac{1}{2} AL\varepsilon fB \quad (28)$$

Magnetic potential energy in Magnetostrictive rod is

$$V_m = \frac{1}{2} ALBH = \frac{1}{2} AL(-F\varepsilon + bB)H$$

$$= -\frac{1}{2} ALBf\varepsilon + \frac{1}{2} ALgB^2 \quad (29)$$

150

Magnetic external work done for N turn coil with coil current I is

$$W_m = INBA \qquad (30)$$

Mechanical External work done is

$$W_E = F\varepsilon L \qquad (31)$$

Total potential energy of the system is equal to $T_p = -V_e - V_m + W_m + W_e$, which in expanded form becomes

$$T_p = -\frac{1}{2}ALE\,\varepsilon^2 + ALfB - \frac{1}{2}ALgB^2 + INBA + F\varepsilon L \qquad (32)$$

Using Hamilton's Principle, we will get two equation for B and ε as

$$-ALE\,\varepsilon + ALfB + FL = 0 \qquad (33)$$

$$ALE\varepsilon f + ALgB + INA = 0 \qquad (34)$$

Dividing by Volume, AL Equations (33) and (34) will become

$$E\varepsilon - fB = \frac{F}{A} \qquad (35)$$

$$-f\varepsilon + gB = \frac{F}{A} \qquad (36)$$

Eliminating B from Equation (35) and substituting this in Equation (36), the stress-strain relationship for the Magnetostrictive material can be obtained as

$$B = \frac{E\varepsilon - F/A}{f} \qquad (37)$$

$$\varepsilon = \frac{F/A + INf(gL)}{E - f^2/g} \qquad (38)$$

Assuming $E\!*$ as the magnetically free elastic modulus

$$E^* = E - f^2/g - Q \qquad (39)$$

Total strain for applied coil current I and tensile force, F will be

$$\varepsilon = \frac{INf}{(gLE^*)} + \frac{F}{AE^*} \qquad (40)$$

Here $E = Q^*$ is the elastic modulus for a magnetically stiffened rod and $Q = E^*$ is for magnetically flexible rod. Magnetically stiffened means that the magnetic flux $B=0$ in side the rod as the rod is wound by short circuited coils. Magnetically flexible means the rod is free from any coil. E-Q relation can be obtained form Equation (27). To model the one-dimensional nonlinear Magnetostrictive stress-strain and magnetic field-magnetic flux relationships, Equation (25) and Equation (26) can be written as,

$$E(\varepsilon) - fB = \sigma \qquad (41)$$

$$-f\varepsilon + g(B) = \frac{IN}{L} \qquad (42)$$

Where f is the real parameter of scalar value, and ε -E (ε), B - g (B) are two real parameter nonlinear curves. The basic advantage of this model is that only two nonlinear curves are required for representing nonlinearity reported in different stress

151

levels. As opposed to this approach, in straight forward polynomial representation of magnetostriction [Krishnamurty et. al (1999)] one requires single nonlinear curve for every stress level. To get the coefficients of two nonlinear curves and the value of real parameter f, experimental data from Etrema manual [Butler (1988)] is used. From strain, applied coil current and stress level available in the manual, these coefficients are evaluated. Considering modulus elasticity as 30GPa, and f as 75.3x106 m/Amp as an initial guess, the values of magnetic flux density, B are calculated from Equation (41). Similarly from Equation (42), values of $g(B)$ are evaluated. From these B and $g(B)$ values, curves of $B-g(B)$ is computed.

This curve is used to get mechanical relationship. Here, the value of B is computed from $B-g(B)$ relationship. From this value of B, using Equation (41), values of $E(\varepsilon)$ is calculated. From the $E(\varepsilon)$ and ε values, the mechanical nonlinear curve of $\varepsilon-E(\varepsilon)$ relationship is computed. In summery, first the magnetic nonlinear curve is evaluated from mechanical nonlinear curve and mechanical nonlinear curve is evaluated from magnetic nonlinear curve with the help of experimental data given in Etrema manual [Butler (1988)]. This iteration will continue till both mechanical curve and magnetic curve converges. Thus, with the help of experimental data given in Etrema manual [Butler (1988)] and Equations (41) and (42), nonlinear mechanical and magnetic relationship is evaluated. Initial values of modulus of elasticity and f are computed on trial and error basis.

For sensor device, where coil current is assumed as zero, strain and the value of magnetic flux due to the application of stress is given by

$$B = \frac{E(\varepsilon) - \sigma}{f} \tag{43}$$

$$\varepsilon = \frac{g(B)}{f} \tag{44}$$

The magnetic field is approximated by a sixth order polynomial of magnetic flux density and the modulus is approximated by a sixth order polynomial of the strain, which is given by

$$g(B) = c_5 * B^5 + .. + c_1 * B + c_0 \tag{45}$$

$$B = d_5 * g(B)^5 + .. + d_1 * g(B) + d_0$$

$$E(\varepsilon) = a_6 * \varepsilon^6 + .. + a_1 * \varepsilon + a_0 \tag{46}$$

$$\varepsilon = b_6 * E(\varepsilon)^6 + .. + b_1 * E(\varepsilon) + b_0$$

These curves are shown in Figure 7. Coefficients of these polynomials curves are given in Table-1, where unit of B is tesla, $g(B)$ is Amp/m and $E(\varepsilon)$ is Pa. The value of magneto-mechanical coupling parameter (f) is 75.3x106 m/amp, which is reciprocal of 13.3x10$-$9 amp/m.

On the basis of these two curves given in Equation (45) and Equation (46) and parameter (f), strain, magnetostriction vs. applied magnetic field for different stress level are plotted in Figure 8.

The experimental data of strain-magnetic field relationships for different stress level is almost matching with this model. Similarly strain-compressive force and

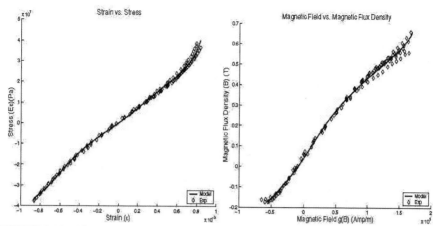

FIGURE 7. Non Linear Stress-Strain curves (left) and Magnetic flux and magnetic field curves (right)

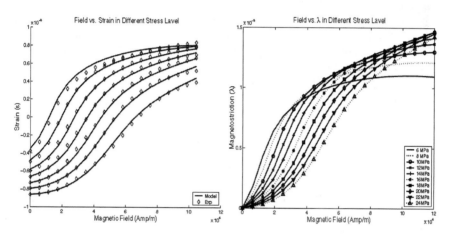

FIGURE 8. Nonlinear Strain-Magnetic field (left) & Magnetostriction-magnetic field(right)

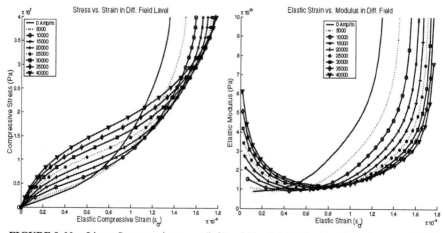

FIGURE 9. Non Linear Stress-strain curves (left) and Elastic Modulus-elastic strain curves (right)

TABLE 1: Coefficients for sixth order polynomial.

	c	d	a	b
6→	0	0	4.5419e+28	1.5853e-50
5→	-2.1687e+06	-1.9526e-27	7.6602e+25	-6.1288e-44
4→	1.5211e+06	1.1589e-21	-4.2662e+22	-1.0355e-34
3→	3.5828e+05	-2.0047e-16	-6.6788e+19	-4.0508e-27
2→	-2.1062e+05	2.0096e-12	2.3911e+16	1.0806e-27
1→	2.2754e+05	4.7789e-06	1.2539e+13	2.7977e-11
0→	-8.8129e+03	4.4239e-02	3.3893e+10	-1.9704e-05

elastic modulus for different magnetic field level is plotted in Figure 9. Elastic modulus is initially decreases and then increases for each magnetic field level, which is also reported in Butler (1988).

The Magnetic Flux, the strain , the stress and the coil current is computed as follows. As two nonlinear curves are related in these relationships, the calculation of magnetic flux and strain from stress and coil current is an iterative procedure. Initially, the value of magnetic flux, B is assumed a certain value. From $B-g(B)$ curve, the value of $g(B)$ is evaluated. From Equation (36) the value of strain is evaluated considering magnetic field as coil current times coil turn per unit length of actuator. Using this strain, from the $\varepsilon -E(\varepsilon)$ curve the value of $E(\varepsilon)$ can be found. From Equation (35), the value of B can be determined. If this B value is not same as assumed, this iteration will be continued until the value converges.

ANN MODEL

To avoid iterative procedure mentioned in nonlinear model, one three layer artificial neural network is developed, which gives direct nonlinear mapping from magnetic field and stress to magnetic flux density and strain. Standard logistic function $y = 1/(1+e^{-1.7159v})$ is used in hidden layer as activation function with linear output layer. Input (stress and magnetic field) and output (strain and magnetic flux density) data is normalized for better performance of network.

$$\sigma_n = \frac{(\sigma - \sigma_{mean})}{(\max|\sigma| - \sigma_{mean})} \tag{47}$$

$$H_n = \frac{(H - H_{mean})}{(\max|H| - H_{mean})} \tag{48}$$

$$\varepsilon_n = \frac{(\varepsilon - \varepsilon_{mean})}{(\max|\varepsilon| - \varepsilon_{mean}} \tag{49}$$

$$B_n = \frac{(B - B_{mean})}{(\max|B| - B_{mean})} \tag{50}$$

σ_n, the normalized stress is calculated using Equation (47). The value of σ_{mean} and $(max|\sigma| - \sigma_{mean})$ are -1.57966×10^7 and 0.830345×10^8 Pa respectively. Similarly H^n is normalized magnetic field, is calculated from Equation (48). The value of H_{mean} and $(max|H| - H_{mean})$ in Equation (48) are both 750 Oe. Normalized strain ε_n is the output of network, from which strain is calculated using Equation (49). The value of ε_{mean} and $(max|\varepsilon| - \varepsilon_{mean})$ are 0.203245×10^{-03} and 0.106504×10^{-02} respectively. In Equation (50) the value of B_{mean} is 0.385126 tesla and the value of $(max|B| - B_{mean})$ is 0.319818 and 0.499656 tesla respectively.

To train this network some training and validation samples are generated through iterative process stated earlier. Weight and bias parameter of the trained network is given in Table-2 and Table-3. Different validation studies are also carried out.

TABLE 2. Connection between input layer and hidden layer.

Input Layer Neurons	Hidden Node 1	Hidden Node 2	Hidden Node 3	Hidden Node 4
H_n	4.6187	-2.1241	-.15066	.38726
σ_n	1.2866	-1.3195	-.43318	.031765
Input Bias	2.2483	-.25721	-.61579	.088719

TABLE 3. Connection between hidden layer and output layer.

Output Layer Neurons	ε_n	B_n
Hidden Node 1	.63447	.67409
Hidden Node 2	-.53313	-.23172
Hidden Node 3	-.69546	-.071654
Hidden Node 4	.48235	2.1900
Output Bias Node	-.29930	-1.5467

COMPARISON BETWEEN DIFFERENT COUPLED MODELS

Comparative study of different models is done taking a Magnetostrictive rod with varying magnetic field and stress level. Three different stress levels (6.9 MPa, 15.1 MPa and 24.1MPa) are taken to compute the total strain and magnetic flux density in the rod for varying magnetic field level and shown in Figure 10. In the left Figure, total strain is shown according to linear, polynomial, ANN and experimental approaches. Both polynomial and ANN approach is showing close result with the experimental data throughout the magnetic field range. As in the case of linear model, results are not matching with the experimental data throughout the magnetic field range. However, this model can be used in low magnetic field level for medium stress level and in medium field level for high stress level. For low stress level, linear model can be used on an average sense. In the absence of experimental data (Etrema manual) of magnetic flux density, only computational results are shown in right figure. Magnetic flux density is shown according to linear, polynomial and ANN approach. Similar to the strain results, the results of ANN model and polynomial model are in excellent agreement However, the results of linear model is not matching through out the magnetic field range. In linear model, for medium stress level in low magnetic

155

field level magnetic flux density is matching with the nonlinear model. For high and low stress level, linear model can be used in the average sense.

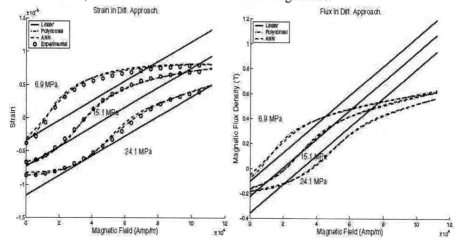

FIGURE 10. Comparison of different models: Strain-Magnetic field curves (left) and Flux density-magnetic field curves (right)

SUMMARY

This article is mainly intended for anhysteretic linear and nonlinear, coupled constitutive relationship of Magnetostrictive material. The coupled model is studied without assuming any direct relationship of magnetic field unlike uncoupled model. In linear-coupled model elastic modulus, permeability and magneto-elastic constant is considered as constant. However, this model cannot predict the highly nonlinear properties of Magnetostrictive material. In the nonlinear coupled model, nonlinearity is decoupled in the magnetic and the mechanical domain using two nonlinear curves for stress-strain and magnetic flux density and magnetic field intensity. In this model, the computation of magnetostriction requires the value of magnetic flux density, which comes through an iterative process for nonlinearity of curves. To avoid this iterative computation, one three layer artificial neural network is developed, which will give nonlinear mapping from stress level and magnetic field to strain and magnetic flux density. Finally, comparative study of linear, polynomial and ANN approach is done and shown that linear coupled model can predict the constitutive relationships in an averaged sense only. Nonlinear models are shown to predict experimental results exactly throughout the magnetic field range

ACKNOWLEDGEMENTS

Author wishes to acknowledge the support extended by The Boeing Company under its umbrella of Strategic University Alliance with the Indian Institute of Science. Author also would like to deeply thank his former graduate student Dr. Debiprasad Ghosh for performing the numerical simulation addressed in this article.

REFERENCES

1. Anjanappa, M.; Bi, J. (1994): Magnetostrictive mini actuators for Smart structures applications. Smart Mater.Struct., vol. 3, pp. 383–390.
2. Benbouzid, M. E. H.; Kvarnsjo, L.; Engdahl, G. (1995): Dynamic modeling of giant magnetostriction in TERFENOL-d rods by the finite element method. IEEE transactions on magnetics, vol. 31, pp. 1821–1823.
3. Benbouzid, M. E. H.; Reyne, G.; Meunier, G. (1993): Nonlinear finite element modeling of giant magnetostriction. IEEE transactions on magnetics, vol. 29,
4. pp. 2467–2469.
5. Butler, J. L. (1988): Application manual for the design of TERFENOL-D Magnetostrictive transducers. Technical Report TS 2003, Ames Iowa, Edge Technologies Inc.
6. Ghosh, D. P.; Gopalakrishnan, S. (2002): Structural health monitoring in a composite beam using Magnetostrictive material through a new FE formulation. In Proceedings of SPIE vol. 5062 Smart Materials, Structures and Systems, edited by Sangeneni Mohan, B. Dattaguru, S. Gopalakrishnan, (SPIE, Bellingham, WA,
7. 2003), 704-711.
8. Hall, D.; Flatau, A. (1994): One-dimensional analytical constant parameter linear electromagnetic magneto-mechanical models of a cylindrical Magnetostrictive TERFENOL-D transducer. Proc. ICIM94: 2nd Int. Conf. on Intelligent Materials (Williamsburg, VA, 1994), pp. 605–616.
9. KrishnaMurty, A. V.; Anjanappa, M.; Wang, Z.; Chen, X. (1999): Sensing of delaminations in composite laminates using embedded Magnetostrictive particle layers. Journal of Intelligent Material Systems Structures, vol. 10, pp. 825–835.
10. Maugin, G. A. (1985): Nonlinear electromechanical effects and applications World Scientific.
11. Moffett, M. B.; Clark, A. E.; Wun-Fogle, M.; Linberg, J.; Teter, J.P.;McLaughlin, E. A. (1989): Characterization of TERFENOL-D for Magnetostrictive transducers. J. Acoust. Soc. Am., pp. 1448–55.
12. Pelinescu, I.; Balachandran, B. (2001): Analytical study of active control of wave transmission through cylindrical struts. SmartMater. Struct., vol. 10, pp. 121–136.
13. Reddy, J. N.; Barbosa, J. I. (2000): On vibration suppression of Magnetostrictive beams. Smart Mater. Struct., vol. 9, pp. 49–58.
14. RoyMahapatra, D. P.; Gopalakrishnan, S.; Balachandran, B. (2001): Active feedback control of multiple waves in helicopter gearbox support struts. Smart Materials and Structures, vol. 10, pp. 1046–1058.
15. Rumelhart, D.; Hinton, G.; Williams, R. (1986): Learning representations by back propagation error. Nature, vol. 323, pp. 533–536.
16. Saidha, E.; Naik, G. N.; Gopalakrishnan, S. (2003): An experimental investigation of a smart laminated composite beam with Magnetostrictive patch for health monitoring applications. Structural Health Monitoring.
17. Toupin, R. A. (1956): The elastic Dielectric. Journal of Rational Mechanics and Analysis, vol. 5, pp. 849–915.
18. Villery, E. (1865): Change of magnetization by tension and by electric current. Ann. Phys. Chem., vol. 126, pp. 87–122.

SESSION C

POLYMERS IN SMART APPLICATIONS

Issues for consideration in the design of shape memory polymer systems

D. Pandit[b], A.P. Deshpande[a] and S.M. Sivakumar[b]

[a]Department of Chemical Engineering
[b]Department of Applied Mechanics
Indian Institute of Technology Madras
Chennai 600 036. INDIA

Abstract. The paper discusses the important issues that have to be taken into consideration in the design of shape memory polymer (SMP) based devices. The study involves identification of performance parameters that link the material design and analysis to design of these devices. With the aim of understanding the dependence of the shape memory effect on parameters that affect it in terms of molecular makeup and thermo-mechanical loading conditions, the response of a shape memory polymer is studied. An attempt has been made here to develop a simple mechanistic model that simulates essential features of a 1-D SMP behavior with reasonable association with the performance parameters. The simulations show that the essential features of the response can be captured by this simple model.

Keywords: shape memory polymers, thermo-mechanical behavior, cross-link, constitutive model, glass transition

INTRODUCTION

In this paper, we discuss the salient features of the thermo mechanical behavior of shape memory polymers (SMPs) that are important for the device designers. Modeling framework, identification of parameters related to the SMP behavior and quantification of performance parameters that are directly relevant to the design of the SMP based devices are also discussed.

Shape memory polymers are an emerging class of soft rubbery polymeric materials that exhibit a shape memory effect that is recoverable over a large strain regime. Lower in stress levels and larger in strain regimes than the shape memory alloys, light weight and biocompatible, they provide a complementary role to the shape memory alloys thus offering the potential to be employed over a wide range of suitable applications. Such applications may be found in smart fabrics, heat –shrinkable tubes for electronics or films for packaging, self deployable sun sails in space craft, self disassembling mobile phones, intelligent medical devices and implants for minimally invasive surgery to name a few [1].

In general, there are two temperatures (say, one termed high temperature and the other low temperature) at which this class of polymers show different properties. At the high temperature state, they can be deformed to any temporary shape due to its rubbery nature. The shape it is deformed to can be retained upon cooling to the low temperature state. The retained shape can be kept indefinitely at this low temperature.

CP1029, *Smart Devices: Modeling of Material Systems, An International Workshop*
edited by S. M. Sivakumar, V. Buravalla, and A. R. Srinivasa
©2008 American Institute of Physics 978-0-7354-0553-0/08/$23.00

This process of shape retention is called 'programming'. Upon heating to the 'high temperature', the shape that it previously had at the high temperature (the permanent or the reference shape) can be recovered. This process is called 'recovery' [1]. This property that is associated with a one way shape memory effect (essentially remembering the high temperature shape) opens up the possibility of creating complex shape changes by simply actuating the material at different locations of a component thus offering scope for a 'distributed actuation'. This has triggered various kinds of applications such as gripping applications, morphing of surfaces with reconfiguration capability, programmable sealing applications, and energy absorption with healing capability [2].

In order to efficiently employ the SMPs in the above application devices, there are currently no standard design procedures available. A thorough understanding of the robust response of these materials under complex force/displacement and temperature protocols is essential to execute the task of designing for a particular application at hand. Thus, the task of constitutive modeling of these materials assumes prime importance. This task of modeling involves, firstly, the identification of the parameter set that serves to characterize the type of response expected in common in SMPs and secondly, to propose appropriate evolution of the parameters that define the change of state of the material. With this, it is possible to compare the materials, using these parameters / properties, for a specific shape memory application. Parameters chosen with a direct link to the physical basis of the SMP function helps in easy comparison. Unlike the usual thermo-mechanical modeling, modeling of these polymers involves the consideration of large deformations, especially, when it comes to shape locking, the thermal effects to model the temperature controlled processes, viscous effects that may affect the performance considerably and the 'plasticity' involved in shape locking.

A simple mechanistic approach to the modeling framework that is thermodynamically consistent is proposed in this work. It consists of a network of the salient thermo mechanical elements that model the basic mechanisms which form the fundamental nature of the SMP material. It also renders the possibility of introducing molecular mechanisms based forms of material functions.

The paper is arranged as follows: With a view to understanding the behavior of SMPs and the identification of performance parameters, the core shape memory effect observed in polymers is described. The general mechanisms involved in exhibiting this shape memory behavior and characteristics of a typical SMP are presented in a subsequent section followed by a discussion on quantification of performance parameters. Finally, modeling the 1-D behavior of the material and the model results are presented.

Shape Memory Effect in polymers

The shape memory effect observed in SMPs can be understood from the following thermo-mechanical process cycle:

After conventionally preparing the material above a certain transition temperature it is isothermally deformed at the same temperature. Keeping the deformation constant by constraining the specimen, it is cooled to a low temperature (T_1) below the transition temperature. Then unloading is done at (T_l),where the specimen shows up the temporary shape as the deformation induced at the high temperature (T_h) got locked up. Unloading is followed by heating to (T_h) under no stress condition and it is observed that the specimen shows up the parent shape with what the thermo mechanical cycle started (Figure 1)

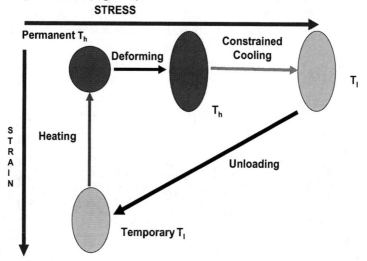

FIGURE 1. This shows a typical thermo-mechanical cycle of loading that demonstrates shape memory effect in certain polymers. The horizontal axis is stress axis and the vertical downward axis is strain axis. The heating and cooling are indicated appropriately. The items in blue are in the 'low-temperature' state and that in red are in the 'high-temperature' state. The process here shows an ideal behavior of typical SMPs.

The Mechanisms involved in the shape memory effect

Polymers are long chain molecules with a large number of repeating units. In general polymers have a highly disordered phase called the amorphous region and a relatively organized region called the crystalline phase.

SMPs are polymer networks, which are equipped with net points and temperature sensitive switching segments in the amorphous region. The net points which can be of chemical (covalent bonds) and/or physical (intermolecular interaction) nature determine the permanent shape of the polymer network along with the crystalline regions. The net points, which are regions of cross-linking, can be of physical nature in polymers whose morphology consists of at least two segregated domains as seen in block copolymers. Here, domains corresponding to highest thermal transition temperature ($T_{permanent}$) act as net points which are the hard segment, while chain segments in domains with the second highest thermal transition T_{trans} act as molecular switches. These switching domains can either be formed by chain segments driving the entropic elastic behavior or by side chains.

In the case of covalent net point networks there is only one transition temperature T_{trans} due to physical interactions. Above the band of T_{trans} the SMPs exhibit entropic

163

elasticity behavior and below T_{trans} band of temperature, they exhibit glassy or solid like behavior.

Entropic elasticity phenomenon occurs due to decrease in entropy of the material in the amorphous region manifested by external loading which straightens the molecular chains. Since all the molecules are under constant thermal motion they strike the straightened chains back into kinked state which is entropic ally more favorable. Similar analogy of entropic elasticity may be obtained from concept of ideal gas. Solid like behavior occurs due to bond stretching.

Figure 2 shows a simple cartoon of the molecular mechanism of shape memory effect in SMPs.

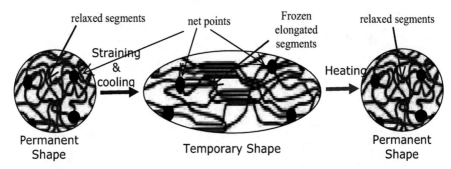

FIGURE 2. The polymer networks with net points and temperature sensitive segments in amorphous regions at different thermo-mechanical states. The permanent shape is preserved by the physical net points of the network while the temporary shape at the low temperature (below the transition temperature) can be arbitrarily set.

Characteristics of the shape memory polymers

The transition temperature may be either the glass transition temperature (T_g) due to the presence of the secondary physical interactions (vitrification) or the melting temperature (T_m) due to the presence of crystallites (crystallization) in a co-polymer or a polymer blend.

If $T_{trans} = T_g$, then the transition from and to the rubbery state to the glassy solid like state takes place over a wide range of temperature. On the other hand, if $T_{trans} = T_m$, then this change occurs over a narrower temperature range. Viscous effects dominate in the band of the temperature about the transition temperature, especially, T_g. Also, at T_l, the stiffness of the material is around two orders of magnitude stiffer than its rubbery state at T_h.

The crosslink density boosts the stiffness at any given temperature and prevents permanent slipping of segments. The volume fraction of amorphous region influences the shape memory phenomenon, since this effect is associated only with the amorphous region.

At small strains, the material generally shows a linear stress-strain relationship both at T_l & T_h.

The qualitative response of a shape memory polymer under a typical thermo mechanical loading program is presented on the stress-strain-temperature axes in Figure 3. The loading program for which the response is presented is the same as that presented in Figure 1. During the initial stretching process, (curve 1 in the figure), the stress varies linearly with strain. It can be seen that for a strain of the order of 8%, the stress is less than 1MPa indicating a low stiffness material at this high temperature state. In the next stage of the loading program, the temperature is decreased below the glass transition temperature keeping the strain fixed, it can be seen that the stress increases with decrease in temperature asymptotically reaching a constant slope equal to the product of elastic modulus and thermal expansion coefficient at the glassy phase (curve 2). Release of the fixed strain at this stage sees the material going stress free at the low temperature condition (curve 3). Notice that, unlike the high temperature state, the strains are small for the stress relaxed indicating a relatively high stiffness of the material at the low temperature condition. In the final stage, upon heating under the stress free condition, the strain that is frozen is released (curve 4) with a rapid degree of release in the glass transition temperature range. The material almost reaches the zero strain state that it was originally in before the loading program was applied.

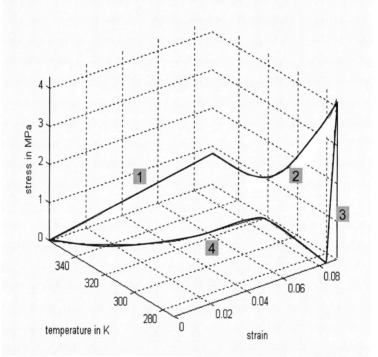

FIGURE 3. Response of a shape memory polymer under a thermo-mechanical loading cycle. The almost linear curve 1 represents the response under stretching at high temperature. The response under cooling upon fixing the strain is shown in curve 2. There is a steady rise in stress due to thermal contraction. After the release of stress at the low temperature (curve 3), the response under heating back is shown by curve 4. With the initial thermal expansion, there is a rapid release of strain near the glass-transition temperature (330K) finally reaching the original state of zero strain and zero stress condition.

165

Quantification of performance of SMPs

The stiffnesses of the shape memory polymer both at T_l & T_h are important from a designer's point of view along with the value of the transition temperature and the bandwidth of the transition region. In Table 1, typical values of the different parameters for a shape memory polymer useful for the designer are presented.

The SMPs will, in general, have to go through a number of thermo-mechanical cycles in practical applications. For quantifying in such applications, strain fixity rate and strain recovery rates are introduced.

Let 'n' denote the number of cycles, R_f denote the strain fixity rate or the ability to fix a mechanical deformation ε_{pre}, resulting in a temporary shape $\varepsilon_{fix}(n)$ & let R_r denotes the strain recovery rate which quantifies the ability to restore the mechanical deformation of the permanent shape $\varepsilon_p(n)$ after n^{th} cycle.

Then, we have

$$R_f(n) = \frac{\varepsilon_{fix}(n)}{\varepsilon_{pre}}$$ 1.

$$R_r(n) = \frac{\varepsilon_{pre} - \varepsilon_p(n)}{\varepsilon_{pre} - \varepsilon_p(n-1)}$$ 2.

where $\varepsilon_p(0) = 0$.

Clearly, ideally both R_f and R_r should be 100%

Sometimes speed of recovery of temporary strain imposed may play an important role.

This is defined by, speed of recovery,

$$V_r = 0.8 R_f \varepsilon_{pre} \frac{\dot{T}}{(T_h - T_l)}$$ 3.

Clearly high value of V_r is desirable most of the times.

TABLE 1. Typical magnitudes of parameters for shape memory polymers

Parameter	Typical values
T_g or T_m	298K to 373K
T_l & T_h	253K to 313K
Maximum stress	2 to 10 MPa
Maximum Strain(ε_{pre})	2 to 100%
Strain fixity rate(R_f)	90 to 100%
Strain recovery rate (R_r)	90 to 100%
Rate of temperature loading ($\dot{T}$)	± 1 to ± 5 K per minute
Speed of recovery (V_r)	0.05 to 0.5 per minute

Constitutive Modeling of SMPs

The design of SMP based devices demands thorough understanding and constitutive modeling of thermo mechanical shape memory (SM) cycle. Constitutive models facilitate the prediction of recoverable stress and strain levels under varying degrees of constraint, as will be invariably experienced in SMP applications.

Though various approaches to modeling the polymer behavior have been attempted using the concepts such as orientation element [3], multi phase crystallinity [4], visco elasticity [5], and viscoplasticity [6], only a very few of these models can be successfully employed to capture the full shape memory behavior of polymers.

In [3], a four state orientation based model is proposed for the prediction of time required to respond to external heating by increase in stress of a constrained amorphous polycarbonate material. Though the model successfully simulates the induction time for the material it is unsuitable for use in non-isothermal processes which will be inevitably experienced by an SMP in the SM cycle.

The viscoelastic model presented in [5], is applicable to crystalline cross linked polymers. It takes account of time dependent phenomenon using modified Voigt model. The model parameters are functions of cross link density and temperature.

The model proposed by Tobushi, et al. [6] incorporates the rate effect and simulates the full SM cycle but suffers from the drawback of mathematical inconsistencies in the constitutive equations derived and presented drawing errors due to this.

The model in [6] which is mathematically rigorous is again inconsistent with second law of thermodynamics, and seriously undermines the actual material behavior by introducing multiphase crystallinity in an amorphous polymer system where no crystalline regions exist.

Under these circumstances there is a need for a model which is simple to understand and implement, captures the salient features of SM effect and is consistent with the physics of the problem. A simple mechanistic model with a bare minimum number of parameters is attempted in the following.

A Simple Mechanistic Model

A form for Helmholtz function is assumed based on the fact of linear elastic behavior of the material under small strain condition. From Clausius – Duhem statement of second law of thermodynamics a form for uniaxial stress component and driving force for plastic dissipation is derived. The model incorporates thermo elastic behavior along with a variation of stiffness over the temperature range relevant to shape memory effect. The model may be physically perceived to be an entropic elasticity spring (B) connected in parallel to the series connection of a glassy spring (A) with an entanglement friction element with a yield stress, σ_y which quantifies the resistance to intermolecular relative displacement (Figure 4).

Thus, Helmholtz potential (ψ) per unit volume can be written as,

$$\psi(T, \varepsilon, \varepsilon_p) = \frac{1}{2} B(\varepsilon - \varepsilon_{th})^2 + \frac{1}{2} A(\varepsilon - \varepsilon_{th} - \varepsilon_p)^2 + \psi_{th} \qquad 4.$$

Where $\quad T$ = the temperature of the material in Kelvin

$\qquad \varepsilon$ = Total strain

167

ε_p = Locked up plastic strain

ε_{th} = Thermal strain

$B = B\,(T)$ denotes the entropic elasticity stiffness

$A = A(T)$ denotes the glassy phase stiffness

$\Psi_{th} = \Psi_{th}\,(T)$ denotes thermal contribution towards Ψ

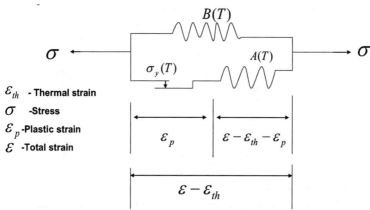

FIGURE 4. A simple mechanistic model for simulating the response of a shape memory polymer under thermo-mechanical loading. The spring $B(T)$ represents the entropic elasticity while $A(T)$ represents the bond-related elasticity. $\sigma_y\,(T)$ is the threshold stress that holds in place the locked up strain. This value is very low at the high temperature state so that strain doesn't at that state.

The constitutive equation is obtained by employing the Clausius – Duhem inequality statement of second law of thermodynamics

$$0 \le \sigma\dot{\varepsilon} - (\dot{\psi} + \eta\dot{T})\qquad\qquad 5.$$

where σ = total stress and η = entropy density.

The constitutive equation is obtained as

$$\sigma = B(\varepsilon - \varepsilon_{th}) + A(\varepsilon - \varepsilon_{th} - \varepsilon_p)$$

$$\varepsilon = \frac{\sigma + A\varepsilon_p}{A + B} + \varepsilon_{th}\qquad\qquad 6.$$

Since $A(T)$ and $\sigma_y\,(T)$ are dominant in the glassy state and $B(T)$ plays a key role in the rubbery state. The model is suitable for amorphous phase, small strain, and no radiation conditions. A piece wise linear form for these material parameters may suitably be evaluated from the shape memory effect experiments.

The model is essentially proposed to simulate the following most important processes of a shape memory cycle:

1. Deformation at high temperature to a prestrain value.

2. Constraining the pre strain followed by cooling to the characteristic low temperature where the material behaves like a glassy substance.

3. Unloading at this low temperature.

4. This is followed by heating under no stress condition.

Experimental data available for an amorphous, epoxy polymer system (CTD DP5.1) [4] was simulated using a MATLAB code written for the model. The simulation results for the typical thermo-mechanical loading program described an earlier section in this article are shown in Figure 5. As can be seen, the trends of prediction from this simplistic mechanistic model seem satisfactory from the design point of view.

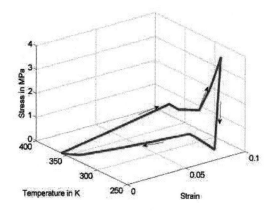

FIGURE 5. Simulation of the response of a shape memory polymer under a thermo-mechanical cycle of loading. The nature of the model ensures full recovery of stress and strain at the original state after the thermo-mechanical loading cycle is complete.

CONCLUSIONS

The model proposed here has the greatest advantage in its 'simple to understand' nature. It captures the physics of the problem when it comes to the question of thermodynamics and intermolecular interactions. It simulates to a reasonable degree of accuracy the shape memory response of amorphous SMPs. Further refinement in terms of introduction of rate effects, thermal recovery effects and finite deformations may be necessary to make it a good designer's tool.

ACKNOWLEDGEMENTS

The authors gratefully acknowledge the help extended by Ms.R.Dhanya and Mr.V.Anand.

REFERENCES

[1].Lendlein, A., and Kelch, S., (2002) "Shape memory polymers," Agnew. Chem. Int. Ed. Vol.41, pp.2034-2057.
[2] Yakacki, C.M., Shandas, R., Lanning, C., Rech, B., Eckstein, A., Gall, K., (2007). Unconstrained recovery characterization of shape memory polymer networks for cardiovascular applications. Biomaterials 28 (14), 2255–2263.
[3] M.Trznadel; (1986) Thermally stimulated shrinkage forces in oriented polymers: Induction time, Polymer, Vol.27, pp.871-876.

[4] Y.Liu, K.Gall, M.L. Dunn, A. R. Greenberg, J. Diani; (2006) Thermo mechanics of shape memory polymers: uniaxial experiments and constitutive modeling; International Journal of Plasticity. Vol.22, pp.279-313.

[5] H. A. Khonakdar, S. H. Jafari, Sorour Rasouli, J.Morshedian, H. Abedini; (2007) Investigation & modeling of temperature dependence recovery of shape memory cross linked polyethylene; Macromol. Theory Simul. Vol.16, pp.43-52.

[6] H. Tobushi, T. Hashimoto, S. Hayashi, E. Yamada; (1997) Thermo mechanical constitutive modeling in shape memory polymer of polyurethane series; Journal of Intelligent Material Systems and Structures , Vol.8, pp.711-718.

Design and Development of a Partially Compliant 4-bar using IPMC for Work Volume Generation

Dibakar Bandopadhya and Bishakh Bhattacharya

Department of Mechanical Engineering, IIT Kanpur, Kanpur – 208016, India

Abstract: This paper illustrates an interesting result on application of Ionic Polymer Metal Composite (IPMC) as a partially compliant smart actuator. A four-bar mechanism is designed and developed in which the conventional rigid link rocker is replaced with an IPMC link to increase the work volume. Simulations are based on the effective length of IPMC and are obtained following Euler-Bernoulli approach for each input voltage. It is demonstrated that controlling the bending of IPMC, work volume of the mechanism can be increased. An experimental set up of four-bar mechanism is developed and experiment is conducted to validate the simulation results. The results prove that a definite work volume is possible by controlling the bending of IPMC.

Keywords: Partially Compliant Mechanism, 4-Bar Mechanism, Ionic Polymer Metal Composite, Work Volume Generation

INTRODUCTION

Traditionally, kinematic chains of mechanisms are designed considering rigid links and assuming negligible elastic deformation at the joints. For a conventional rigid four-bar mechanism, input torque is given to the crank and the follower generates motion based on the link-lengths of the mechanism. Elastic deformation of the links was first considered by Crosseley [1] for varying path generation. Now-a-days, mechanisms that derive their path due to elastic deformation of the joints or links are classified as compliant mechanisms (Midha [2-4], Ananthasuresh [5], Saggere [6], etc.). Saggere et. al. [6] have considered a four-bar mechanism for segmented path generation problem, where the coupler is a flexible link and it undergoes a prescribed shape change along with rigid body motion. However, the compliant mechanism is passive in nature, i.e. external force is required to activate the change in the link-shape. With the advent of smart materials, it has become possible to design links that can change its shape by the application of an actuating voltage.

A variety of smart materials are available today such as shape memory alloys, piezo-ceramics, electro-active polymers etc., which can be used either for driving the mechanism, or for active link shape control. Sitti [7], for example, has developed a piezoelectric actuator driven four-bar mechanism for actuation of a micro insect wing thorax.

In the quest for advanced smart materials, polymer labeled as EAPs (Electro-active Polymer) are being considered in application both for sensing and actuation. Electro-

CP1029, *Smart Devices: Modeling of Material Systems, An International Workshop*
edited by S. M. Sivakumar, V. Buravalla, and A. R. Srinivasa

active Polymers (EAP) are a class of active smart materials that are divided into two categories, electronic, driven by electric field, and ionic, driven by diffusion of ions. Ionic Polymer Metal Composite (IPMC) has both fixed anions and mobile cations [8]. In hydrated state, the cations diffuse towards the electrode on the surface of the material under an applied electric field. Inside the polymer structure, anions are interconnected as clusters to provide channels for cations to flow towards the electrode.

This motion of ions causes the structure to bend towards the anode. The high strain generation by IPMC has attracted attention for its use as mechanical actuator in applications requiring large motion but force of small magnitude [9]. Also, unlike PZT actuators, IPMCs require only a small voltage for actuation, usually less than 5 V. IPMCs are generally made of nafion, a perfluorinated polymer electrolyte, sandwiched between platinum (or gold) electrodes on both sides [10]. Large deformation actuation of IPMC with low actuation voltage bestow them the capability of use as remote arm manipulator (RAM) and as flexible low mass robotic arm propellers for swimming robots. There are other possible examples where this device can be used, such as in vibration control [11], adjustable reflectors and micro path generators. The objectives of this paper are

(i)   Design and development of a partially compliant 4-bar mechanism, introducing IPMC as rocker for varying path generation

(ii)   To calculate the effective length of bent IPMC-rocker for each path under input voltages following the Euler-Bernoulli approach

(iii)   Demonstration of how a work volume can be generated through bending of rocker

In the present study we have used IPMC which produces large strains hence resulting in very large work volume being generated at the tip of the rocker. The three conventional tasks for which four-bar mechanisms have been synthesized are path generation, force and motion generation. This paper concerns a conventional task of path generation by a compliant four-bar mechanism in which the rocker is an ionic polymer metal composite. The effective length of this rocker can be changed by the application of a voltage and hence it behaves like a link with variable length. It is well known that in a Grashof four-bar linkage the tip of the rocker generates a curve for one revolution of the crank. With the change in the length of the rocker, several curves can be generated by the rocker. This in effect generates a work volume inside which the tip of the rocker can be manipulated. Section 2 deals with the bending characteristics of IPMC, where, relationships of bending moment and tip force are established with tip deflection. Section 3 presents the basic analysis of a conventional four-bar linkage and it is shown how a work volume can be generated by changing the length of the rocker. Section 4 discusses the control of IPMC bending and effective length calculation for each input voltage. Simulation results and discussions are given in section 5 while experiment set up and results are discussed in section 6. Finally conclusion is drawn in section 7.

PROPERTIES OF IPMC

Strip of IPMC of size 40x10x0.5 (mm^3) has been tested in cantilever mode. Copper strips are used at one end and voltage is applied quasi-statically from a DC power supply (0-60V, 0-10A, supplied by Elnova Ltd., New Delhi) and subsequently bending of IPMC is recorded through graph paper for each input voltage.

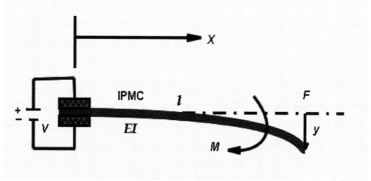

Figure 1. Bending configuration of IPMC

For a cantilevered bender, an expression for the mechanical stiffness can be derived by considering the quasi-static relationship between applied force and deflection for an Euler-Bernoulli beam. Due to application of voltage there will be distributed bending moment which will cause the IPMC to bend. Now for an analogy if we consider that the application of an external force F causes the same amount of deflection, then one can write the equation as

$$M = F(l - x) \tag{1}$$

Where, x is the distance from the supported end along the length of the IPMC of length l. For small deflection, the bending moment and the deflection are related through

$$\frac{d^2 y}{dx^2} = \frac{M}{EI} \tag{2}$$

Where, E is the modulus of elasticity of IPMC and I is the area moment of inertia. Substituting Eq. (1) into (2) one can get

$$\frac{d^2 y}{dx^2} = \frac{F}{EI}(l - x) \tag{3}$$

Integrating Eq. (3) and putting boundary conditions for a cantilevered link $y|_{x=0} = 0$

and $\left.\dfrac{dy}{dx}\right|_{x=0} = 0$, and after simplification, the Y-deflection of IPMC is obtained as:

$$y = \frac{12F}{Ewh^3} \times \frac{l^3}{3} = \frac{4Fl^3}{Ewh^3} \tag{4}$$

Where, w is the width and h is the thickness of IPMC. Therefore, force generated at the tip of the IPMC is obtained as

$$F = \frac{yEwh^3}{4l^3} \tag{5}$$

Again, bending moment generated can be expressed as $M = F \times l$, therefore, maximum bending moment for each deflection is obtained as:

$$M = \frac{yEwh^3}{4l^2} \tag{6}$$

The bending moment generated depends significantly on the moisture contents and thus movement of ions within IPMC from the backbone materials. The metallization of the top and bottom surface of IPMC acts as electrodes. These are quite thin in the order of 2-5 μm and hence do not affect the modulus of elasticity of the IPMC significantly. Through experiment it is found that the relation between tip deflection and applied voltage remains almost linear for both increasing and decreasing voltage. Therefore, the experimental data has been plotted with linear curve fitting approximation $(Y = mV + b)$ as shown in Fig. 2 with a quality factor (r^2) of 0.9951. Fig. 3 shows the bending moment for each input voltage. The input voltage and y-deflection can be approximated as

$$Y = 5.5V + 1.0833 \qquad 0.5 \leq V \leq 4.5 \tag{7}$$

where, m and b are given by,

$$m = \frac{\sum\limits_{i=1}^{c} V_i \sum\limits_{i=1}^{c} Y_i - \sum\limits_{i=1}^{c} V_i Y_i}{\left\{\sum\limits_{i=1}^{c} V_i\right\}^2 - c \sum\limits_{i=1}^{c} V_i^2} \qquad c = 9 \tag{8}$$

$$b = \frac{\left[\sum\limits_{i=1}^{c} y_i - m \sum\limits_{i=1}^{c} V_i\right]}{c} \tag{9}$$

and the quality factor of the approximation is given by:

$$r^2 = m^2 \frac{\left[\sum\limits_{i=1}^{c} V_i^2 - \frac{1}{c}\left(\sum\limits_{i=1}^{n} V_i\right)^2\right]}{\left[\sum\limits_{i=1}^{n} Y_i^2 - \frac{1}{c}\left(\sum\limits_{i=1}^{n} Y_i\right)^2\right]} \tag{10}$$

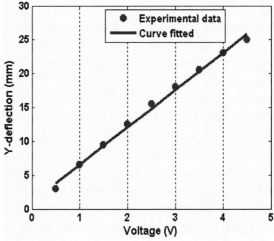

Figure 2. Curve fitted along the experimentally obtained voltage-deflection data

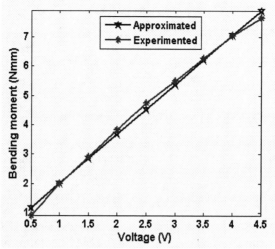

Figure 3. Variation of bending moment with input voltage

OVERVIEW OF PATH GENERATION BY A RIGID ROCKER

In this section a brief overview of path generation by a rigid rocker of a conventional 4-bar mechanism is discussed.

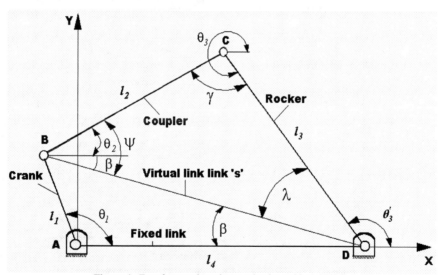

Figure 4. Four-bar crank-rocker mechanism with rigid links

Consider the crank rocker mechanism as shown in Fig. 4 with conventional rigid rocker. To obtain the analytical solution, virtual link-'s' is joined and then cosine rule is used for finding the angles specified. Thus, from the geometry,

$$s = \sqrt{l_1^2 + l_4^2 - 2l_1 l_4 \cos\theta_1} \tag{11}$$

175

$$\beta = cos^{-1} \frac{l_4^2 + s^2 - l_1^2}{2l_4 s}$$

(12)

$$\psi = cos^{-1} \frac{l_2^2 + s^2 - l_3^2}{2l_2 s}$$

(13)

$$\lambda = cos^{-1} \frac{l_3^2 + s^2 - l_2^2}{2l_3 s}$$

(14)

There are usually two values of λ corresponding to each value of θ_1. For crank angle θ_1 in the range of $0 \le \theta_1 \le \pi$, $\theta_2 = \psi - \beta$ and $\theta_3' = \pi - \lambda - \beta$. However, if the crank angle θ_1 is in the range $\pi \le \theta_1 \le 2\pi$ then $\theta_2 = \psi + \beta$ and $\theta_3' = \pi - \lambda + \beta, \theta_3 = \theta_3' + \pi$. For one revolution of the crank, the tip position of the rocker will generate an arc of a circle of radius equal to the length of rocker. A MATLAB program is developed for the synthesis of the rocker curve generation such that the Grashof's criterion is satisfied. As an example for a four-bar system of $l_1 = 15mm, l_2 = 40mm$, $l_3 = 40mm, l_4 = 40mm$.

The coordinates of the joint C as shown in Fig. 5 are $x_3 = l_4 + l_3 \cos\theta_3', y_3 = l_3 \sin\theta_3'$ which are clearly functions of l_3 and θ_3'.

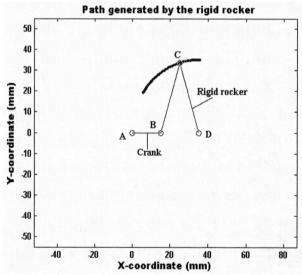

Figure 5. Path generated by the crank-rocker mechanism with rigid links

Path generation by a variable rocker length

In this section, a rigid rocker is replaced by a variable length rocker which can be fabricated by using smart actuators like piezo stacks as a limb as shown in Fig. 6. In this case several curves are generated for different lengths of the PZT-rocker, giving rise to a work volume as shown in Fig. 7. Thus the position of tip point of rocker i.e. (x_3, y_3) is

176

calculated for every angular displacement of crank. The change in rocker length may be initiated by applying voltage on the piezo stack as shown in Fig. 6. The relationship between input voltage and resulting strain may be expressed as $\varepsilon_{11} = d_{33}\dfrac{V}{t}$, where, d_{33} is the piezoelectric modulus, V is the input voltage and t is the thickness of PZT stack.

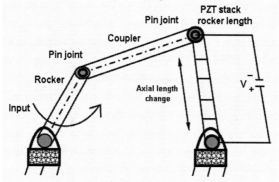

Figure 6. Change of PZT stack rocker length due to applied voltage

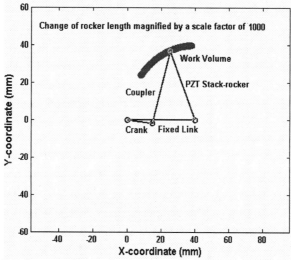

Figure 7. Path generated by varying the PZT stack rocker length by voltage variation

A simulation of work volume generated has been done in MATLAB that computes the positions of the rocker tip based on the crank angle and rocker link lengths with the parameters $d_{33} = 600 \times 10^{-12} \, m \, / \, V$ (soft PZT), initial input 100V varied up to 1KV and thickness (here axial length of PZT stack) is 40 mm. Input is given as angular displacement of crank in increments of *0.0126* radian and the rocker curves are obtained due to change of rocker/PZT stack length. The length of change of PZT stack due to each input voltage is shown in Fig. 8.

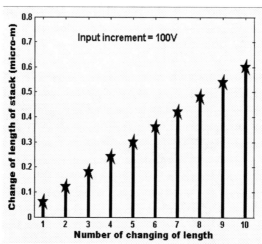

Figure 8. Change in length of PZT rocker due to input voltage

In the Fig. 5 it is shown, with a rigid rocker link, for every revolution of the crank a fixed path can be generated at the rocker tip-point. However, in Fig. 7, it is demonstrated that variables path can be generated at the rocker tip by changing the length of the PZT-stack rocker by applying input voltage. It is evident that such a method of length change by changing applied voltage is only suitable for micro-scale positioning. A larger work volume would demand the introduction of a smart and compliant member. IPMC based link would be an ideal candidate for this purpose.

Effective Length Calculation of IPMC Based on Euler-Bernoulli Approach

In order to obtain a desired path, it is required to position the tip of the rocker at the desired location for a particular instant of rotation of the crank. This requires that the length of the rocker is to be controlled accurately as shown in Fig. 9. A simple method has been proposed to control the deflection of the IPMC based rocker. However, the rocker is connected with the fixed link through pin joint; therefore, there will be no contribution to the bending of the rocker due to dynamic affects.

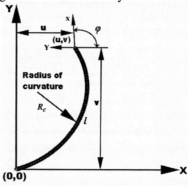

Figure 9. Bending of configuration of IPMC for an input voltage V

Tip position of the IPMC is obtained by applying Euler-Bernoulli equation for an input voltage of V. Considering uniform bending moment generated throughout IPMC then

$$\frac{d\phi}{ds} = \frac{M}{EI} \qquad (15)$$

Using equation (15) horizontal component of tip position can be obtained as:

$$\frac{M}{EI} = \frac{d\phi}{ds} = \frac{d\phi}{dx} cos\,\phi \Rightarrow \int_0^u dx = \frac{EI}{M} \int_0^\phi cos\,\phi.d\phi \qquad (16)$$

Therefore,

$$u = Q\,sin\,\phi \qquad (17)$$

Where, $Q = \frac{EI}{M}$ and ϕ is the tip angle of the IPMC. Similarly vertical deflection of IPMC is obtained as:

$$\int_0^v dy = Q \int_0^\phi sin\,\phi.d\phi = Q(1 - cos\,\phi) \qquad (18)$$

Therefore,

$$v = Q(1 - cos\,\phi) \qquad (19)$$

Therefore, effective length of IPMC is calculated as:

$$l_e = \sqrt{u^2 + v^2} \qquad (20)$$

Fig. 10 is showing the decrement of IPMC-rocker length for each input voltage.

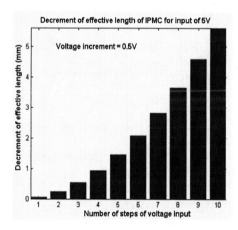

Figure 10. Decrement of effective length of IPMC due to varying input voltage up to 5V

SIMULATION RESULTS

In this section path generation is obtained for the tip of the rocker made of with IPMC. A program has been developed in MATLAB for path generation taking into account of the compliance of IPMC. Coordinates of the points P_1, P_2, P_3 and P_4 are given in terms of link length and variable angle. Considering point P_1 at the origin, the coordinates of the other points are given by: $P_4(x_4 = l_4, y_4 = 0)$, $P_2(x_2 = l_1 \cos\theta_1, y_2 = l_1 \sin\theta_1)$ and $P_3(x_3 = l_4 + l_e \cos\theta_3', y_3 = l_e \sin\theta_3')$. Table 1 lists the properties of the 4-bar mechanism used for simulation. A motor torque of 10 Nmm is used for the simulation. Fig. 11 shows the work volume generated for an input voltage of 5V.

Table 1. Properties of the 4-bar mechanism used for simulation

Crank length (l_1) = 15 mm	Fixed length (l_4) = 40mm
Coupler length (l_2) = 40 mm	No of steps (n) =500
Rocker length (l_3) = 40 mm	Increment (θ_1) = 0.0126 rad

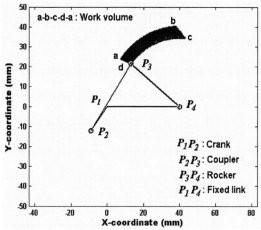

Figure 11. Work volume generated due to varying path generated by rocker for an input of 5V

EXPERIMENTAL RESULTS

An experimental setup of a partially compliant crank-rocker mechanism is built (see Fig. 12) to prove the feasibility of the proposed method. The setup consists of rigid links made up of light Perspex material and the rocker is of ionic polymer metal composite (IPMC). Low torque capacity DC motor (supplied by KTA Japan) is used to provide input torque to the crank. Crank and coupler form revolute joints, while one end of IPMC strip is pinned to the frame and other end forms a fixed joint with coupler. Length of crank is 15mm while coupler and fixed frame are each of 40mm length. The sizes of the IPMC strip are 40 mm x 10 mm x 1 mm and a set of two copper strips are placed at the joint end of the IPMC actuator. The necessary inputs and control configuration are shown in the Fig. 13. PC interfacing is carried out through DAQ-PCI-6251 and NI-DAQmx supplied by National Instrumentation and amplifier of gain 10 is used.

Experiment is conducted to generate a path for 360^0 revolution of crank in 10 seconds with the motor attached to the crank. Fig. 14 shows the experimentally obtained work volume for maximum input of 4.5V while Fig. 15 shows the experimentally obtained bending of IPMC-rocker. It is observed that experimental results slightly differ from theoretical, it is because of bending characteristic of IPMC greatly depends on the water contents and thus its ions movement which gradually degrades due to dehydration of IPMC.

Figure 12. A Partially Compliant four-bar mechanism

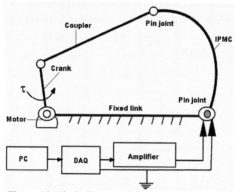

Figure 13. Block diagram of the experimental set up

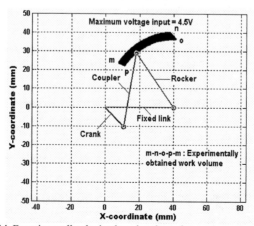

Figure 14. Experimentally obtained work volume for maximum input of 4.5V

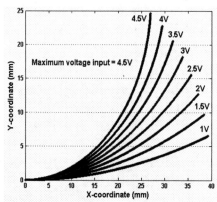

Figure 15. Experimentally obtained bent rocker and its tip position for maximum voltage input of 4.5V

CONCLUSION

In this paper, a partially compliant four-bar crank-rocker mechanism has been used successfully for variable path generation in which the rocker is made of an IPMC, whose length can be changed by application of different voltages. Using the proposed mechanism the tip of the rocker can be manipulated inside its work volume to follow different paths. The experimental result verifies the proposed design. It is observed that increasing the IPMC-rocker length work volume generated by the mechanism can be increased. This mechanism has the potential for application in micro robotics and as compliant mechanisms.

REFERENCES

1. R.H. Burns and F.R.E.Crossley, 1993 "Kinetostatic Synthesis of Flexible Link Mechanisms", ASME paper no: 68-MECH-36.
2. A. Midha, 1993. "Chapter-9: Elastic Mechanisms", Modern Kinematics – The Development in the last forty years, A.G. Erdman (Ed) John Wiley and Sons, Inc.
3. L.L. Howel and A. Midha, 1994. "A loop closure theory for the application and synthesis of compliant mechanisms", ASME J. of Mechanical Design, 118(1), 121-125
4. A. Midha, T. Norton and L.L. Howel, 1992 "On the nomenclature, classification and abstractions of compliant mechanisms", ASME J. of Mechanical Design, 116 (1), 270-279
5. G. K. Ananthasuresh, 1994. *A new design paradigm in micro-electromechanical systems & Investigations on compliant mechanisms* PhD dissertation, Univ. of Michigan.
6. L. Saggere and S. Kota, 2001. "Synthesis of planer, Compliant Four-bar Mechanisms for Compliant Segment Motion Generation", ASME Journal of Mechanical Design, 123, 535-541.
7. Metin Sitti, 2003. "Piezo-electrically actuated four-bar mechanism with two flexible links for micromechanical flying insect thorax", IEEE/ASME Transactions on Mechatronics, 8(1), 26-36.
8. M. Shahinpoor, K. J. Kim, 2001. "Ionic polymer metal composites Fundamentals I", Smart. Mater. Struct. 10, 819-833.
9. Kim K J and Shahinpoor M 2003 "Ionic Polymer Metal Composites –II Manufacturing Technique", Smart Materials and Structure 12 65-69
10. M. Shahinpoor, K. J. Kim, 2004. "Ionic polymer-metal composites: III.Modeling and simulation as biomimetic sensors, actuators, transducers and artificial muscles. Smart Mater. Struct. 13 1362-1388
11. Bandopadhya, D., Bhattacharya, B. and Dutta, A. 2008. Active Vibration Control Strategy for a Single Link Flexible Manipulator Using Ionic Polymer Metal Composite, *Jl. of Intelligent Mat. Systems and Struct.* **19**, 487 - 496.

A nonlinear generalized continuum approach for electro-mechanical coupling

S. Skatulla*, A. Arockiarajan† and C. Sansour*

*School of Civil Engineering, The University of Nottingham, Nottingham, UK
†Department of Applied Mechanics, Indian Institute of Technology Madras, India

Abstract. Electro-active polymers (EAP) are "smart materials" whose mechanical properties may be changed significantly by the application of electric field. Hence, these materials can serve as actuators in electro-mechanical systems, artificial muscles, etc. In this paper, we provide a generalized continuum framework basis for the characterization of the nonlinear electroelastic properties of these materials. This approach introduces new strain and stress measures which lead to the formulation of a corresponding generalized variational principle. The theory is then completed by Dirichlet boundary conditions for the displacement field and the electric potential and then derivatives normal to the boundary. The basic idea behind this generalized continuum framework is the consideration of a micro-and a macro-space which together span the generalized space. All quantities including the constitutive law for the electro-mechanically coupled nonlinear hyperelasticity are defined in the generalized space. Numerical examples are presented to demonstrate the numerical accuracy of the implemented formulation using the mesh free method.

Keywords: Generalized continua, electro-mechanical coupling, strain gradient theories, size-scale effects, multiscale modelling, meshfree methods, oriented material behaviour

INTRODUCTION

Within the last decade, material science has undergone a distinct evolution: from the use of inert structured materials to materials built for a particular function, and then to the so-called smart materials with pronounced recognition, discrimination and reaction capabilities. The sensing and actuating functions of smart materials are explored to develop smart structures. Passive smart structures respond to external changes in a useful manner without assistance, whereas active smart structures have a feedback loop, which is composed of sensors and actuators together with external macro-scale control and logic circuits. Electro-active polymers (EAP) are one among the smart materials of interest, being very attractive, since they sustain large strains, and are easily processed [1]. Hence, they are considered as good substitutes for the low strain piezoceramics and piezopolymers in many applications with improved performance. EAP are characterized by electromechanical coupling effects, e.g. a mechanical deformation can be induced by the application of electric field loading and, in particular, exhibit large displacement in response to electrical stimulation enabling great potential for many engineering fields.

CP1029, *Smart Devices: Modeling of Material Systems, An International Workshop*
edited by S. M. Sivakumar, V. Buravalla, and A. R. Srinivasa
©2008 American Institute of Physics 978-0-7354-0553-0/08/$23.00

A significant area of application of these active polymers is found in biomimetics - the practice of taking up ideas and concepts from nature and implementing them in engineering and design, since its characteristics are similar to biological muscles. An example is the index finger of the hand of a robot, wherein currently being driven by conventional motors serving as a baseline, these materials are substituted as effective actuators [2]. Enhancement of the performance of EAP will require advancement in related computational chemistry models, comprehensive material science, nonlinear electro-mechanics, and improved material processing techniques. Among these, it is important to study the nonlinear effects in the electro-mechanical coupled problems related to the EAP.

When it comes to dealing with smaller structures, scale effects may play an important role. This kind of phenomena, however, are not neatly addressed by classical continuum theories, where the variation of strain is considered to be only substantial within a so-called *representative volume element* (RVE) to which the external loading corresponds. This concept is based on the presumption that the RVE is large enough compared to the size of the micro-constituents such as granular, crystalline or other heterogeneous aggregates, in the case of EAP molecule chains. Thus, the behaviour of the single constituents and their interaction with each other can be neglected. It also implies that a large number of micro-constituents respond to an external stimuli as a whole and mean values of material properties can be taken which represent the entire body. In the small-scale structures though, the external loading interrelates to sub-continua which refer to the micro-constituents. That is, the dimensions of the body under consideration are not very much larger than the characteristic length of the material of this body, e.g. interatomic spacing in a crystal, the particle size of a granular material, or distance between the molecule chains of polymers. Here, the generalized continuum theories can provide a remedy, as they can be formulated to account for the micro-structural influence on the macroscopic material behaviour.

In this contribution, a strain gradient approach is proposed, where the micro-structure is modelled in a very general manner. That is to incorporate the size and the dimension of the micro-continuum into an electro-mechanically coupled continuum model (inter-action between the electric fields and the mechanical deformation). The approach is based on a theory introduced in [3, 4], where a generalized continuum is formulated as to consist of a macro-and a micro-continuum. Accordingly, the generalized deformation is composed by a macro- and micro-component. The dimension of the micro-continuum or the number of degrees of freedom additional to those needed for a classical contin-uum, respectively, may be freely chosen depending on the accuracy of the description of a physical property, but must be finite. In this work however, the dimension of the micro-space is restricted to be maximal three-dimensional. The approach considers a ge-ometrically exact description of finite deformation within the macro-continuum, but as a first step linearizes the deformation within the micro-continuum. In general, however, this ansatz also allows for the formulation of nonlinear micro-deformation. The elec-tric potential is assumed be constant within the micro-space. Based on the approach of a generalized deformation, new non-linear strain and corresponding stress measures are defined and, taking up the idea of an electro-mechanically coupled stored energy density function [5], a generalized electro-mechanically coupled variational principle is formu-lated. For modelling purposes, an moving least square (MLS) approximation scheme is chosen, because it provides the flexibility in terms of continuity and consistency needed by this generalized formulation. Finally, some numerical examples are solved based on the developed theory.

The plan of this paper is as follows: In section 2 the theory of the generalized continuum is outlined followed by section 3 which briefly addresses the basics of electro-mechanics. Subsequently, in section 4 an electro-mechanically coupled variational formulation is proposed comprising the previously introduced generalized continuum and electro-mechanical coupling. The applicability of the generalized electro-mechanical approach with MLS is illustrated in section 5 by various applications within the domain of hyperelasticity.

GENERALIZED DEFORMATION AND STRAIN

The generalized continuum framework is based the on mathematical concept of a fibre bundle (see e.g. Choquet-Bruhat et al. [6]), where in the simplest case the generalized space is constructed as the Cartesian product of a macro space $\mathscr{B} \subset \mathbb{E}(3)$ and a micro space $\mathscr{S}$ which we write as $\mathscr{G} := \mathscr{B} \times \mathscr{S}$. This definition assumes an additive structure of $\mathscr{G}$ which implies that the integration over the macro- and the micro-continuum can be performed separately. The macro-space $\mathscr{B}$ is parameterized by the curvilinear coordinates ϑ^i, $i = 1, 2, 3$ and the micro-space $\mathscr{S}$ by the curvilinear coordinates ζ^α. Here, and in what follows, Greek indices take the values $1, \ldots$ or n. The dimension of $\mathscr{S}$ denoted by n is arbitrary, but finite. Furthermore, we want to exclude that the dimension and topology of the micro-space is dependent on ϑ^i. Each material point $\tilde{\mathbf{X}} \in \mathscr{G}$ is related to its spatial placement $\tilde{\mathbf{x}} \in \mathscr{G}_t$ at time $t \in \mathbb{R}$ by the mapping $\tilde{\varphi}(t) : \mathscr{G} \longrightarrow \mathscr{G}_t$. For convenience but without loss of generality we identify $\mathscr{G}$ with the undeformed reference configuration at a fixed time t_0 in what follows.

The tangent space $\mathscr{T}\mathscr{G}$ in the reference configuration is defined by the pair $(\tilde{\mathbf{G}}_i \times \mathbf{I}_\alpha)$ given by

$$\tilde{\mathbf{G}}_i = \frac{\partial \tilde{\mathbf{X}}}{\partial \vartheta^i} \qquad \text{and} \qquad \mathbf{I}_\alpha = \frac{\partial \tilde{\mathbf{X}}}{\partial \zeta^\alpha}, \tag{1}$$

where the corresponding dual contra-variant vectors are denoted by $\tilde{\mathbf{G}}^i$ and $\mathbf{I}^\alpha$, respectively. A corresponding tangent space in the current configuration $\mathscr{T}\mathscr{G}_t$ is spanned by the pair $(\tilde{\mathbf{g}}_i \times \mathbf{i}_\alpha)$ given by

$$\tilde{\mathbf{g}}_i = \frac{\partial \tilde{\mathbf{x}}}{\partial \vartheta^i} \qquad \text{and} \qquad \mathbf{i}_\alpha = \frac{\partial \tilde{\mathbf{x}}}{\partial \zeta^\alpha}. \tag{2}$$

Now, we choose the placement vector $\tilde{\mathbf{x}}$ of a material point P $(\tilde{\mathbf{X}} \in \mathscr{G})$ to be the sum of its position in the macro-continuum $\mathbf{x} \in \mathscr{B}_t$ and in the micro-continuum $\xi \in \mathscr{S}_t$ as follows

$$\tilde{\mathbf{x}}\left(\vartheta^k, \zeta^\beta, t\right) = \mathbf{x}\left(\vartheta^k, t\right) + \xi\left(\vartheta^k, \zeta^\beta, t\right). \tag{3}$$

Thereby, the macro-placement vector $\mathbf{x}$ defines the centerpoint of the micro-coordinate system such that the micro-placement ξ is assumed to be relative to the macro-

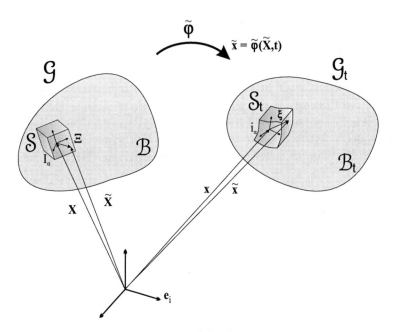

FIGURE 1. configuration spaces

placement. The micro-motion can be assumed to be linear in ζ^α considering it to be very small compared to the macro-motion, a simplification also suggested by Eringen [7], and we arrive at

$$\tilde{\mathbf{x}} = \mathbf{x}\left(\vartheta^k, t\right) + \zeta^\alpha \mathbf{a}_\alpha\left(\vartheta^k, t\right). \tag{4}$$

The vector functions $\mathbf{a}_\alpha\left(\vartheta^k, t\right)$ can be viewed as the directors of the micro-continuum with their corresponding micro-coordinates ζ^α. The number of which must be chosen according to a specific topology of the micro-space as well as certain physical properties of a material due to its intrinsic structure.

In order to avoid the incorporation of additional degrees of freedom, other than the displacement degrees of freedom, the micro-motion is assumed to be dependent on the macroscopic spatial derivatives of the macroscopic placement vector $\mathbf{x}_{,\alpha}$, i.e. the directors of the micro-continuum $\mathbf{a}_\alpha$ are defined as follows

$$\mathbf{a}_\alpha\left(\vartheta^k, t\right) = \frac{\partial \mathbf{x}}{\partial \vartheta^\alpha} = \mathbf{x}_{,\alpha}\left(\vartheta^k, t\right). \tag{5}$$

Then, the generalized deformation field (4) takes the following form

$$\tilde{\mathbf{x}}\left(\vartheta^k, \zeta^\beta, t\right) = \mathbf{x}\left(\vartheta^k, t\right) + \zeta^\alpha \mathbf{x}_{,\alpha}\left(\vartheta^k, t\right). \tag{6}$$

Note, even if the micro-continuum $\mathcal{S}$ is defined by the gradient of macroscopic placement vector, it is important to realize that the dimension of the micro-space does not have to coincide with dimension of the macro-space, but must not be larger than three.

186

Taking the derivatives of $\tilde{\mathbf{x}}$ with respect to ϑ^i and ζ^α the generalized deformation gradient tensor can be expressed as follows

$$\tilde{\mathbf{F}}\left(\vartheta^k,\zeta^\beta,t\right) = \left[\mathbf{x}_{,i}\left(\vartheta^k,t\right) + \zeta^\alpha \mathbf{x}_{,\alpha i}\left(\vartheta^k,t\right)\right] \otimes \tilde{\mathbf{G}}^i + \mathbf{x}_{,\alpha}\left(\vartheta^k,t\right) \otimes \mathbf{I}^\alpha. \qquad (7)$$

In order to formulate generalized strain measures based on (7) we proceed correspondingly to the definition of the classical right *Cauchy-Green* deformation tensor, thus, its generalized equivalent is expressed as

$$\tilde{\mathbf{C}} = \tilde{\mathbf{F}}^T \tilde{\mathbf{F}}. \qquad (8)$$

Neglecting higher order terms in ζ^α and extracting only the dominant part of $\tilde{\mathbf{C}}$ we arrive at

$$\tilde{\mathbf{C}} = \left(\mathbf{x}_{,k} \cdot \mathbf{x}_{,l} + \zeta^\alpha \left(\mathbf{x}_{,k} \cdot \mathbf{x}_{,\alpha l} + \mathbf{x}_{,\alpha k} \cdot \mathbf{x}_{,l}\right)\right) \tilde{\mathbf{G}}^k \otimes \tilde{\mathbf{G}}^l = \mathbf{C} + \zeta^\alpha \mathbf{K}_\alpha, \qquad (9)$$

where $\mathbf{C}$ represents the conventional right *Cauchy-Green* deformation tensor. The scalar products of vectors are denoted by a dot.

In summary, as the deformation of the micro-continuum depends on the gradient of the macro-placement vector, it undergoes rotation, stretch and shear and we find an analogy to the so-called *micromorph continuum*[7]. However, no additional degrees of freedom are involved which go beyond those needed in classical mechanics. Any generalized strain measure derived from (6) can only be linear with respect to the micro-coordinate ζ^α, as the generalized deformation is linear in ζ^α. The micro-deformation ξ is obviously dependent on the macro deformation gradient $\mathrm{Grad}\,\mathbf{x}$, that is, the generalized deformation $\tilde{\mathbf{x}}$ is completely described by the unknown macro-displacement field $\mathbf{u}$ and its first order derivatives.

In the following we want to confine ourselves to the quasi-static case and consider no time dependency.

BASIC RELATIONS OF ELECTRO-MECHANICS

This section is not meant to be a comprehensive introduction in electro-mechanics but rather to present the basic electric fields in the proposed generalized continuum framework. The generalised space $\mathscr{G}$ is indicated by a swung dash. For further details on nonlinear electroelasticity we refer [8][9][10]. First, the electric field $\tilde{\mathbf{e}}$ is defined by

$$\tilde{\mathbf{e}} = -\frac{\partial \tilde{\phi}}{\partial \theta^i}\,\tilde{\mathbf{G}}_i, \qquad (10)$$

where $\tilde{\phi}$ denotes the electric potential. Then the electric displacement vector is expressed as

$$\tilde{\mathbf{d}} = \varepsilon_0 \tilde{J} \tilde{\mathbf{C}}^{-1} \tilde{\mathbf{e}} + \tilde{\mathbf{p}}, \qquad (11)$$

where the constant ε_0 denotes the vacuum electric permittivity, the generalized Jacobian $\tilde{J} = \sqrt{\det \tilde{\mathbf{C}}}$ and $\tilde{\mathbf{p}}$ denotes the spontaneous polarisation. As $\tilde{\phi}$ is assumed to be constant within the microspace $\mathscr{S}$ the governing equations for electric field and displacement be formulated for the macrospace as follows

$$\mathrm{Curl}\,\mathbf{e} = \mathbf{0} \quad \text{on } \partial\mathscr{B}, \qquad \mathrm{Div}\,\mathbf{d} = 0 \quad \text{in } \mathscr{B}. \tag{12}$$

For compressible neo-Hookian material the stored energy function can be considered as

$$
\tilde{\rho}_0\,\psi\left(\tilde{\mathbf{C}},\tilde{\mathbf{e}}\right) = \frac{\mu}{2}\left(\tilde{\mathbf{C}}:\mathbf{1}-3\right) - \mu\ln\tilde{J} + \frac{\lambda}{2}\left(\ln\tilde{J}\right)^2 + c_1\,\mathbf{1}:\tilde{\mathbf{e}}\otimes\tilde{\mathbf{e}} + c_2\,\tilde{\mathbf{C}}:(\tilde{\mathbf{e}}\otimes\tilde{\mathbf{e}})
$$
$$
-\frac{1}{2}\varepsilon_0\tilde{J}\tilde{\mathbf{C}}^{-1}:(\tilde{\mathbf{e}}\otimes\tilde{\mathbf{e}}) \tag{13}
$$

where λ and μ are the Lamé constants, and c_1 and c_2 are material constants which need to be determined experimentally [5]. The double dot operator (:) denotes the scalar product of tensors. Note that for problems involving large deformations the last term in (13) can be considered to be negligible. Alternatively, the mechanical part of the stored energy function (13) may be substituted with a linear elastic constitutive law such as the *Saint-Venant-Kirchhoff model* and we have

$$
\tilde{\rho}_0\,\psi\left(\tilde{\mathbf{C}},\tilde{\mathbf{e}}\right) = \frac{1}{8}\overset{4}{\mathbb{C}}\left(\tilde{\mathbf{C}}-\mathbf{1}\right):\left(\tilde{\mathbf{C}}-\mathbf{1}\right) + c_1\,\mathbf{1}:\tilde{\mathbf{e}}\otimes\tilde{\mathbf{e}} + c_2\,\tilde{\mathbf{C}}:(\tilde{\mathbf{e}}\otimes\tilde{\mathbf{e}})
$$
$$
-\frac{1}{2}\varepsilon_0\tilde{J}\tilde{\mathbf{C}}^{-1}:(\tilde{\mathbf{e}}\otimes\tilde{\mathbf{e}}) \tag{14}
$$

where $\overset{4}{\mathbb{C}}$ represents the wellknown fourth order constitutive tensor containing the elastic moduli. In this work however, we replace the mechanical part of (13) with a non-linear statistically based hyperelastic material law [11, 12]. This model has been originally developed to describe nearly incompressible behaviour of rubber material.

Now, with the spontaneous polarisation is given by

$$\tilde{\mathbf{p}} = -\rho_0\frac{\partial\psi\left(\tilde{\mathbf{C}},\tilde{\mathbf{e}}\right)}{\partial\tilde{\mathbf{e}}}, \tag{15}$$

and the electric displacement (11) can be expressed as

$$\tilde{\mathbf{d}} = \varepsilon_0\tilde{J}\tilde{\mathbf{C}}^{-1}\tilde{\mathbf{e}} - \rho_0\frac{\partial\psi\left(\tilde{\mathbf{C}},\tilde{\mathbf{e}}\right)}{\partial\tilde{\mathbf{e}}}. \tag{16}$$

GENERALIZED ELECTRO-MECHANICALLY COUPLED VARIATIONAL PRINCIPLE

Leaning on a classical electro-mechanically coupled variational formulation based on the conventional right *Cauchy-Green deformation tensor* and the electric field $\mathbf{e}$ an adjacent approach is to establish a generalized variational principle based on the generalized right *Cauchy-Green* deformation tensor $\tilde{\mathbf{C}}$. For this let us consider a non-linear boundary value problem in the domain $\mathscr{B} \times \mathscr{S}$ with the boundary $\partial\mathscr{B} \times \mathscr{S}$. Accordingly, Dirichlet boundary conditions are prescribed on $\partial\mathscr{B}_D \times \mathscr{S} \subset \partial\mathscr{B} \times \mathscr{S}$ denoted by $\tilde{\mathbf{h}}_{\tilde{\mathbf{u}}}^{(\mathbf{n})}$ where $\mathbf{n}$ defines the normal vector on $\partial\mathscr{B}$. Neumann boundary conditions are prescribed on $\partial\mathscr{B}_N \times \mathscr{S} = \partial\mathscr{B} \times \mathscr{S} \setminus \partial\mathscr{B}_D \times \mathscr{S}$ denoted by $\tilde{\mathbf{t}}^{(\mathbf{n})}$.

Now, let $\mathscr{W}_{ext}$ define the external virtual work in the Lagrangian form as follows

$$
\mathscr{W}_{ext}(\mathbf{u}) = \int_{\mathscr{B}} \mathbf{b} \cdot \delta\mathbf{u}\, dV + \int_{\mathscr{B}} \mathbf{l}^{\alpha} \cdot \delta\frac{\partial\mathbf{u}}{\partial\vartheta^{\alpha}}\, dV
$$
$$
+ \int_{\partial\mathscr{B}_N} \mathbf{t}^{(\mathbf{n})} \cdot \delta\mathbf{u}\, dA + \int_{\partial\mathscr{B}_N} \mathbf{q}^{(\mathbf{n})\alpha} \cdot \delta\frac{\partial\mathbf{u}}{\partial\vartheta^{\alpha}}\, dA \qquad (17)
$$

where the external body force and body couples

$$
\mathbf{b}\left(\vartheta^k\right) = \int_{\mathscr{S}} \tilde{\rho}_0\, \tilde{\mathbf{b}}\left(\vartheta^k, \zeta^{\beta}\right)\, dS, \qquad \mathbf{l}^{\alpha}\left(\vartheta^k\right) = \int_{\mathscr{S}} \zeta^{\alpha}\, \tilde{\rho}_0\, \tilde{\mathbf{b}}\left(\vartheta^k, \zeta^{\beta}\right)\, dS \qquad (18)
$$

are acting on $\mathscr{B} \times \mathscr{S}$, and the external traction and surface couples

$$
\mathbf{t}^{(\mathbf{n})}(\eta^r) = \int_{\mathscr{S}} \tilde{\rho}_0\, \tilde{\mathbf{t}}^{(\mathbf{n})}\left(\eta^r, \zeta^{\beta}\right)\, dS, \qquad \mathbf{q}^{(\mathbf{n})\alpha}(\eta^r) = \int_{\mathscr{S}} \zeta^{\alpha}\, \tilde{\rho}_0\, \tilde{\mathbf{t}}^{(\mathbf{n})}\left(\eta^r, \zeta^{\beta}\right)\, dS \qquad (19)
$$

are acting on $\partial\mathscr{B}_N \times \mathscr{S}$. Note that η^r, $r = 1,2$ is the coordinate chart on $\partial\mathscr{B}$. Furthermore, dV is a volume element of the macroscopic domain $\mathscr{B}$, whereas dA is a surface element of its corresponding boundary $\partial\mathscr{B}$, and accordingly, dS is a volume element of the microscopic domain $\mathscr{S}$. The density of the generalized space in the reference configuration is denoted by $\tilde{\rho}_0$ and the density of the macro-space is consequently expressed by

$$
\rho_0 = \int_{\mathscr{S}} \tilde{\rho}_0\, dS. \qquad (20)
$$

Furthermore, we assume now that the body under consideration $\mathscr{B} \times \mathscr{S}$ is hyperelastic and possesses an elastic potential Ψ represented by the stored strain energy per unit undeformed volume $\tilde{\rho}_0\, \psi(\tilde{\mathbf{C}}, \tilde{\mathbf{e}})$.

For the static case and considering only mechanical and electric processes the *first law of thermodynamics* provides the following variational statement

$$
\int_{\mathscr{B}} \int_{\mathscr{S}} \left\{ \tilde{\rho}_0 \frac{\partial\psi}{\partial\tilde{\mathbf{C}}} : \delta\tilde{\mathbf{C}} + \tilde{\rho}_0 \frac{\partial\psi}{\partial\tilde{\mathbf{e}}} \cdot \delta\tilde{\mathbf{e}} \right\} dS\, dV - \mathscr{W}_{ext} = 0. \qquad (21)
$$

Integrating over the microspace $\mathscr{S}$ we further simplify the variational formulation as follows

$$
\int_{\mathscr{B}} \left\{ \frac{1}{2} (\mathbf{S} : \delta\mathbf{C} + \mathbf{M}^{\alpha} : \delta\mathbf{K}_{\alpha}) - \mathbf{d} \cdot \delta\mathbf{e} \right\} dV - \mathscr{W}_{ext} = 0, \qquad (22)
$$

where

$$
\mathbf{S} = \int_{\mathscr{S}} 2\tilde{\rho}_0 \frac{\partial\psi}{\partial\tilde{\mathbf{C}}}\, dS, \qquad \mathbf{M}^{\alpha} = \int_{\mathscr{S}} 2\xi^{\alpha}\tilde{\rho}_0 \frac{\partial\psi}{\partial\tilde{\mathbf{C}}}\, dS, \qquad \mathbf{d} = -\int_{\mathscr{S}} \tilde{\rho}_0 \frac{\partial\psi}{\partial\tilde{\mathbf{e}}}\, dS.
$$

The generalized variational principle is supplemented by Dirichlet boundary conditions for the displacement field, its derivatives normal to the boundary and for the electric potential

$$\mathbf{u} = \mathbf{h}_u \quad \text{on } \partial\mathscr{B}_D^u, \tag{23}$$

$$\frac{\partial \mathbf{u}}{\partial \vartheta^\alpha} = \mathbf{h}_\gamma^\alpha \quad \text{on } \partial\mathscr{B}_D^\gamma, \tag{24}$$

$$\phi = h_\phi \quad \text{on } \partial\mathscr{B}_D^\phi. \tag{25}$$

which represent the microscopic average of the generalized displacement field, its gradient and the electric potential at points $\mathbf{X} \in \partial\mathscr{B}_D$. Note that the domain $\mathscr{B}$ has the boundary $\partial\mathscr{B}$ which consists of the Dirichlet boundary $\partial\mathscr{B}_D \subset \partial\mathscr{B}$ and the Neumann boundary $\partial\mathscr{B}_N = \partial\mathscr{B} \setminus \partial\mathscr{B}_D$.

NUMERICAL EXAMPLES

The following examples within this section are modelled making use of the MLS-approximation scheme. The MLS-approximation functions have to meet the continuity and consistency requirements of the proposed variational formulation (22) which is continuity of $C^1(\Omega)$ and consistency of order 2 due the incorporated strain gradients. This is ensured by choosing a cubic weight function and second order basis polynomial. For further details the reader is referred to [13, 14, 15, 16]. The numerical integration over the micro-continuum $\mathscr{S}$ is carried out with the help of the *Gauss quadrature*, the order of which has to be second according to the used basis polynomial.

As previously mentioned, the mechanical part of the stored energy density function (13) is replaced with a non-linear statistically based hyperelastic material law [11, 12]. which makes use of three constants, the shear modulus C_R, the bulk modulus κ and parameter N which addresses the the limited extensibility of the macromolecular network structure of the polymer material.

All applications throughout this section share the same problem configuration which is a plate with a hole at its centre subjected to electric charge loading as depicted in Fig. 2. The meshfree particle distribution corresponds to the illustrated mesh. However, only one quarter of the plate is modelled applying the appropriate symmetry conditions and using 3 equally spaced layers of 65 particles each.

In graphs Fig. 3 and Fig. 4 the classical solution is denoted by a red line and the generalized solutions are modelled assuming a two-dimensional micro-space with the directors (5) defined as $\mathbf{a}_1 = \mathbf{x}_{,1} = \mathbf{x}_{,x}$ and $\mathbf{a}_2 = \mathbf{x}_{,2} = \mathbf{x}_{,y}$. Three different scaling levels are investigated firstly, choosing the internal length scale parameters $l = l_x = l_y = 5.0$ denoted by the blue line, secondly $l = l_x = l_y = 20.0$ denoted by the green line, and thirdly, $l = l_x = l_y = 40.0$ denoted by the purple line. The transition to the classical solution is achieved by setting the internal length scale parameters to very small values $l = l_x = l_y \leq 10^{-3}$. For this case the strain gradient contribution is negligible small and

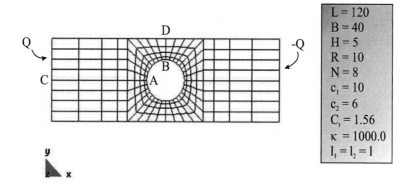

L = 120	
B = 40	
H = 5	
R = 10	
N = 8	
c_1 = 10	
c_2 = 6	
C_r = 1.56	
κ = 1000.0	
$l_1 = l_2 = 1$	

FIGURE 2. problem configuration

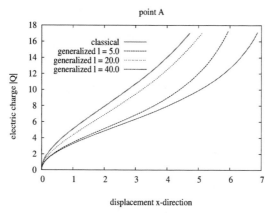

FIGURE 3. electric charge versus displacement at point A in x-direction simulating different magnitudes for the internal length scale parameters

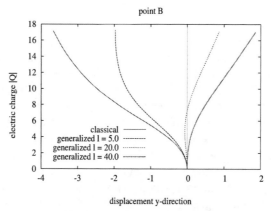

FIGURE 4. electric charge versus displacement at point B in y-direction simulating different magnitudes for the internal length scale parameters

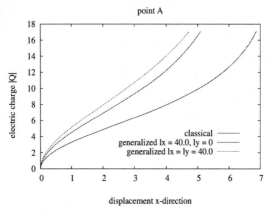

FIGURE 5. electric charge versus displacement at point A in x-direction simulating one and two-dimensional microspaces

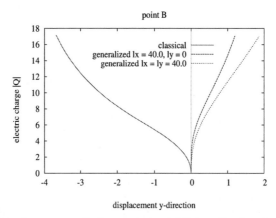

FIGURE 6. electric charge versus displacement at point B in y-direction simulating one and two-dimensional microspaces

the conventional part of (22) is dominant. Note that the size of the microspace is specified by the internal length scale parameters. In graph Fig. 3 the displacement in x-direction versus electric charge loading at point A is displayed, where the closure of the hole is increasingly less with larger values for the internal length scale parameters. The graph Fig. 4 illustrates displacement in y-direction versus electric charge loading at point B. Here, an interesting effect can be observed that, in contrast to the classical result where the hole is contracting, for the two highest scaling levels the diameter of the hole is actually expanding during the simulation.

Next, the effect of a one-dimensional and a two dimensional micro-space having the same magnitude for the internal length scale parameters is studied and compared with the classical result. In graphs Fig. 5 and Fig. 6 the classical solution is again denoted by a red line, the generalized solution for the one-dimensional micro-continuum

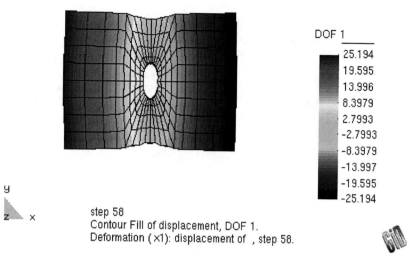

step 58
Contour Fill of displacement, DOF 1.
Deformation (×1): displacement of , step 58.

FIGURE 7. deformed configuration at $u_x = 20$ at point C simulated with the classical approach

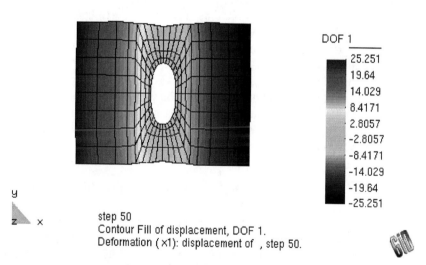

DOF 1

step 50
Contour Fill of displacement, DOF 1.
Deformation (×1): displacement of , step 50.

FIGURE 8. deformed configuration at $u_x = 20$ at point C simulated with the generalized approach using $l_x = 40.0$ and $l_y = 0.0$

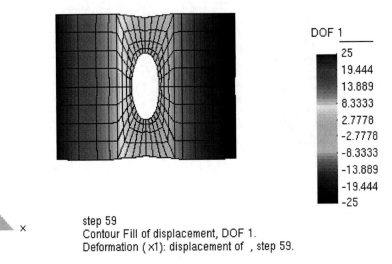

DOF 1
25
19.444
13.889
8.3333
2.7778
-2.7778
-8.3333
-13.889
-19.444
-25

y
z   x

step 59
Contour Fill of displacement, DOF 1.
Deformation (×1): displacement of , step 59.

FIGURE 9. deformed configuration at $u_x = 20$ at point C simulated with the generalized approach using $l_x = 40.0$ and $l_y = 40.0$

with $l = l_x = 40$ and $l_y = 0$ is denoted by the blue line, and for the two-dimensional micro-continuum with $l = l_x = l_y = 40.0$ denoted by the green line. These two graphs are accompanied by corresponding deformed configurations with conture plots of the displacement in x-direction as shown in Fig. 7, Fig. 8, and Fig. 9. In graph Fig. 5 the displacement in x-direction versus electric charge loading at point A is depicted, where the closure of the hole is increasingly less starting with the classical solution, followed by the one-dimensional micro-continuum and then the two-dimensional micro-space. The graph Fig. 6 displaying the displacement in y-direction versus electric charge loading at point B shows that the one-dimensional micro-space is less expanding than for the two-dimensional case.

CONCLUSIONS

Electroactive polymers (EAP) are of increasing importance in engineering. Understanding the behaviour of this type of materials is a challenge for engineers. A nonlinear hyperelastic constitutive law with scale effects in a generalized continuum setting is employed to simulate the nonlinear coupling behavior that EAPs exhibit under electrical stimuli. Numerical examples show the significant impact of scale effects. The generalized continuum formulation, mesh free method implementation and the numerical examples presented in this paper set the initial step towards a more realistic modelling of EAP, in particular and, of smart materials, in general. The future work will focus on deriving a robust internal potential energy function which provides further realistic modelling of electro-mechanically coupled problems.

REFERENCES

1. K. Ren, S. Liu, M. Lin, Y. Wang, and Q. Zhang, *Sensors and Actuators A: Physical* (2008), doi:10.1016/j.sna.2007.10.083.
2. Y. Bar-Cohen, *Proceedings of the SPIE Smart Structures and Materials Symposium* **4695**, 4695–02 (2002).
3. C. Sansour, *Journal de Physique IV Proceedings* **8**, 341–348 (1998).
4. C. Sansour, and S. Skatulla, *Geomechanics and Geoengineering* **2**, 3–15 (2007).
5. D. Vu, P. Steinmann, and G. Possart, *International Journal for Numerical Methods in Engineering* **70**, 685–704 (2007).
6. Y. Choquet-Bruhat, C. DeWitt-Morette, and M. Dillard-Bleick, *Analysis, Manifolds and Physics Part I*, North-Holland, Amsterdam, 1982.
7. A. Eringen, *Microcontinuum field theories I: Foundations and Solids*, Springer, New York, 1999.
8. G. Maugin, *Continuum mechanics of electromagnetic solids*, North-Holland, Amsterdam, 1988.
9. P. Voltairas, D. Fotiadis, and C. Massalas, "A theoretical study of the hyperelasticity of electro-gels," in *Proceedings of the Royal Society, Mathematical Physical and Engineering Sciences*, 2003, vol. 459, pp. 2121–2130.
10. A. Dorfmann, and R. Ogden, *Acta Mechanica* **174**, 167–183 (2005).
11. E. Arruda, and M. Boyce, *Journal of the Mechanics and Physics of Solids* **41**, 389–412 (1998).
12. C. Sansour, S. Feih, and W. Wagner, *International Journal for Computer-Aided Engineering and Software* **20**, 875–895 (2003).
13. P. Lancaster, and K. Salkauskas, *Mathematics of Computations* **37**, 141–158 (1981).
14. B. Nayroles, G. Touzot, and P. Villon, *Computational Mechanics* **10**, 307–318 (1992).
15. T. Belytschko, L. Gu, and Y. Lu, *Modelling and Simulation in Materials Science and Engineering* **2**, 519–534 (1994).
16. W. Liu, and Y. Chen, *Journal of Numerical Methods in Fluids* **21**, 901–931 (1995).

SESSION D1

FERROELECTRIC MATERIALS—NONLINEAR BEHAVIOR

Experimental Studies on the Non-linear Ferroelectric and Ferroelastic Properties of Piezoceramics

Dayu Zhou and Marc Kamlah

Forschungszentrum Karlsruhe, Institut fuer Materialforschung II,
Hermann-von-Helmholtz-Platz 1, D-76344 Eggenstein-Leopoldshafen, Germany

Abstract. We summarize some of the experimental work on the constitutive behaviour of Soft-PZT piezoceramics carried out at the Research Center Karlsruhe during the recent years. In particular, multi-axial, non-proportional polarization rotation tests are presented for a commercial soft PZT material under purely electric field loading. In addition to the polarization measurement along the field loading direction, the normal strain responses in all three coordinate directions were monitored simultaneously using a strain gauge technique. Based on a series of polarization and strain versus electric field curves, switching (domain reorientation threshold) surfaces were constructed in the bi-axial electric field plane using the conventional offset method. The experimental data can be used to examine the existing switching criteria in phenomenological models for the non-linear constitutive behavior of piezoceramics.

INTRODUCTION

As one of the main candidates of intelligent materials, ferroelectric piezoceramics, e.g. lead zirconate titanate (PZT), are presently being used increasingly for low-frequency (off-resonance) actuator applications, including noise controls, active vibration suppression, ultra-high-precision positioning, and especially fuel injection valves in new generation common rail diesel and gasoline engines [1,2]. In practice, these piezo-actuators are required to fulfill their functions in a severe working environment with high reliability over the entire lifetime of service. For instance, in order to obtain a large force or strain output, the active material has to be subjected to a high driving electric field. The tensile strength of piezoceramics is very low. Therefore, the actuators are normally designed to operate under a significant compressive stress preload. In addition, their complicated structures give rise to local stress and/or electric field concentration around the inhomogeneities, including processing defects, micro-cracks, inner electrodes, etc.

As one of the main candidates of intelligent materials, ferroelectric piezoceramics, e.g. lead zirconate titanate (PZT), are presently being used increasingly for low-frequency (off-resonance) actuator applications, including noise controls, active vibration suppression, ultra-high-precision positioning, and especially fuel injection valves in new generation common rail diesel and gasoline engines [1,2]. In practice, these piezo-actuators are required to fulfill their functions in a severe working environment with high reliability over the entire lifetime of service. For instance, in order to obtain a large force or strain output, the active material has to be subjected to a high driving electric field. The tensile strength of piezoceramics is very low. Therefore, the actuators are normally designed to operate under a significant compressive stress preload. In addition, their complicated structures give rise to local stress and/or electric field concentration around the inhomogeneities, including processing defects, micro-cracks, inner electrodes, etc.

There exists a significant amount of experimental data available for ferroelectric piezoceramics under uni-axial electric field and mechanical stress loading conditions [5 -9]. These measurements are essential for the basic test of the accuracy of those constitutive theories. In order to further calibrate and validate them in more general loading scenarios, a detailed knowledge of the material response, especially, the domain switching condition under multi-axial loading is desirable. Several experimental efforts have been made for this purpose. In the work of Chen and Lynch [10], multi-axial strain measurements were performed on thin-walled 8/65/35 PLZT tubular specimens upon simultaneous applications of axial load, internal pressure, and external pressure. The results were used to construct a domain switching surface in the principal stress space, which was found to obey the Tresca criterion for a composition in the unpoled state. Fett et al. [9] first reported the non-symmetric behavior of soft PZT under uni-axial compressive and tensile stress loading. In combination with torsion tests on thin-walled tubes made by the same material, they found that the Drucker-Prager yielding criterion was in good agreement with the experimental results [11].

These aforementioned multi-axial works were based on purely mechanical tests, i.e. the electrical loading was not taken into consideration. Huber et al. [12] reported their electrical polarization rotation tests and proportional, coaxial electromechanical tests for BaTiO$_3$, as well as soft and hard PZT. Due to the lack of strain measurements, the switching surfaces constructed in the bi-axial electric fields (E_3, E_1) and stress-electric field (E_3, σ_3) planes can only predict the critical load level for the onset of irreversible polarization changes. However, this approach gave us an incomplete picture. For instance, no net polarization can be induced by purely mechanical load in the proportional electromechanical test, therefore the corresponding coercive stress appears to be infinitive and the switching surfaces become open. Recently, we reported a more detailed set of proportional electromechanical measurements for PIC151 soft PZT [13]. The main advance of this work is the additional measurement of the strain responses in three perpendicular directions, from which closed and thus, more realistic switching surfaces were obtained in the (E_3, σ_3) plane. It was found that the switching surface corresponding to an offset of 1% remnant strain was close to the Tresca and Drucker-Prager yield condition.

In this paper, we present the experimental results of a series of systematic polarization rotation tests for PIC151 soft PZT subjected to bi-axial electric field loading. In addition to the polarization measurement along the field loading direction, the normal strain responses along three axes of the local coordinates were monitored simultaneously. Using the conventional offset method, domain switching surfaces for the onset of irreversible polarization and strain changes were mapped out in the (E_3, E_2) plane and compared with the results of an initially unpoled material.

EXPERIMENTAL PROCEDURE

Measurements were performed using the polycrystalline PIC151 bulk ceramics (PI Ceramic, Lederhose, Germany). This material is a $Pb(Ni_{1/3}Sb_{2/3})O_3–PbTiO_3–PbZrO_3$ ternary phase system formed in the vicinity of the morphotropic phase boundary of PZT in the tetragonal range. An Sb^{5+} dopant acts as a donor to make the material"soft". Hence, the material may be considered a "soft" PZT. Detailed physical properties of this material can be found on the internet: www.piceramic.de.

Previous work has been carried out to measure the uni-axial polarization and strain response of initially unpoled samples under purely electric field loading [7]. The experimental results are listing in Table. 1 and were used as guide values for subsequent tests of this work. E_3^{sat} is the peak value of applied field to produce saturated polarization and "butterfly" strain hysteresis curves, from which the remnant polarization P_3^r, the remnant strains S_3^r and S_1^r (S_2^r), as well as the coercive field E_3^c were read directly.

Table. 1. Preliminary measurement results of PIC151 soft PZT.

Material	Saturation field E_3^{sat} (kV/mm)	Remnant polarization P_3^r (C/m²)	Coercive field E_3^c (kV/mm)	Remnant strain in longitudinal direction S_3^r (‰)	Remnant strains in transverse directions S_1^r (S_2^r) (‰)
PIC151 soft PZT	2.0	0.342	1.0	2.474	-1.167

Multi-axial polarization rotation tests were preformed using a procedure similar to the one described by Huber et al [12]. Plates of dimension 48mm × 7mm × 20mm were first poled by the manufacturer. The exact poling procedure is proprietary to PI Ceramic, but is known to be performed at room temperature by a applying an electric field of maximum strength 2.5kV/mm. As shown in Fig. 1, rectangular blocks of uniform size 15mm × 5mm × 5mm were cut from the large parent-plates with their long axes inclined at a set of angles (from 0° to 180°, in steps of 15°) to the initial poling (remnant polarization) direction. To preserve the remnant pre-polarization, extra caution was taken to minimize the depolarization effects induced by thermal heating and mechanical load during the cutting process. After cutting, the samples were lightly polished followed by painting a thin layer of silver paint on the top and bottom 5mm × 5mm surfaces to allow for subsequent application of an electric field at an angle (0°-180°) to the direction of remnant pre-polarization (P_3^r).

As seen in Figs. 1 & 2(a), the global and local Cartesian coordinate systems are introduced to describe the loading conditions in a simple manner. The parent-plate was first fully poled by applying E_3 along the 3-axis of the global coordinates, resulting in

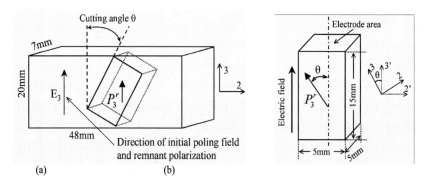

FIGURE 1. (a) Cutting out of a rectangular block of pre-poled material at an angle θ. (b) Applying an electric field (along the 3'-axis of the local coordinates) to the block at an angle θ to the original poling direction (along the 3-axis of the global coordinates).

a remnant polarization of P_3^r. After cutting and electroding, an electric field E was applied subsequently along the 3'-axis of the local coordinates, which is inclined at an angle θ with respect to the 3-axis and P_3^r. The reloading field E is therefore lying in the biaxial $E_3 - E_2$ plane, with components of $E\cos\theta$ and $E\sin\theta$ along the 3-axis and 2-axis of the global coordinate system, respectively. The 1 and 1' axes of these two coordinates are not shown. Note that they are coinciding and always perpendicular to the remnant polarization P_3^r.

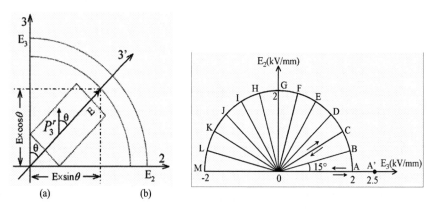

FIGURE 2. (a) Applying an electric field at an angle θ to the remnant polarization P_3^r in the $E_3 - E_2$ plane. (b) Loading paths for original poling operation and subsequent bi-axial electrical loading.

The loading paths for the polarization rotation tests are shown in Fig. 2(b). The original poling operation for the parent-plate corresponds to a loading along path *OA'* (from 0 to 2.5kV/mm) followed by an unloading. The angled-cut specimens were installed into a carefully designed test fixture described elsewhere [8]. A ramp-shaped electric field was applied to each specimen along one of the loading paths from OA to OM, with a maximum strength of 2kV/mm and a loading rate of 0.08kV.mm^{-1}.s^{-1}.

Polarization changes in the field loading direction (along the 3'-axis) were measured using Sawyer-Tower method. A high-input-resistance electrometer (6517A, Keithley Instruments, Cleveland, OH) was used to monitor the voltage fluctuations over a 10μF reference capacitor connected in series with the specimen. The normal strain responses along all three axes of the local coordinates were monitored simultaneously using strain gauge technique. A pair of strain gauges attached on two neighboring 15mm × 5mm surfaces was used to measure the longitudinal strain along the field loading direction (3'-axis). After the experiment, the strain data recorded were averaged to plot the curves. Transverse strains along the 2'-axis and 1'-axis were monitored by mounting strain gauges on the remaining two 15mm × 5mm surfaces, respectively.

During experiment, the specimen was immersed into an insulated electric liquid (FC-40, 3M, St. Paul, MN) to prevent arcing. A computer equipped with a data acquisition board and running DASYLAB software (Dasytec, Amherst, NH) was used to digitally record the measurement signals and to control a bipolar high-voltage power supply (HCB 15–30 000, F. u. G., Rosenheim, Germany) to provide high electric field.

RESULTS AND DISCUSSION

Fig. 3 shows the polarization and strain measurement results of pre-poled specimens subjected to an electric field at different angles to the original poling direction. For the sake of clarity, the results are given in steps of 45°. The measurements of an initially un-poled specimen are also given for comparison.

The case of $\theta = 0°$ corresponds to reloading an electric field parallel to the initial poling direction (or P_3^r). Almost linear dielectric and piezoelectric responses can be observed in the whole courses of the polarization and strains verses electric field curves, indicating that an approximately saturated domain orientation state had been induced by the original poling operation and remained quite well after cutting and specimen preparation. It is evident that, with an increase in the angle between the remnant polarization and the reloading field, non-linear changes in the polarization and strains For the particular cases of $\theta = 0°$ and 180°, the transverse strains along 1' and 2' axes of the local coordinates are identical. However, for a certain loading angle between 0° and 180°, different domain orientation states with respect to these two local directions cause the transverse strains tracing different routes to develop. For the representative instance of $\theta = 90°$, domain-switching gives rise to more non-linear strain change along the 2'-axis; by contrast, the slight non-linear deformation along 1'-axis is mainly due to the variation of piezoelectric coefficient during the polarization rotation process. For the particular cases of $\theta = 0°$ and 180°, the transverse strains along 1' and 2' axes of the local coordinates are identical. However, for a certain loading angle between 0° and 180°, different domain orientation states with respect to these two local directions cause the transverse strains tracing different routes to develop. For the representative instance of $\theta = 90°$, domain-switching gives rise to more non-linear strain change along the 2'-axis; by contrast, the slight non-linear

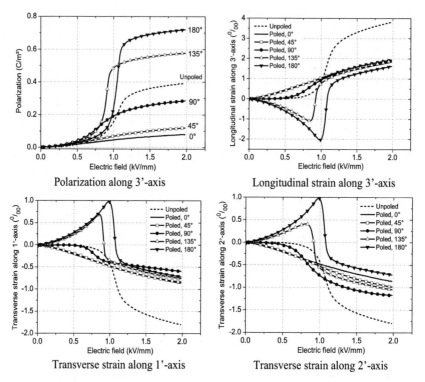

FIGURE 3. Polarization and strain measurement results of pre-poled samples subjected to an electric field at different angles to the original poling direction. The dashed curves correspond to the measurement of an initially unpoled material.

deformation along 1'-axis is mainly due to the variation of piezoelectric coefficient during the polarization rotation process.

The critical loading condition required for the irreversible polarization and strain to reach a certain level was determined using the conventional offset method. Straight lines were constructed parallel to the initial linear dielectric and piezoelectric portions of the polarization and strain curves at some specific offsets, e.g. 1% of P_r ($0.331 C/m^2$, half of the polarization change at $\theta = 180°$) and S_3^r (-2.474‰). The intersections of the parallel lines and the measured curves correspond to a series of electric field values, which are used to construct switching (yielding) surfaces in polar form versus the loading angle. Fig. 4 shows the offsets (from 1% to 40%) switching surfaces attained by polarization and longitudinal strain measurements in the re-loading field direction. The data obtained from loading along paths *OA-OM* form the upper half ($0° \leq \theta \leq 180°$) of the switching surfaces. The lower half of each surface is mapped out from symmetry. At angles close to zero, the 10%, 20%, and 40% offsets in polarization as well as the 40% offset in strain are not reached due to the saturation effects, consequently the corresponding switching surfaces are open.

Switching surfaces attained using small offset values are more suitable for determination of the critical loading condition required for the onset of incipient domain switching and, are therefore more informative for constitutive models. In Fig.

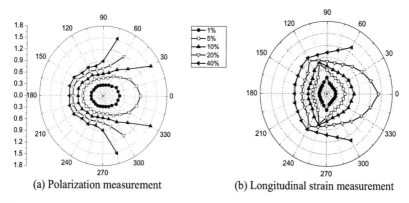

(a) Polarization measurement　　　　　(b) Longitudinal strain measurement

Figure 4. Domain switching surfaces obtained from polarization (a) and strain (b) measurements in the field loading direction. The radial axis represents the strength of the electric field in kV/mm units and, the polar axis shows the angle θ between the remnant polarization and the reloading electric field. The curves labeled with different legends correspond to the offsets of 1%, 5%, 10%, 20%, and 40% of the remnant polarization and remnant strain, respectively.

5,

the 1% offset switching surfaces obtained from polarization and strain measurements of initially unpoled and pre-poled specimens are plotted in the (E_3, E_2) plane. The fact of isotropic domain distribution results the yielding surfaces of initially unpoled specimen in circular shape centred at E = 0, indicating the switching (yielding) condition is independent of the loading path. An elliptical switching surface is obtained from the polarization measurements of pre-poled specimens. A shrinkage in radius can be clearly observed in $\pm E_2$ directions. That is, at θ = 90° (or 270°) and its neighbourhood, a 1% offset in polarization is induced at lower critical field levels than that needed at θ = 0° and 180°. As seen in Fig. 3, the P-E curves measured at θ = 90° and 135° exhibit apparently earlier non-linear increase in polarization in comparison with the P-E curve of θ = 180°. In fact, this phenomenon does exist in the range of 45° < θ < 135°. Huber et al. [14]. attributed their similar experimental observations for PZT-5H to the presence of residual compressive stresses in the remnant polarization direction, which resulted from mismatch effects during the original poling process. With the assistance of the compressive stress, a relatively low electric field is needed to initiate domain switching towards the 90° direction. On the switching surface from the strain measurement of pre-poled specimens, the abnormal data points observed at 90° and its neighbourhood are due to the insensitivity of strain gauges to the shear. The data recorded around 0° and 180° are more reliable for prediction of the change in irreversible strain.

By comparing Figs. 5(a) and 5(b), it is interesting to find the difference in size of the switching surfaces obtained from polarization and strain measurements. For the unpoled specimen, the critical electric field required for a 1% offset in polarization is much lower than that needed for a 1% offset in strain. This phenomenon may be accounted for in the following way: At the early stage of electric field loading, the unpoled material mainly undergoes 180° domain switching, which results in irreversible change in polarization, but provides no contribution to the strain. An

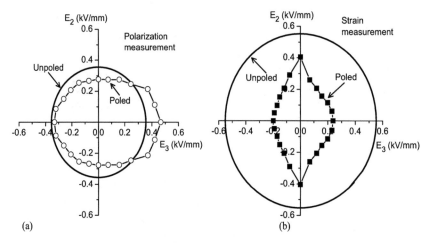

FIGURE 5: Comparison of 1% offset domain switching surfaces of pre-poled and initially unpoled specimens in the bi-axial (E_3, E_2) plane. The results correspond to polarization (a) and strain (b) measurements along the field loading direction.

evidence for this explanation is that, the pre-poled 180° specimen needs nearly the same value of critical field to attain a 1% offset in polarization. Strain measurement for pre-poled specimens results in a significant decrease in size of the switching surface.

For specimens in poled state, the experimental results provide some evidences of a translation of the switching surfaces in the $+E_3$ direction. The distance of translation depends on the selection of polarization and strain measurement and the usage of offset values measuring the extent of switching. For the 1% offset switching surfaces, polarization measurements make the switching surface centred at $E_3 = 0.0666$ kV/mm; whereas, a slight shift of 0.0198 kV/mm is attained from the strain measurements. In phenomenological models, the feature of translation of the switching surfaces can be characterized approximately by incorporating an appropriate kinematic hardening law in the switching function. A switching function based on the result of polarization measurement is tentatively given by

$$\sqrt{(E_3 - E_3^B)^2 + C(E_2)^2} = E_0,$$

where E_0 is the critical field required for the irreversible polarization to attain a certain value, E_3^B is the back-field describing a translation of the switching surface in $+E_3$ direction, and C is a shape factor scaling the distortion of the switching surface in $\pm E_2$ directions. It should be kept in mind that all these parameters are depending on the extent of change in irreversible polarization, i.e. the loading history. For the case of a 1% offset in polarization, the measurements give $E_0 = 0.399$kV/mm, $E_3^B = 0.0666$kV/mm, and C = 1.965.

This work provides a solid experimental database for the development of multi-axial constitutive laws for ferroelectric ceramic materials. It also opens challenging

206

questions that need to be addressed, for instance, the significant difference in size of the switching surfaces obtained from polarization and strain measurements of pre-poled samples. In addition, further consideration of the electrical boundary condition during the domain switching process will be of great value for deep understanding of the evolution of the internal electric field and its influence on the experimental results.

CONCLUSION

A systematic set of simultaneous polarization and strain measurements has been performed for pre-poled soft PZT material re-loaded by an electric field at an angle to the original poling direction. The anisotropic domain distribution results in significantly different non-linear responses measured at different angles. Domain switching (yielding) surfaces were obtained in the bi-axial electric field space using the conventional offset method. Switching surfaces obtained by polarization and strain measurements were found to differ from each other in size. In contrast to the circular yielding surfaces of isotropic unpoled specimen, the switching surface obtained from polarization measurement for pre-poled specimens exhibits a distortion in shape and a translation in the pre-poling direction. In the case of the strain measurement, a significant apparent softening with respect to the unpoled state was observed. The experimental results are valuable for testing the qualitative accuracy of currently available phenomenological constitutive theories.

REFERENCES

1.   C. Schuh, Th. Steinkopff, A. Wolff, and K. Lubitz, "Piezoceramic multilayer actuators for fuel injection systems in automotive area", pp. 165-75 in Proceedings of SPIE, Vol. **3992**, *Smart Structures and Materials 2000: Active Materials: Behavior and Mechanics*, Edited by C. S. Lynch. SPIE, Newport Beach, CA, 2000.

2.   R. Müller-Fiedler and V. Knoblauch, "Reliability aspects of microsensors and micromechatronic actuators for automotive applications", Microelectron. Reliab., **43**[7] 1085-1097 (2003).

3.   M. Kamlah, "Ferroelectric and ferroelastic piezoceramics-modeling of electromechanical hysteresis phenomena", Continuum Mech Thermodyn., **13** [4] 219-68 (2001).

4.   C. M. Landis, "Non-linear constitutive modeling of ferroelectrics", Curr. Opin. Solid St. M., **8**, 59-69 (2004).

5.   H. Cao and A. G. Evans, "Nonlinear Deformation of Ferroelectric Ceramics", J. Am. Ceram. Soc., **76** [4] 890-96 (1993).

6.   C. S. Lynch, "The effect of uniaxial stress on the electro-mechanical response of 8/65/35 PLZT", Acta Mater., **44** [10] 4137-48 (1996).

7.   D.Y. Zhou, M. Kamlah, and D. Munz, "Effects of Uniaxial Prestress on the Ferroelectric Hysteretic Response of Soft PZT", J. Eur. Ceram. Soc., **25**[4] 425-32 (2005).

8.   D.Y. Zhou, M. Kamlah, and D. Munz, "Effects of Bias Electric Fields on the Non-linear Ferroelastic Behavior of Soft Lead Zirconate Titanate Piezoceramics", J. Am. Ceram. Soc., **88** [4] 867–874 (2005).

9.   T. Fett, S. Müller, D. Munz, and G. Thun, "Nonsymmetry in the deformation behavior of PZT", J. Mater. Sci. Lett., **17**, 261-65 (1998).

10.   W. Chen and C.S. Lynch, "Multiaxial Constitutive Behavior of Ferroelectric Materials", J. Eng. Mater. Technol., **123**, 169-175 (2001).

11.   T. Fett, D. Munz, and G. Thun, "Multiaxial deformation behavior of PZT from torsion tests", J. Am. Ceram. Soc., **86**, 427, (2003).

12.   J. E. Huber, J. Shieh, and N. A. Fleck, "Multiaxial response of hard and soft ferroelectrics under stress and electric field", pp. 133-42 in Proceedings of SPIE, Vol. **4699**, *Smart Structures and Materials 2002: Active Materials: Behavior and Mechanics*, edited by C. S. Lynch. SPIE, San Diego, CA, 2002.

13.   D.Y. Zhou, Z.G. Wang, and M. Kamlah, "Experimental investigation of domain switching criterion for soft lead zirconate titanate piezoceramics under coaxial proportional electromechanical loading", J. Appl. Phys., **97**[8] Art. No. 084105 (2005).

14.   J. E. Huber and N. A. Fleck, "Multiaxial electrical switching of a ferroelectric: theory versus experiment", J. Mech. Phys. Solids, 49, 785-811 (2001).

Nonlinear piezoelectric effects – towards physics-based computational modelling of micro-cracking, fatigue, and switching

A. Menzel[*,†], J. Utzinger[**] and A. Arockiarajan[‡]

[*]*Dortmund University of Technology, Faculty of Mechanical Engineering, Group of Mechanics and Machine Dynamics, Leonhard-Euler-Str. 5, D-44221 Dortmund, Germany*
[†]*Lund University, Division of Solid Mechanics, P.O. Box 118 SE-221 00 Lund, Sweden*
[**]*University of Kaiserslautern, Department of Mechanical and Process Engineering, Chair of Applied Mechanics, P.O. Box 3049, D-67653 Kaiserslautern, Germany*
[‡]*University of Nottingham, School of Civil Engineering, University Park, NG7 2RD, UK*

Abstract. Piezoelectric ceramics – as one widely commercialised group of smart materials – exhibit a great potential for various engineering applications. Their high-frequency capabilities are in particular attractive for actuator and sensor devices, which nowadays are present in daily-life-technologies such as cellular phones, fuel injection systems, and so forth.
At high loading levels however, severely nonlinear behaviour of these materials is observed which, from the control point of view, must be further investigated in order to be able to precisely account for such effects within the design of intelligent systems. The reasons for these nonlinearities are manifold and, even investigated within the last decades, not fully understood. Nevertheless, two important sources for these observations are so-called micro-cracking, together with fatigue phenomena, as well as switching or rather phase transformations. Accordingly, the main goal of this contribution is to study these effects by means of developing related constitutive models that can be embedded into iterative algorithmic schemes such as the finite element method.
One the one hand, the grain-structure of a piezoceramic specimen will be modelled via the direct incorporation of the grain-boundaries as so-called interface elements. The underlying cohesive-like constitutive law of this layer includes both degrees of freedom of the surrounding bulk material – or rather the jumps in these fields – namely displacements and the electric potential. Based on the resulting traction-separation-type relations, micro-cracking is directly accounted for on this micro-level. Moreover, the constitutive law of the interfacial layer is supplemented by additional variables that enable the formulation of fatigue under cyclic loading conditions. On the other hand, phase transformations – modelled in terms of an energy-based switching criterion – are discussed and embedded into an iterative finite element context. Symmetry relations of the underlying unit cells are directly included so that the switching model accounts for the micro-mechanical properties of the piezoelectric materials of interest. At this stage, representative numerical simulations of poly-crystalline specimens are based on straightforward averaging-techniques, while the combination of the developed micro-cracking model (grain boundaries) with the proposed switching model (bulk) constitutes future research.

Keywords: ferroelectricity, fatigue, switching, grain boundaries, interface, finite element method
PACS: 77.22.Gm, 77.22.Ej, 77.80.Dj, 77.80.Fm

CP1029, *Smart Devices: Modeling of Material Systems, An International Workshop*
edited by S. M. Sivakumar, V. Buravalla, and A. R. Srinivasa
©2008 American Institute of Physics 978-0-7354-0553-0/08/\$23.00

INTRODUCTION

By definition, ferroelectric materials convert electrical energy into mechanical energy (actuator), respectively mechanical energy into electrical energy (sensor), so that related balance equations result in so-called coupled problems. At high electro-mechanical loading conditions such piezoelectric materials exhibit highly nonlinear behaviour stemming from, for instance, changes in spontaneous polarisation directions – so-called switching – or micro-cracking and fatigue phenomena; see the monographs by Smith [23] and Lupascu [19] as well as the overview article by Kamlah [12] and references cited therein. While the nonlinear response due to phase transformations has been studied and modelled by various authors and research groups – among those see, for example, Landis [17], Huber [9], Belov and Kreher [5], and our own investigations reported in [2, 3] – while micro-cracking and related fatigue seems, at least from the computational mechanics point of view, by now have been studied by a smaller community; see Fang and Liu [7], Arias et al. [1], Westram et al. [26], and Utzinger et al. [25]. The modelling of polycrystalline piezoelectric materials is, per se, of multi-scale-type so that formulations proposed in the literature refer to rather different scales of observation. In view of the switching criterion reflecting the onset of phase transformations, we will adopt an ansatz that is motivated by the physics of the underlying unit cells and which dates back to the contributions by Hwang et al. [10] and McMeeking and Hwang [20]. In addition to this phase transition, special emphasis will be placed on so-called intergranular effects in the context of fatigue. Two different types are thereby distinguished, namely embrittlement along the grain boundaries and direct crack growth – both being relevant for low as well as high cycle fatigue. Practically speaking, along the grain boundaries – the grains themselves being modelled by means of coupled finite elements – we will introduce so-called interface elements supplemented with a damage-like electromechanically coupled constitutive law such that low cycle fatigue phenomena are nicely captured. Nowadays, iterative finite element techniques are successfully developed for nonlinear coupled problems; in the present context, the reader is referred to Gabbert et al. [8], Kamlah and Böhle [13], Landis [16], Kim and Jiang [14], Schröder and Gross [22], Klinkel [15] as well as to [4]. An overview on the implementation of interface elements is provided by Schellekens and de Borst [21]. In view of the application of inelastic constitutive relations, capable of cyclic deformations, to cohesive-type laws, see also Utzinger et al. [24].

The paper is organised as follows: first, fundamental relations as essential balance and constitutive equations as well as the basic finite element scheme are reviewed. To be specific, tetragonal material symmetry will later on be approximated via transversely isotropic relations. Next, discrete micro-cracking together with fatigue-related constitutive equations are discussed. The subsequent finite element example is directly referred to a micrograph take from the literature. Thereafter, the domain switching model adopted is addressed. Several numerical examples underline that the model is able to produce simulation results which are in good agreement with experimental data. Finally, a short summary concludes the paper.

MODELLING-RELATIONS FOR PIEZOELECTRIC CONTINUA

The subsequent section reviews general relations relevant for the modelling of piezo-electric continua. For the sake of clarity, we here restrict ourselves to the linear case – nonlinear effects being addressed later on – and thereby also introduce the notation applied in this contribution.

Fundamental balance equations

Let the configuration of a body B of interest, which is assumed to undergo small deformations, be denoted by $\mathscr{B} \subset \mathbb{R}^3$. Material points of B are characterised by position vectors $\mathbf{x} \in \mathbb{R}^3$ and the essential degrees of freedom for the coupled problem at hand are the displacements $\mathbf{u} \in \mathbb{R}^3$ as well as the electric potential $\phi \in \mathbb{R}$. When neglecting acceleration and source terms, the local format of the balance of linear momentum and Gauß' law read

$$\nabla \cdot \sigma = \mathbf{0}, \qquad \nabla \cdot \mathbf{D} = 0 \qquad \text{in } \mathscr{B} \tag{1}$$

together with the boundary conditions $\mathbf{u} = \mathbf{u}^p$ on $\partial \mathscr{B}_u$, $\mathbf{t} \, (= \sigma \cdot \mathbf{n}_\sigma) = \mathbf{t}^p$ on $\partial \mathscr{B}_\sigma$, $\phi = \phi^p$ on $\partial \mathscr{B}_\phi$, and $q \, (= -\mathbf{D} \cdot \mathbf{n}_D) = q^p$ on $\partial \mathscr{B}_D$. The flux terms σ and $\mathbf{D}$ are established as Cauchy stresses and dielectric displacements – the constitutive equations for both being discussed in the following.

Basic constitutive equations

As mentioned above, the type of piezoelectric materials of interest for this study allow characterisation by means of tetragonal unit cells which we approximate via transversely isotropic constitutive equations. The orientation of such a unit cell is determined by a unit vector $\mathbf{a}$ so that the underlying (spontaneous) polarisation vector takes the representation $\mathbf{P}_s = P_s \mathbf{a}$. When making use of well-established potential-relations, such constitutive equations, that specify σ and $\mathbf{D}$ in terms of the additively decomposed strain tensor and the electric field

$$\nabla^{\text{sym}} \mathbf{u} = \varepsilon = \varepsilon_e + \varepsilon_s, \qquad \nabla \phi = -\mathbf{E}, \tag{2}$$

can be defined as derivatives of the electric enthalpy H. We here solely account for linear and quadratic combinations of the arguments $\{\varepsilon_s, \mathbf{E}, \mathbf{a}\}$ which results in

$$\sigma = \mathbf{E} : \varepsilon_e - \mathbf{d}_\varepsilon^{\mathrm{T}} \cdot \mathbf{E}, \qquad \mathbf{D} = \mathbf{d}_\varepsilon : \varepsilon_e + \mathbf{k}_\varepsilon \cdot \mathbf{E} + \mathbf{P}_s, \tag{3}$$

wherein $\mathbf{E}$ is the fourth-order elasticity tensor, $\mathbf{d}_\varepsilon$ denotes the third-order piezoelectric tensor, and $\mathbf{k}_\varepsilon$ characterises the second-order dielectric permittivity. In addition, we assume the spontaneous strains to take the format $\varepsilon_s = \varepsilon_s \left[3 \mathbf{a} \otimes \mathbf{a} - \mathbf{I} \right]/2$ with $\mathbf{I}$ being the second-order identity tensor.

211

Finite element discretisation

The essential degrees of freedom of the coupled problem at hand are ϕ and $\mathbf{u}$ and a straightforward finite element setting can directly be based on the principle of virtual work so that the related weak forms read

$$G_\phi = - \int_{\mathcal{B}} \delta \mathbf{E} \cdot \mathbf{D} \, dv \; - \; \int_{\partial \mathcal{B}_D} \delta \phi \, q \, da \; = \; - G_\phi^{\text{int}} + G_\phi^{\text{ext}} \; = \; 0,$$

$$G_\mathbf{u} = \int_{\mathcal{B}} \delta \varepsilon : \sigma \, dv \; - \; \int_{\partial \mathcal{B}_\sigma} \delta \mathbf{u} \cdot \mathbf{t} \, da \; = \; G_\mathbf{u}^{\text{int}} - G_\mathbf{u}^{\text{ext}} \; = \; 0. \tag{4}$$

Based on standard (iso-parametric) interpolation, the approximated electric potential $\phi^h = \sum_{i=1}^{n_{\text{nod}}} N_i \, \phi_i$ and the approximated displacement field $\mathbf{u}^h = \sum_{i=1}^{n_{\text{nod}}} N_i \mathbf{u}_i$ render $\mathbf{E}^h = - \sum_{i=1}^{n_{\text{nod}}} \phi_i \, \nabla N_i$, respectively $\varepsilon^h = \sum_{i=1}^{n_{\text{nod}}} [\mathbf{u}_i \otimes \nabla N_i]^{\text{sym}}$. The approximation of the corresponding variations or rather test functions as well as of the incremental degrees of freedom follow by analogy so that

$$G_\phi^{h,e} = \int_{\mathcal{B}^e} \sum_{i=1}^{n_{\text{nod}}} \delta \phi_i \; [\nabla N_i \cdot \mathbf{D}] \, dv \; - \; \int_{\partial \mathcal{B}_D^e} \sum_{i=1}^{n_{\text{nod}}} \delta \phi_i \; [N_i q] \, da \; = \; 0,$$

$$G_\mathbf{u}^{h,e} = \int_{\mathcal{B}^e} \sum_{i=1}^{n_{\text{nod}}} \delta \mathbf{u}_i \cdot [\nabla N_i \cdot \sigma] \, dv \; - \; \int_{\partial \mathcal{B}_\sigma^e} \sum_{i=1}^{n_{\text{nod}}} \delta \mathbf{u}_i \cdot [N_i \mathbf{t}] \, da \; = \; 0, \tag{5}$$

together with the incremental relation, whereby the constitutive tensors are understood to remain constant,

$$\Delta G_\phi^{h,e} = - \int_{\mathcal{B}^e} \sum_{i,j=1}^{n_{\text{nod}}} \delta \phi_i \; [\nabla N_i \cdot \mathbf{k}_\varepsilon \cdot \nabla N_j] \, \Delta \phi_j \; + \; \delta \phi_i \; [\nabla N_i \cdot \mathbf{d}_\varepsilon \cdot \nabla N_j] \cdot \Delta \mathbf{u}_j \, dv,$$

$$\Delta G_\mathbf{u}^{h,e} = - \int_{\mathcal{B}^e} \sum_{i,j=1}^{n_{\text{nod}}} \delta \mathbf{u}_i \cdot [\nabla N_i \cdot \mathbf{d}_\varepsilon^{\text{T}} \cdot \nabla N_j] \, \Delta \phi_j \; - \; \delta \mathbf{u}_i \cdot [\nabla N_i \cdot \mathbf{E} \cdot \nabla N_j] \cdot \Delta \mathbf{u}_j \, dv. \tag{6}$$

MICRO-CRACKING AND FATIGUE IN POLYCRYSTALLINE CERAMICS

In this section, the computational modelling of micro-cracking by means of so-called interface elements is discussed. To be specific, such finite elements enable the incorporation of jumps of the degrees of freedom along the underlying grain boundaries. In addition, fatigue phenomena are accounted for via adequate traction-separation-laws with switching effects being neglected at this stage. For conceptual simplicity, we restrict ourselves to the two-dimensional case in this section.

Interfacial constitutive equations

The strain-type and electric-field-like quantities for the interface $\Gamma \subset \mathcal{B}$, or rather the grain boundaries, are the jump in the displacement field $[\![\mathbf{u}]\!]$ and in the electric potential $[\![\phi]\!]$. Moreover, let $\mathbf{n}$ denote the outward unit vector of the interface considered and $\mathbf{m}$ the related unit tangent vector. By analogy with equation (3), the traction vector and dielectric displacement contribution are assumed to take the representations

$$\tau = \mathbf{E}^{\mathrm{ifa}} \cdot [\![\mathbf{u}]\!] + \mathbf{d}^{\mathrm{ifa}}\,[\![\phi]\!] + \tfrac{1}{2}\, p\,[[\mathbf{u}]\!] \cdot \mathbf{n}]^2\,\mathbf{n}, \qquad \Lambda = \mathbf{d}^{\mathrm{ifa}} \cdot [\![\mathbf{u}]\!] - \mathbf{k}^{\mathrm{ifa}}\,[\![\phi]\!], \qquad (7)$$

wherein the penalty-type parameter p is introduced to reduce penetration whenever it occurs. In addition, it is worth to note that the, say, interfacial electric field corresponds to the jump in the electric potential, $\mathrm{E} = -\,[\![\phi]\!]/[\mathrm{length}]$. Furthermore, let the interfacial elastic stiffness decouple with respect to $\mathbf{n}$ and $\mathbf{m}$, i.e. $\mathbf{E}^{\mathrm{ifa}} = \mathrm{E}_{\mathrm{n}}^{\mathrm{ifa}}\,\mathbf{n} \otimes \mathbf{n} + \mathrm{E}_{\mathrm{m}}^{\mathrm{ifa}}\,\mathbf{m} \otimes \mathbf{m}$, while $\mathbf{d}^{\mathrm{ifa}} = \mathrm{d}^{\mathrm{ifa}}\,\mathbf{n}$. To account for fatigue phenomena, we next introduce damage-like evolution equations for $\mathbf{E}^{\mathrm{ifa}}$ and $\mathrm{k}^{\mathrm{ifa}}$. To be specific, use of a scalar-valued variable is made, $d \in [0, 1]$ with $d|_{t_0} = 0$, so that

$$\mathrm{E}_{n,m}^{\mathrm{ifa}} = [1 - d]\,\mathrm{E}_{n,m}^0, \qquad \mathrm{k}^{\mathrm{ifa}} = \mathrm{k}^0 + [\mathrm{k}_\varepsilon - \mathrm{k}^0]\,d, \qquad (8)$$

with k_ε being characteristic in view of the dielectric permittivity of the bulk, for example $\mathrm{k}_\varepsilon = \mathbf{a} \cdot \mathbf{k}_\varepsilon \cdot \mathbf{a}$. The remaining task consists in setting up an evolution equation for d, which we assume to be determined in terms of the representative jump-quantity

$$\delta = \sqrt{\beta_{\mathrm{n}}^2\,[[\mathbf{u}]\!] \cdot \mathbf{n}]^2 + \beta_{\mathrm{m}}^2\,[[\mathbf{u}]\!] \cdot \mathbf{m}]^2 + \beta_\phi^2\,[\![\phi]\!]^2}. \qquad (9)$$

Here, we assume a time-based fatigue formulation capturing low cycle fatigue. To do so, an additional history variable is introduced, namely $\tilde{\delta}$ with initial value $\tilde{\delta}_0 = 0$. By means of this quantity we characterise fatigue via

$$d = \exp(-\beta_d/\tilde{\delta}_{n+1}) \qquad \text{and} \qquad \tilde{\delta}_{n+1} = \tilde{\delta}_n + \langle \delta_{n+1} - \delta_n \rangle, \qquad (10)$$

whereby a finite interval of time $\Delta t = t_{n+1} - t_n > 0$ has been considered and $\langle \bullet \rangle = \tfrac{1}{2}[\bullet + |\bullet|]$.

Finite element discretisation

Similar to equation (4), the weak form of the balance equations – the penalty-type term being omitted – results in

$$G_{[\phi]} = \int_\Gamma \delta[\![\phi]\!]\,\Lambda\,\mathrm{d}s = 0, \qquad G_{[\mathbf{u}]} = \int_\Gamma \delta[\![\mathbf{u}]\!] \cdot \tau\,\mathrm{d}s = 0. \qquad (11)$$

We here do not place further emphasis on finite element interpolations and incremental relations but solely mention that – when making use of the, say, classical $\mathbf{B}$-matrix and

213

having two-node interface elements (*I* and *II*) in mind – the interfacial jumps may be approximated via

$$[\![\xi]\!]^h = \mathbf{B} \cdot [\xi^i, \xi^{ii}, \xi^{iii}, \xi^{iv}] \quad \text{with}$$
$$\mathbf{B} = [-\operatorname{diag}(N_I), -\operatorname{diag}(N_{II}), \operatorname{diag}(N_I), \operatorname{diag}(N_{II})], \tag{12}$$

wherein $\xi = [\mathbf{u}, \phi]$, and *i–iv* are the 'attached' finite element nodes of the ambient bulk elements.

Numerical example

A PZT micrograph, taken from Jaffe et al. [11], is chosen for the subsequent numerical example. The micrograph itself together with the applied discretisation (three-node isoparametric triangular elements with identical shape functions for $\mathbf{u}$ and ϕ) as well as the boundary conditions chosen are displayed in figure 1. For conceptual simplicity, we assume the orientation of the underlying unit cells to coincide with the longitudinal loading direction. The grains are modelled by means of the formulation previously highlighted while interface elements, as discussed in this section, are introduced along all grain boundaries. In addition to the fixed displacement boundary conditions as displayed in figure 1, displacement-driven cyclic longitudinal loading is applied. To be specific, 16 cycles are considered ($C_{max} = 16$) with identical maximum and minimum values of the prescribed degrees of freedom – all of them being positive, so that the specimen is macroscopically loaded under tension. Due to the incorporation of fatigue, representative quantities, such as stresses or the electric field, change from cycle to cycle. This effect is clearly seen in figure 2 and 3, where the projection of σ and $\mathbf{E}$ with respect to the longitudinal loading direction is shown.

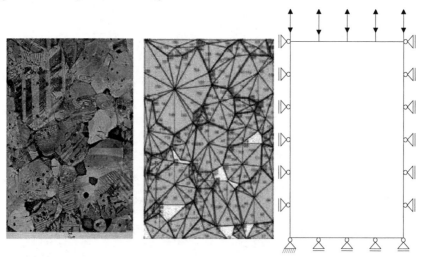

FIGURE 1. Micrograph (left), discretisation (middle), and boundary conditions (right).

FIGURE 2. Mechanical fatigue: longitudinal electric field $10 \times E_y$ [kV] for $C = 1, 10, 16$.

FIGURE 3. Mechanical fatigue: longitudinal stresses σ_{yy} [MPa] for $C = 1, 10, 16$.

SWITCHING EFFECTS IN POLYCRYSTALLINE CERAMICS

Apart from micro-cracking and fatigue phenomena, as discussed in the previous section, one of the main reasons for the overall nonlinear response of piezoelectric materials at high loading levels is so-called switching. In order to account for such reorientation or rather phase transformation of the underlying tetragonal unit cells, it turns out to be useful to introduce an orthonormal frame $\{\mathbf{a}_{1,2,3}\}$ with, for example, $\mathbf{a}_3 = \mathbf{a}$. In view of transitions within the $\langle 1\,0\,0 \rangle$ family, six different orientations must, in general, be taken into account – the initial state, four types of $90°$ domain switching, and $180°$ domain switching. In the following, an energy-based switching criterion is adopted, whereby, for conceptual simplicity, we will not distinguish between domains and grains.

Energy-based switching criterion

First-order phase-transformations are commonly assumed to be driven by the reduction in Gibbs free energy G. In this context, we assume local Reuss bounds for the piezoelectric problem at hand and, by making use of equation (3) together with a Legendre transformation of the underlying electric enthalpy, obtain

$$G = -\sigma : \varepsilon_s - \mathbf{E} \cdot \mathbf{P}_s - \tfrac{1}{2} \mathbf{E} \cdot \mathbf{k}_\sigma \cdot \mathbf{E} - \tfrac{1}{2} \sigma : \mathbf{C} : \sigma - \mathbf{E} \cdot \mathbf{d}_\sigma : \sigma + G_0 \qquad (13)$$

with $\mathbf{C} = \mathbf{E}^{-1}$, $\mathbf{d}_\sigma^T = \mathbf{C} : \mathbf{d}_\varepsilon^T$, and $\mathbf{k}_\sigma = \mathbf{k}_\varepsilon + \mathbf{d}_\varepsilon : \mathbf{d}_\sigma^T$. When considering a finite time interval of interest, $\Delta t = t_{n+1} - t_n > 0$ within which a phase transformation is assumed to be active, the change in Gibbs free energy or rather the switching criterion takes the format

$$\Delta G = \sigma : \Delta \varepsilon_s + \mathbf{E} \cdot \Delta \mathbf{P}_s + \tfrac{1}{2} \mathbf{E} \cdot \Delta \mathbf{k}_\sigma \cdot \mathbf{E} + \tfrac{1}{2} \sigma : \Delta \mathbf{C} : \sigma + \mathbf{E} \cdot \Delta \mathbf{d}_\sigma : \sigma > G^*, \qquad (14)$$

wherein G^* denotes a critical threshold value to initiate switching and $\Delta \bullet = \bullet_{n+1} - \bullet_n$. Practically speaking, the new local polarisation direction

$$\mathbf{a}_{n+1} = \mathbf{a}_{3\,n+1} \in \{\pm \mathbf{a}_1, \pm \mathbf{a}_2, -\mathbf{a}_3\}_n \qquad (15)$$

determines the 'Δ-quantities' in equation (14) – the particular one being realised corresponding to that polarisation direction which renders the highest local reduction in Gibbs free energy.

For polycrystalline ferroelectrics, however, grains and domains strongly interact which may render the local loading levels to appear much higher than reflected by the macroscopic boundary conditions. To phenomenologically incorporate these intergranular effects, we modify the threshold value of the switching criterion, $G^* := G^{\ddagger} P$, whereby $G^{\ddagger} = \mathrm{const}$ and P is of Weibull-type, namely

$$P = \begin{cases} 1 - \exp\left(-q \left[\dfrac{\Delta G}{G^{\ddagger}} \right]^r \right) & \text{if} \quad \Delta G < G^{\ddagger} \\ 1 & \text{if} \quad \Delta G \geq G^{\ddagger} \end{cases} \qquad (16)$$

The parameters $q, r \geq 0$ provide additional freedom to calibrate simulation results with experimental measurements.

Finite element discretisation

The subsequent numerical examples are directly based on the finite element formulation highlighted above. In addition, each finite element (eight-node three-dimensional isoparametric brick elements with identical shape functions for $\mathbf{u}$ and ϕ) is supplemented by an internal-variable-like orthonormal frame $\{\mathbf{a}_{1,2,3}\}$ which we initially generate by random to obtain a macroscopically, say, un-poled and isotropic specimen. Interface elements, as introduced in the previous section, are not accounted for. With this initialisation and suitable boundary conditions in hand, a straightforward staggered iteration technique is applied within each load step to incorporate switching effects:

(i) based on the couple finite element formulation, compute $\mathbf{u}$ and ϕ at fixed $\{\mathbf{a}_{1,2,3}\}$ for given boundary and loading conditions

(ii) based on the switching criterion, compute $\{\mathbf{a}_{1,2,3}\}$ at fixed $\mathbf{u}$ and ϕ

(iii) based on the coupled finite element formulation, recompute $\mathbf{u}$ and ϕ at fixed $\{\mathbf{a}_{1,2,3}\}$ for given boundary and loading conditions

Please note that steps (ii) and (iii) may be repeated until some criterion is met. For the purpose of visualisation, we finally make use of a simple volume-averaging technique combined with a projection onto the loading direction, say $\mathbf{e}$; in other words

$$\mathsf{E} = \frac{1}{\mathscr{V}} \int_{\mathscr{V}} \mathbf{E} \cdot \mathbf{e} \, \mathrm{d}v, \quad \mathsf{e} = \frac{1}{\mathscr{V}} \int_{\mathscr{V}} \mathbf{e} \cdot \boldsymbol{\varepsilon} \cdot \mathbf{e} \, \mathrm{d}v, \quad \mathsf{D} = \frac{1}{\mathscr{V}} \int_{\mathscr{V}} \mathbf{D} \cdot \mathbf{e} \, \mathrm{d}v, \quad \mathsf{S} = \frac{1}{\mathscr{V}} \int_{\mathscr{V}} \mathbf{e} \cdot \boldsymbol{\sigma} \cdot \mathbf{e} \, \mathrm{d}v.$$
(17)

Numerical examples

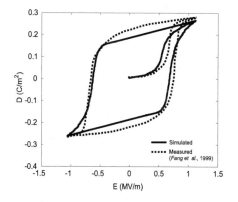
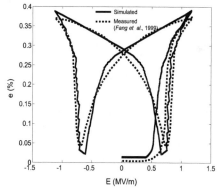

FIGURE 4. Hysteresis and butterfly curve without additional superimposed stresses.

A block-like PZT 51 specimen is considered, totally clamped at its bottom, the side surfaces being free, whereas cyclic electric loading conditions are applied to the top surface. The simulation results are in good agreement with experimental measurements reported by Fang and Li [6] and Lu et al. [18].

Figure 4 shows the classical hysteresis and butterfly loops, whereas the result displayed in figure 5 additionally account for compressive stresses. It is seen that the switching range is enlarged under such external stresses and that the graphs, say, 'flatten'. Finally, figure 6 shows a macroscopic (longitudinal) stress versus strains, respectively dielectric displacements curve. It is clearly visualised in figure 6 that a

217

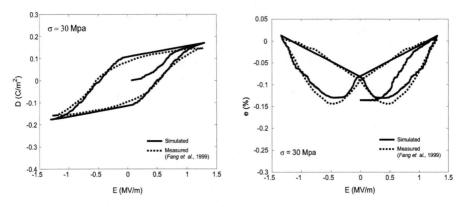

FIGURE 5. Hysteresis and butterfly curve with axial compressive stresses $\sigma = 30$ MPa.

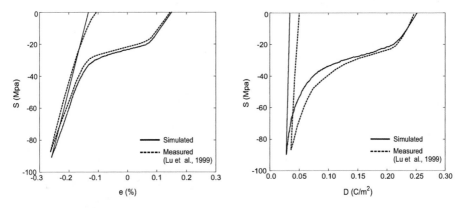

FIGURE 6. Stresses S versus strains e and dielectric displacements D.

compressive stress amplitude 'below' -20 [MPa] initiates switching processes. Saturation is obtained at a level of -90 [MPa] so that further increasing the applied stresses renders entirely linear response.

SUMMARY

The main goal of this contribution was to focus on two sources, and their computational modelling, for the nonlinear response of piezoceramics – micro-cracking combined with fatigue as well as switching effects. Both investigations seem to be promising and provide some potential for future research: the fatigue law should further be elaborated, in particular in view of a sound framework for the degradation of the interfacial dielectric permittivity. However, further experimental studies should be taken into account before enhancing this particular part of the model proposed. A major step for the switching formulation consists in precisely capturing the geometry of domains and grains. This should also render the probability function P to become superfluous. Finally, the combination of micro-cracking, fatigue and switching is of cardinal importance.

ACKNOWLEDGMENTS

Partial financial support for this work was provided by the Swedish Research Council (Vetenskapsrådet) under grant 622-2006-578 and is gratefully acknowledged.

REFERENCES

1. I. Arias, S. Serebrinsky, and M. Ortiz. A phenomenological cohesive model of ferroelectric fatigue. *Acta Mater.*, 54:975–984, 2006.
2. A. Arockiarajan and A. Menzel. On the modelling of rate–dependent domain switching in piezoelectric materials under superimposed stresses. *Comput. Model. Eng. Sci.*, 19(2):163–180, 2007. printed version: 20(1), 55–72.
3. A. Arockiarajan, A. Menzel, B. Delibas, and W. Seemann. Computational modeling of rate–dependent domain switching in piezoelectric materials. *Euro. J. Mech. A/Solids*, 25:950–964, 2006.
4. A. Arockiarajan, A. Menzel, B. Delibas, and W. Seemann. Micromechanical modeling of switching effects in piezoelectric materials – a robust coupled finite element approach. *J. Intel. Mat. Sys. Struct.*, 18:983–999, 2007.
5. A.Y. Belov and W.S. Kreher. Viscoplastic behaviour of perovskite type ferroelectrics. *Mat. Sci. Eng. B*, 118:7–11, 2005.
6. D. Fang and C. Li. Nonlinear electro–mechanical behavior of a soft PZT–51 ferroelectric ceramic. *J. Mat. Sci.*, 34:4001–4010, 1999.
7. D. Fang and B. Liu. Fatigue crack growth in ferroelectric ceramics driven by alternating electric fields. *J. Amer. Ceramic Soc.*, 87:840–846, 2004.
8. U. Gabbert, H. Berger, H. Köppe, and X. Cao. On modelling and analysis of piezoelectric smart structures by the finite element method. *Int. J. Appl. Mech. Eng.*, 5:127–142, 2000.
9. J.E. Huber. Micromechanical modelling of ferroelectrics. *Curr. Opin. Solid State Mater. Sci.*, 9:100–106, 2005.
10. S.C. Hwang, C.S. Lynch, and R.M. McMeeking. Ferroelectric/ferroelastic interactions and a polarization switching model. *Acta Metal. Mater.*, 43(5):2073–2084, 1995.
11. B. Jaffe, W.R. Cook, and H. Jaffe. *Piezoelectric Ceramics*. Academic Press, London, 1971.
12. M. Kamlah. Ferroelectric and ferroelastic piezoceramics – modeling of electromechanical hysteresis phenomena. *Continuum Mech. Thermodyn.*, 13(4):219–268, 2001.
13. M. Kamlah and U. Böhle. Finite element analysis of piezoceramic components taking into account ferroelectric hysteresis behaviour. *Int. J. Solids Struct.*, 38:605–633, 2001.
14. S.J. Kim and Q. Jiang. A finite element model for rate–dependent behavior of ferroelectric ceramics. *Int. J. Solids Struct.*, 39:1015–1030, 2002.
15. S. Klinkel. A phenomenological constitutive model for ferroelastic and ferroelectric hysteresis effects in ferroelectric ceramics. *Int. J. Solids Struct.*, 43:7197–7222, 2006.
16. C.M. Landis. A new finite element formulation for electromechanical boundary value problems. *Int. J. Numer. Methods Engng.*, 55:613–628, 2002.
17. C.M. Landis. Non–linear constitutive modeling of ferroelectrics. *Curr. Opin. Solid State Mater. Sci.*, 8:59–69, 2004.
18. W. Lu, D.-N. Fang, C.Q. Li, and K.C. Hwang. Nonlinear electric–mechanical behaviour and micromechanics modeling of ferroelectric domain evolution. *Acta Mater.*, 47:2913–2926, 1999.
19. D.C. Lupascu. *Fatigue in Ferroelectric Ceramics and Related Issues*. Number 61 in Materials Science. Springer, 2004.
20. R.M. McMeeking and S.C. Hwang. On the potential energy of a piezoelectric inclusion and the criterion for ferroelectric switching. *Ferroelectrics*, 200:151–173, 1997.

21. J.C.J. Schellekens and R. de Borst. On the numerical integration of interface elements. *Int. J. Numer. Methods Engng.*, 36:43–66, 1993.
22. J. Schröder and D. Gross. Invariant formulation of the electromechanical enthalpy function of transversely isotropic piezoelectric materials. *Arch. Appl. Mech.*, 73:533–552, 2004.
23. R.C. Smith. *Smart Material Systems – Model Development*. Frontiers in Applied Mathematics. SIAM, 2005.
24. J. Utzinger, M. Bos, M. Floeck, A. Menzel, E. Kuhl, R. Renz, K. Friedrich, A.K. Schlarb, and P. Steinmann. Computational modelling of thermal impact welded PEEK/steel single lap tensile specimens. *Comput. Mater. Sci.*, 41:287–296, 2008.
25. J. Utzinger, A. Menzel, and P. Steinmann. Investigation of microcracks in ferroelectric materials by application of a grain–boundary–motivated cohesive law. *Proc. Appl. Math. Mech.*, 2007. in press.
26. I. Westram, W.S. Oates, D.C. Lupascu, J. Rödel, and C.S. Lynch. Mechanism of electric fatigue crack growth in Lead Zirconate Titanate. *Acta Mater.*, 55:301–312, 2007.

Studies on grain-boundary effects of ferroelectric polycrystals

K. Jayabal*, A. Arockiarajan*, S. M. Sivakumar* and C. Sansour†

*Department of Applied Mechanics, Indian Institute of Technology Madras, Chennai 600036
†School of Civil Engineering, University of Nottingham, Nottingham NG7 2RD, UK

Abstract. The aim of this paper is to study the nonlinear dissipative effects of ferroelectric polycrystals based on firm thermodynamics principles. The developed micro-mechanically motivated model is embedded into an electromechanically coupled finite element formulation. In this framework, each domain is represented by a single finite element, and initial dipole directions are randomly oriented so that the virgin state of the particular bulk ceramics of interest reflects an un-poled material. Thermodynamically consistent energy criterion based on Gibbs free energy is adopted for the initiation of domain switching processes. The so-called grain-boundary effects, that is the constraint imposed by the surrounding grains on a grain at its boundary is incorporated in this model by means of micro-macromechanically motivated concept. In the expression for the driving force, an additional term is incorporated based on the change in Gibbs free energy of the neighboring grains for the particular switching domain/grain of interest. To study the overall bulk ceramics behavior, straightforward volume averaging techniques are applied. The simulated numerical results show appreciable improvement in modeling the nonlinear response for ferroelectrics subjected to various loading aspects compared with the experimental data from the literature.

Keywords: Ferroelectrics, micromechanical modeling, Domain switching, Finite element method
PACS: URL broken!

INTRODUCTION

An active or "smart" material is often defined as one that gives an unexpected response to an input, for example, an electrical or magnetic response to a mechanical or thermal input. In the field of engineering, active materials of interest are those utilize in modern structural design and intelligent systems. Ferroelectric materials represent an important class of active materials for applications as transducers, actuators, and sensors. Ferroelectric ceramics are widely used in a wide range of applications as MEMS devices, FRAM (ferroelectric random access memories), nanopositioning, active damping, ultrasonics and so forth; see [1]. These materials have unique properties compared to traditional engineering materials and their constitutive behavior involves electromechanical coupling allowing them to serve as smart materials. Under the action of low electric fields or low mechanical stresses, the behavior of the materials of interest is almost linear but exhibits strong nonlinear response under high electric fields or mechan-

CP1029, *Smart Devices: Modeling of Material Systems, An International Workshop*
edited by S. M. Sivakumar, V. Buravalla, and A. R. Srinivasa
©2008 American Institute of Physics 978-0-7354-0553-0/08/$23.00

ical stresses, respectively. Domain switching effects are accepted to be the main source for this highly nonlinear behavior, stemming from reorientation of the underlying polarization directions with respect to the applied loading directions; see the overview article [2]. Accordingly, it is of cardinal importance to account for a realistic constitutive behavior, some aspects of which are addressed in the literature; for an overview, see the excellent review articles [3, 4].

Ferroelectric constitutive models can be, in general, classified as macroscopic (phenomenological) models and microscopic (sub grain) models. Phenomenological models are generally derived within a thermodynamic framework. The state of the material is defined by internal variables at any given time and the evolution of these internal variables is determined by corresponding constitutive laws (evolution equations); refer [5, 6, 7, 8]. In general, remanent strain and remanent polarization are used as internal variables which define the irreversible process at hand. These models use plasticity-based approach in which the electric yield surface and stress yield surface are developed and the macroscopic reversible behavior occurs within these surfaces. During the switching process, the switching surface displays an irreversible change due to variation in the remanent strain and remanent polarization. Subsequently, the switching surface expands or moves depending upon the hardening rule that determines the evolution of the new surfaces; for further formulations we refer the reader to the contributions [9, 10, 11]. The phenomenological models are computationally effective and the implementation of the developed formulation is straight forward. However, these models incorporate a large number of material constants and their experimental determination becomes elaborate task.

The second approach is focused on micromechanical models which account for the internal microstructure and the microscopic switching mechanisms. These models could setup different choices due to particular aspects of the material behavior and the material substructure. These aspects as well as length scales at which the model operates can be included in the framework under consideration. Since this approach includes more physical insight of the material, it is more appropriate than phenomenological models from a modeling point of view, for an extensive overview we refer the reader to the contributions [12, 13, 14, 15]. In general, the underlying assumption is that each grain consists of one single domain or of multiple domains which are allowed to switch when the ferroelectric grain meets some switching criterion. The criterion may be based on the work exerted, the total potential, Gibb's energy, or the internal energy density; see [16, 17]. The driving force of the switching process at each increment of the loading is calculated and checked for the switching criterion and, satisfying the criterion, the existing domains are converted to another set of domains that are favorable to the external forcing fields. The macroscopic material behavior is obtained by microscopic averaging over the individual grains or domains. Though the method is computationally costly in comparison with phenomenological models, which prevents it from being as popular, it has specific advantages over the latter models. In particular, the material constants can be related to the microstructure of the ferroelectric and can be easily assigned a clear physical meaning as well as are simpler to obtain.

The present paper deals with a thermodynamically consistent micromechanical model using a new domain switching criterion that includes grain boundary effects. The main features of the present work are:

- A thermodynamically motivated dissipation maximization-based formulation, including domain switching mechanisms under electromechanical coupled loading, is formulated.

- Boundary effects between the grains are captured via a microscopic approach by incorporating additional term which account for the change in Gibbs free energy in the neighboring domains at particular switching domain under consideration. This term additionally contributes to the threshold for the onset of domain switching.

- In the modeling of perovskite crystallites with tetragonal microstructure, the material parameters depend on random orientations of the unit-cell, i.e. material constants will change with respect to domain switching.

- The developed framework is embedded into a coupled finite element formulation, whereby a straightforward staggered iteration scheme is applied to solve the nonlinear coupled problem for each and every loading / time step.

FUNDAMENTAL EQUATIONS

The configuration of a ferroelectric body of interest is denoted by $\mathscr{B} \subset \mathbb{R}^3$, its boundary $\partial \mathscr{B}$ and positions of material points are characterized by means of $x \in \mathbb{R}^3$. The governing mechanical and electrostatic equations for the bulk material, the balance of linear momentum and Gauss' law, read

$$\rho \ddot{x} = \nabla \cdot \sigma + b \quad \text{in} \quad \mathscr{B} \quad \text{and} \quad 0 = \nabla \cdot D - q \quad \text{in} \quad \mathscr{B}, \quad (1)$$

where b and q are source terms representing the mechanical body force and the electric charge density, respectively. σ and D denote respectively the Cauchy stress tensor and the electric displacement. ∇ - nabla operator (defines the gradient), : - double contraction, $\cdot$ - single contraction, if applied to vectors it results in the classical scalar product. As essential degrees of freedom we introduce the displacement field $u \in \mathbb{R}^3$ and the electric potential $\phi \in \mathbb{R}$, the corresponding boundary conditions either prescribe the displacement u on the boundary $\partial \mathscr{B}_u$ (essential boundary conditions) or the tractions $\sigma \cdot n_\sigma$ on the boundary $\partial \mathscr{B}_\sigma$ (natural boundary conditions)

$$u = u^p \quad \text{on} \quad \partial \mathscr{B}_u \quad \text{or} \quad t = t^p = \sigma \cdot n_\sigma \quad \text{on} \quad \partial \mathscr{B}_\sigma. \quad (2)$$

In addition, corresponding boundary conditions either prescribe the electric potential ϕ on the boundary $\partial \mathscr{B}_\phi$ (essential boundary conditions) or the surface charge density $D \cdot n_D$ on the boundary $\partial \mathscr{B}_D$ (natural boundary conditions)

$$\phi = \phi^p \quad \text{on} \quad \partial \mathscr{B}_\phi \quad \text{or} \quad q = q^p = -D \cdot n_D \quad \text{on} \quad \partial \mathscr{B}_D, \quad (3)$$

with $\partial \mathscr{B}_u \cup \partial \mathscr{B}_\sigma = \partial \mathscr{B}_\phi \cup \partial \mathscr{B}_D = \partial \mathscr{B}$ as well as $\partial \mathscr{B}_u \cap \partial \mathscr{B}_\sigma = \partial \mathscr{B}_\phi \cap \partial \mathscr{B}_D = \emptyset$, $n_{\sigma,D}$ are outward unit vectors defined with respect to the surfaces $\partial \mathscr{B}_\sigma$, $\partial \mathscr{B}_D$. The kinematic relation (linkage between strain tensor and displacements; between electric field and potential) in the bulk ceramic are given by

$$\varepsilon = \tfrac{1}{2}(\nabla \otimes u + u \otimes \nabla) \qquad \text{and} \qquad E = -\nabla\phi. \tag{4}$$

The classical linear ansatz for the modeling of piezoelectric materials in the piezoelectric phase reads

$$D = \mathbf{d} : \varepsilon + k \cdot E \qquad \text{and} \qquad \sigma = \mathsf{C} : \varepsilon - \mathbf{d}^{\mathrm{T}} \cdot E. \tag{5}$$

wherein $\mathsf{C} \in \mathbb{R}^{3\times3\times3\times3}$ is the elastic stiffness, which exhibits the well known major and minor symmetries, $k \in \mathbb{R}^{3\times3}$ is the dielectric permittivity and $\mathbf{d} \in \mathbb{R}^{3\times3\times3}$ is the piezoelectric tensor, which is symmetric with respect to the first two indices.

THERMODYNAMICAL FORMULATION

Ferroelectric materials exhibit electrical, mechanical, and thermal coupled phenomena, which can be described based on a thermodynamical approach to model the material behavior. To lay the ground for developing the constitutive description for multi domain crystallite, we recall the first and second laws of thermodynamics for deformable dielectric materials.

$$\rho\dot{u} = \sigma : \dot{\varepsilon} + E \cdot \dot{D} + \rho r - \operatorname{div} q, \tag{6}$$

where u denote the internal energy density, r the heat generated per unit mass, and q the heat flux on the boundary. The second law of thermodynamics postulates that the rate of entropy is always greater than or equal to the external entropy supply rate which is related to conduction and radiation of heat as

$$\dot{\eta} \geq \int_{\mathscr{B}} \rho \frac{r}{\theta} \, \mathrm{d}v - \int_{\partial\mathscr{B}} \frac{q}{\theta} \cdot n \, \mathrm{d}a. \tag{7}$$

The local form of these equations read

$$\rho\dot{\eta} \geq \frac{\rho r}{\theta} - \operatorname{div}\frac{q}{\theta} \qquad \Rightarrow \qquad \rho\theta\dot{\eta} \geq \rho r - \operatorname{div}q + \frac{1}{\theta}q \cdot \operatorname{Grad}\theta. \tag{8}$$

By replacing ρr in Eq. (8) by the expression resulting from Eq. (6), the following relation is obtained:

$$\rho(\theta\dot{\eta} - \dot{u}) + \sigma : \dot{\varepsilon} + E \cdot \dot{D} - \frac{1}{\theta}q \cdot \operatorname{Grad}\theta \geq 0. \tag{9}$$

Now, it is necessary to introduce different variables, in specific, those which account for the history dependence and the associated dissipation. The state variables can be split into two parts such as the observable variables and the internal variables. The electric field E, the total strain ε and the temperature θ are observable variables. The current state of the ferroelectrics depends on the load history, which could be represented using internal variables which are associated with dissipation. When ferroelectric materials are subjected to higher electromechanical loading and unloading, the material exhibits

224

at zero electric field and /or at zero mechanical stress some non-vanishing electrical displacement and strain which are denoted as remanent electric displacement (polarization) and remanent strain. Due to this unrecoverable values, the electric displacement and total strain can be assumed to be decomposable in a recoverable part $\{\bullet^r\}$ and an irrecoverable part $\{\bullet^i\}$:

$$D = D^r + D^i \qquad \text{and} \qquad \varepsilon = \varepsilon^r + \varepsilon^i. \tag{10}$$

At the microscopic level, the unit cell of the considered ferroelectric material possesses cubic symmetry for operating temperatures above the Curie temperature. This phase is called paraelectric whereby, the relative atom positions in the crystal lattice give rise to a vanishing net dipole moment, i.e. no piezoelectric effect. However, below the Curie temperature, the symmetry of the material might switch from simple cubic structures to tetragonal or rhombohedral arrangements. The new phase is denoted as the ferroelectric phase, where relative atom positions in the crystal lattice might change which gives rise to nonzero net dipole moment known as spontaneous polarization (vectors). During such phase transitions, the movements of atoms apparently result in lattice distortions which consequently cause mechanical strain, so-called spontaneous strain. For poly-crystalline materials at virgin or unpoled state, the polarization vector field is randomly oriented in general. Practically speaking, each grain is divided into domains of different polarization orientation. Since the direction of the polarization is equally distributed in the virgin state, the net polarization and the strain of the bulk material vanishes at the macro level. If the material is subjected to external loads either electrical (a prescribed electric field) and/ or mechanical (traction), the lattice structure undergoes a recoverable change in polarization and strain within the existing domain type. When the loads exceed a certain limit, the unit cell may reorient in a way that the polarization vectors within the grains align according to this loading direction. This change in domain type in ferroelectrics is referred to as domain switching and is irrecoverable in nature. As the external loading increases, a specific domain type is favored at the expense of others. Domain switching initiates at the boundaries in general, on the surface, and propagates inside or it may nucleate at various regions in the material and expand to the whole region.

In ferroelectric crystal structure, the remanent quantities represent macroscopic averages of the microscopic spontaneous polarization and strain. Based on this argument, in this micromechanical model the remanent polarization and strain are assumed to be equivalent to the spontaneous polarization P^s and strain ε^s. The so-called non-linear constitutive equations of the ferroelectric material additionally incorporate a spontaneous polarization vector and a spontaneous strain tensor. The extension of Eq. (5) consequently reads

$$D = \mathbf{d} : [\varepsilon - \varepsilon^s] + k \cdot E + P^s \qquad \text{and} \qquad \sigma = \quad \mathbf{C} : [\varepsilon - \varepsilon^s] - \mathbf{d}^T \cdot E. \tag{11}$$

The dissipation can be derived based on the existence of a thermodynamic potential. Inserting the state variables for D and ε from Eq. (10) in Eq. (9) results in

$$\rho(\theta\dot{\eta} - \dot{u}) + \sigma : \dot{\varepsilon}^r + E \cdot \dot{D}^r + \sigma : \dot{\varepsilon}^i + E \cdot \dot{D}^i - \frac{1}{\theta} q \cdot \text{Grad}\,\theta \geq 0. \tag{12}$$

225

Upon the application of a Legendre transformation, Gibbs free energy density G for the ferroelectric materials consequently results in

$$G = u - \theta\eta - \frac{\sigma : \varepsilon^{\mathrm{r}}}{\rho} - \frac{E \cdot D^{\mathrm{r}}}{\rho} \Rightarrow \dot{G} = \dot{u} - \theta\dot{\eta} - \dot{\theta}\eta - \frac{\sigma : \dot{\varepsilon}^{\mathrm{r}}}{\rho} - \frac{\dot{\sigma} : \varepsilon^{\mathrm{r}}}{\rho} - \frac{E \cdot \dot{D}^{\mathrm{r}}}{\rho} - \frac{\dot{E} \cdot D^{\mathrm{r}}}{\rho}.$$
(13)

Substitution of Eq. (13) in Eq. (12) yields,

$$-\rho\dot{G} - \rho\dot{\theta}\eta - \dot{\sigma} : \varepsilon^{\mathrm{r}} - \dot{E} \cdot D^{\mathrm{r}} + \sigma : \dot{\varepsilon}^{\mathrm{i}} + E \cdot \dot{D}^{\mathrm{i}} - \frac{1}{\theta} q \cdot \mathrm{Grad}\,\theta \geq 0.$$
(14)

The Gibbs free energy for a single crystal can be expressed as the function of observable state variables E, σ, θ and internal variables ζ^{α} as

$$G = G(\sigma, E, \theta, \zeta^{\alpha}),$$
(15)

where ζ^{α} - the internal variable for the particular domain type α among the six distinct domain structures. Here, we neglected the interface contributions between the domains, Eq. (15) results in

$$\dot{G} = \frac{\partial G}{\partial \sigma} : \dot{\sigma} + \frac{\partial G}{\partial E} \cdot \dot{E} + \frac{\partial G}{\partial \theta}\dot{\theta} + \frac{\partial G}{\partial \zeta^{\alpha}}\dot{\zeta}^{\alpha},$$
(16)

and the resulting dissipation inequality, Eq. (14), reads

$$-\left[\rho\frac{\partial G}{\partial \sigma} + \varepsilon^{\mathrm{r}}\right]\dot{\sigma} - \left[\rho\frac{\partial G}{\partial E} + D^{\mathrm{r}}\right]\dot{E} - \left[\rho\frac{\partial G}{\partial \theta} - \rho\eta\right]\dot{\theta}$$
$$-\rho\frac{\partial G}{\partial \zeta^{\alpha}}\dot{\zeta}^{\alpha} + \sigma : \dot{\varepsilon}^{\mathrm{i}} + E \cdot \dot{D}^{\mathrm{i}} - \frac{1}{\theta} q \cdot \mathrm{Grad}\,\theta \geq 0.$$
(17)

Since Eq. (17) holds for all admissible processes and σ, E, and θ are controllable variables, and thus following classical arguments of thermodynamics we arrive at the following mechanical, electrical and thermal constitutive relations between the arguments and their conjugates:

$$\varepsilon^{\mathrm{r}} = -\rho\frac{\partial G}{\partial \sigma}, \qquad\qquad D^{\mathrm{r}} = -\rho\frac{\partial G}{\partial E}, \qquad\qquad \eta = -\frac{\partial G}{\partial \theta},$$
(18)

which leave us with the reduced dissipation inequality

$$\sigma : \dot{\varepsilon}^{\mathrm{i}} + E\dot{D}^{\mathrm{i}} - \rho\frac{\partial G}{\partial \zeta^{\alpha}}\dot{\zeta}^{\alpha} - \frac{1}{\theta} q \cdot \mathrm{Grad}\,\theta \geq 0.$$
(19)

Assuming isothermal processes and homogeneous temperature fields, the heat flux term can be eliminated which results in

$$\sigma : \dot{\varepsilon}^{\mathrm{i}} + E \cdot \dot{D}^{\mathrm{i}} - \rho\frac{\partial G}{\partial \zeta^{\alpha}}\dot{\zeta}^{\alpha} \geq 0.$$
(20)

We assume that the evolution of the irreversible strain (ε^i) and polarization (D^i) can be expressed as a function of ζ^α as

$$\dot{\varepsilon}^i = \varepsilon^i(\dot{\zeta}^\alpha) = \varepsilon^* \dot{\zeta}^\alpha \qquad \text{and} \qquad \dot{D}^i = D^i(\dot{\zeta}^\alpha) = D^* \dot{\zeta}^\alpha, \qquad (21)$$

where, ε^*, D^* are transformation strain and electrical displacement. Using this assumption, the generalized dissipation inequality for the ferroelectrics can be expressed as

$$\sigma : \varepsilon^* \dot{\zeta}^\alpha + E \cdot D^* \dot{\zeta}^\alpha - \rho \frac{\partial G}{\partial \zeta^\alpha} \dot{\zeta}^\alpha \geq 0 \Rightarrow \left[\sigma : \varepsilon^* + E \cdot D^* - \rho \frac{\partial G}{\partial \zeta^\alpha} \right] \dot{\zeta}^\alpha \geq 0. \quad (22)$$

which is the product of driving force (shown within the parenthesis in the above equation), and the rate at which the volume fraction of the particular domain type evolves.

SWITCHING CRITERION

In this section, the generalized thermodynamic aspects are narrowed down towards the microscopic level of ferroelectric material behavior. The dissipation potential in Eq. (22) defines the thermodynamic driving force for transformation ($f^{drv} = \sigma : \varepsilon^* + E \cdot D^* - \rho \frac{\partial G}{\partial \zeta^\alpha}$). Within the above context it is reasonable to identify ζ^α with the volume fraction of a particular domain type and $\dot{\zeta}^\alpha$ with the rate at which this domain evolves. Upon application of external loads and on reaching a critical level, the underlying unit-cell or domain switches from one state to another possible state. The phase transformation refers to a switching process in which change in spontaneous polarization and strain occurs. Assume the present domain type β switches to other domain type α in the unit-cell, the transformation strain and displacement or polarization yields

$$\varepsilon^* = \Delta\varepsilon^s_{\beta\to\alpha} = \varepsilon^s_\alpha - \varepsilon^s_\beta \qquad \text{and} \qquad D^* = \Delta P^s_{\beta\to\alpha} = P^s_\alpha - P^s_\beta. \quad (23)$$

The reversible bulk Gibb's energy density G of this crystal can be expressed as (assume only two domain types α and β)

$$G \quad = \quad \zeta^\alpha G^\alpha + \zeta^\beta G^\beta \quad = \quad G^\beta + \Delta G^{\alpha-\beta} \zeta^\alpha, \qquad (24)$$

where

$$\Delta G^{\alpha-\beta} = G^\alpha - G^\beta = - \left[\frac{1}{2}\sigma : (C_\alpha - C_\beta) : \sigma + \frac{1}{2}E \cdot (k_\alpha - k_\beta) \cdot E + E \cdot (d_\alpha - d_\beta) : \sigma \right] \quad (25)$$

Here, ζ^α and ζ^β refer to the volume fractions of the domains α and β respectively, and the constraint holds $\zeta^\alpha + \zeta^\beta = 1$. Inserting Eq. (23) and the derivative of Eq. (24) in Eq. (22) yields the inequality

$$[\sigma : \Delta\varepsilon^s_{\beta\to\alpha} + E \cdot \Delta P^s_{\beta\to\alpha} +$$
$$\tfrac{1}{2}\sigma : (C_\alpha - C_\beta) : \sigma + \tfrac{1}{2}E \cdot (k_\alpha - k_\beta) \cdot E + E \cdot (d_\alpha - d_\beta) : \sigma]\dot{\zeta}^\alpha \geq 0 \quad (26)$$

227

which identifies the thermodynamic force f^{drv} responsible for the phase transformation or domain switching from β phase to α phase within a unit-cell of ferroelectrics. For domain switching to occur, the energy dissipated by domain wall motion upon transforming a unit volume of crystal from one state to the next state must attain a maximum value defined by f^{crit}. Accordingly, the thermodynamic driving force (f^{drv}) must satisfy the following inequality at all times.

$$f^{drv} \dot{\zeta}^\alpha \le f^{crit} \dot{\zeta}^\alpha. \tag{27}$$

Hence domain switching occurs for the driving force of the system attaining the critical value (f^{crit}). Within the $\langle 1\,0\,0 \rangle$ family, six different state or orientations of the particular unit cell of interest come into the picture – the initial state, four types of 90^0 domain switching, and 180^0 domain switching. In the case of spontaneous strain (mechanical state), solely four $90°$ strain switching states have to be considered. For spontaneous polarization (electrical state), four 90^0 switching states together with one $180°$ switching state are considered. In the sequel, we review the adopted energy-based switching criterion due to phase transformations taking place in a particular time interval $\Delta t = t_{n+1} - t_n > 0$ of interest. Experimental evidence shows that the switching thresholds are varying throughout the loading. The present work incorporates the critical energy barrier in which the underlying mechanisms for 90^0 switchings and 180^0 switchings are different, which gives different critical values for their occurrence. Assuming the electrical loading contributes to the direct 180^0 switching and applying the conditions in Eqs. (26) and (27) result in the following critical value for 180^0 switching:

$$f_{180}^{crit} = E \cdot \Delta P_{\beta \to \alpha}^s + \frac{1}{2} E \cdot (k_\alpha - k_\beta) \cdot E \quad = \quad 2E_c P_0 + 0 \quad = \quad 2E_c P_0, \tag{28}$$

where E_c - the coercive electric field; the change in spontaneous polarization $\Delta P_{\beta \to \alpha}^s = 2P^s = 2P_0$; P_0 - the spontaneous polarization; the dielectric constant $(k_\alpha - k_\beta)$ between two states is unchanged for 180^0 switching. Likewise the critical value for 90^0 switching is

$$f_{90}^{crit} = E \cdot \Delta P_{\beta \to \alpha}^s + \frac{1}{2} E \cdot (k_\alpha - k_\beta) \cdot E \quad = \quad E_c P_0 + \frac{1}{2} E_c^2 (k_{33} - k_{11}). \tag{29}$$

Regarding the ferroelastic switching case, it exhibits only 90^0 switching due to the application of pure mechanical stress, which yields

$$f_{90ela}^{crit} = \sigma : \Delta \varepsilon_{\beta \to \alpha}^s + \frac{1}{2} \sigma : (C_\alpha - C_\beta) : \sigma \quad = \quad \frac{3}{2} \sigma_c \varepsilon_0 + \frac{1}{2} \sigma_c^2 (C_{3333} - C_{1111}), \tag{30}$$

where σ_c - the coercive stress; $\varepsilon^s = \varepsilon_0$ - the spontaneous strain; C_{3333}, C_{1111} - the compliance constants.

GRAIN BOUNDARY EFFECTS

Polycrystalline microstructures often give rise to local loading levels such that individual grains and domains, possibly possessing different polarization directions, interact with the ambient material given rise to so-called grain boundary effects or intergranular effects. Accordingly, domain switching might occur at macroscopic loading levels below the established coercive values so that nonlinear responses are observed even within small loading ranges. Moreover, since we are considering one crystal per element, the inhomogeneity causes a non-smooth variations which amplify the grain boundary-like effect. In order to take these effects into account the proposed formulation incorporates an additional term, which is based on the change in Gibbs free energy in the neighboring domains at particular switching domain under consideration, is added to the threshold for the onset of domain switching

$$\left[f^{drv} = \left(1 - \frac{p}{2}\right) f_0^{drv} + \left(\frac{p_1}{2}\right) f_{i-e}^{drv} + \left(\frac{p_2}{2}\right) f_{i-m}^{drv} + \left(\frac{p_3}{2}\right) f_{i-em}^{drv} \right] \leq f^{crit}, \quad (31)$$

with $p = \frac{p_1 + p_2 + p_3}{3}$, $0 < p \leq 1$, p_1, p_2, p_3 - the material dependent parameters ; f_0^{drv} - the driving force of the switching domain; $f_{i-e}^{drv}, f_{i-m}^{drv}, f_{i-em}^{drv}$ - the electrical, mechanical and electromechanical energies of neighboring or surrounding domains of the switching domain the definition of which is given below:

$$
\begin{aligned}
f_0^{drv} &= \sigma : \Delta\varepsilon_{\beta\rightarrow\alpha}^s + E \cdot \Delta P_{\beta\rightarrow\alpha}^s + \Delta G_{\alpha-\beta} \\
f_{i-e}^{drv} &= \frac{1}{n} \sum_{i=1}^{n} \left[E^{(i)} \cdot \Delta P_{\beta\rightarrow\alpha}^{s(i)} + \frac{1}{2} E^{(i)} \cdot \left(k_\alpha - k_\beta\right)^{(i)} \cdot E^{(i)} \right] \\
f_{i-m}^{drv} &= \frac{1}{n} \sum_{i=1}^{n} \left[\sigma^{(i)} : \Delta\varepsilon_{\beta\rightarrow\alpha}^{s(i)} + \frac{1}{2} \sigma^{(i)} : \left(C_\alpha - C_\beta\right)^{(i)} : \sigma^{(i)} \right] \\
f_{i-em}^{drv} &= \frac{1}{n} \sum_{i=1}^{n} \left[E^{(i)} \cdot \left(d_\alpha - d_\beta\right)^{(i)} : \sigma^{(i)} \right],
\end{aligned}
\quad (32)
$$

where $\Delta G_{\alpha-\beta} = \frac{1}{2}\sigma : \left(C_\alpha - C_\beta\right) : \sigma + \frac{1}{2} E \cdot \left(k_\alpha - k_\beta\right) \cdot E + E \cdot \left(d_\alpha - d_\beta\right) : \sigma$; $\{\bullet\}^{(i)}$ indicates the electric field, stress and material constants for the surrounding domains; n - no. of surrounding domains. In this model, eight surrounding domains are considered in general ($n = 8$). However it will vary in some specific cases; for a domain at the corner of the specimen, n will be 3; if the switching domain is at an edge, the number of surrounding domains n would be 5. Substituting Eq. (32) in Eq. (31) yields the new expression for the driving force

$$
\begin{aligned}
f^{drv} = &\left(1 - \frac{p_1 + p_2 + p_3}{6}\right) f_0^{drv} \\
&+ \left(\frac{p_1}{2}\right) \frac{1}{n} \sum_{i=1}^{n} \left[E^{(i)} \cdot \Delta P_{\beta\rightarrow\alpha}^{s(i)} + \frac{1}{2} E^{(i)} \cdot \left(k_\alpha - k_\beta\right)^{(i)} \cdot E^{(i)} \right] \\
&+ \left(\frac{p_2}{2}\right) \frac{1}{n} \sum_{i=1}^{n} \left[\sigma^{(i)} : \Delta\varepsilon_{\beta\rightarrow\alpha}^{s(i)} + \frac{1}{2} \sigma^{(i)} : \left(C_\alpha - C_\beta\right)^{(i)} : \sigma^{(i)} \right] \\
&+ \left(\frac{p_3}{2}\right) \frac{1}{n} \sum_{i=1}^{n} \left[E^{(i)} \cdot \left(d_\alpha - d_\beta\right)^{(i)} : \sigma^{(i)} \right] \leq f^{crit}.
\end{aligned}
\quad (33)
$$

229

ALGORITHMIC SETTING

The subsequent numerical examples are based on a three-dimensional finite element framework wherein the above elaborated switching model is embedded. Both, the spontaneous polarization P^s as well as the spontaneous strain ε^s are thereby introduced as so-called internal variables stored at the integration point at the element level. We do not distinguish between grains and domains but rather attach an individual unit-cell orientation, as represented by $\{m_{1,2,3}\}$ to one single (eight-node brick) finite element. For the computation of the initial unit-cell orientation (m), which also determines the initial polarization vector $(P^s = P^s m)$ as well as the initial spontaneous strains $(\varepsilon^s = \varepsilon^s[3M - I]/2)$, for $\varepsilon^s > 0$, orientations are applied by using Eulerian angles $(\Phi, \Psi \in (0, 2\pi), \sin(\Theta - \frac{\pi}{2}) \in [-1, 1])$. In this numerical setup, the applied switching criterion refers to quantities averaged over single finite element, viz

$$f_e^{drv}(\langle \sigma_n \rangle_e, \langle E_n \rangle_e, m_n, m_{n+1}) \leq f^{crit} \tag{34}$$

wherein $\langle \bullet \rangle_e = [\mathcal{V}_e]^{-1} \int_{\mathcal{B}_e} \bullet \, dv$ for $\mathcal{V}_e$ denoting the volume of a particular finite element $\mathcal{B}_e$. To compare the simulated results with experimental data and for the purpose of visualization, projection with respect to the macroscopic loading direction e are performed, namely

$$S = \langle \sigma : (e \otimes e) \rangle, \quad e = \langle \varepsilon : (e \otimes e) \rangle, \quad D = \langle D \cdot e \rangle, \quad E = \langle E \cdot e \rangle \tag{35}$$

with the global volume-averaging $\langle \circ \rangle = [\mathcal{V}]^{-1} \int_{\mathcal{B}} \circ \, dv$, for $\mathcal{V}$ characterizing the volume of the entire body $\mathcal{B}$. The developed rate-independent evolution is solved by using a simple staggered iteration technique for each load step to evaluate the switching effects. The following steps are performed:

1. based on the coupled finite element formulation, compute u and ϕ for the initial orientations.

2. While fixing the u and ϕ, calculate the energy barrier for switching based on Eqs. (33), (34) for all the elements, 5 possibilities for each element (unit-cell), and identify the number of elements n_{tot} that exceed the critical value based on the largest energy reduction. Since an element with a higher energy reduction is more likely to switch than an element with a lower energy reduction, the n_{tot} elements are prioritized based on their maximum local Gibbs free energy with the one possessing the highest maximum energy reduction ranked first at the top of the list. These elements will be referred to as switching elements.

3. For the first switching element, recompute u and ϕ for updated unit-cell orientations for given boundary and loading conditions.

4. Repeat step (2) and (3) in the sequence of the second, the third, ... and finally the n_{tot} switching elements to obtain a fully equilibrated state.

5. Finally at the global level, the volume averaged S, e, E and D are computed based on Reuss approximation.

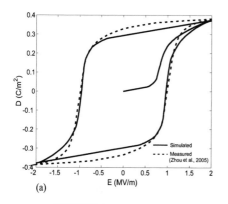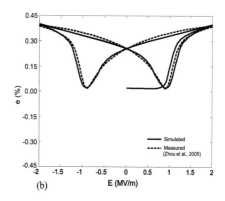

FIGURE 1. Hysteresis and butterfly curves – cyclic uniaxial electrical loading

Even though, we use a simple staggered iteration technique, we are able to predict the important insights of the nonlinearity. However, the art of the paper is focused on micromechanical modeling and incorporation of grain boundary effects. In the future, the algorithmic setup could be improved in the aspects of robustness and computational efficiency.

NUMERICAL EXAMPLES

All numerical examples studied in the following refer to a block-like $10 \times 10 \times 10$ specimen, whereby the discretization is performed with $10 \times 10 \times 10$ eight node bricks (Q1Q1). In view of material parameters, representative PIC 151 values have been adopted from the literature: $k_{11} = 0.0198$ [μ F/m], $k_{33} = 0.024$ [μ F/m], $d_{33} = 0.45 \times 10^{-9}$ [m/V], $d_{31} = -0.21 \times 10^{-9}$ [m/V], $d_{15} = 0.58 \times 10^{-9}$ [m/V], $\lambda = 17.48$ [GPa], $\mu_L = 2.46$ [GPa], $\mu_T = 11.65$ [GPa], $\alpha = 1.67$ [GPa], $\beta = -20.74$ [GPa], $\varepsilon^s = 2.75 \times 10^{-3}$, $P^s = P_0 = 0.3$ [C/m²], $E_0 = 0.7$ [MV/m], $\sigma_0 = 75$ [MPa], $p_1 = 0.7$, $p_2 = 0.5$ and $p_3 = 0.9$.

The boundary condition for the specimen under cyclic electrical loading is represented by a prescribed electric potential at the top surface ϕ_{top}^p and a zero electric potential at the bottom surface ϕ_{bot}^p with a linear variation between them. In addition compressive stresses (prestress) are uniformly applied to the top surface, t_{top}^p, and, the bottom surface is clamped such that only the direction in which the traction is applied is set to be zero.

The simulation starts with solely electrical cyclic loading without any prescribed mechanical stresses. The starting point for the first cycle is at zero potential for the macroscopically unpoled ceramic, whereby randomly oriented polarization vectors, spontaneous strains, and material constants are incorporated, respectively. The electric field is applied incrementally until +2.0 [MV/m] and unloaded to zero field. Subsequently, the field is applied in the reverse direction to −2.0 [MV/m] and once again, the load is applied until +2.0 [MV/m] as a cyclic loading. Classical hysteresis and butterfly curves are showed in Fig.1, i.e. the electric displacement D and the total strain e are monitored versus the electric field E. The macroscopic behavior exhibits linear response starting from

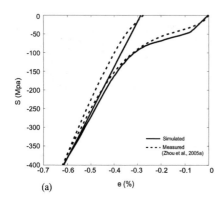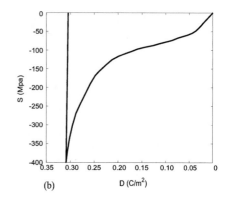

FIGURE 2. Stress versus strain and versus electric displacement curves – uniaxial mechanical loading

the origin to the field nearer to 0.7 [MV/m], while the remanent strain remains zero. The electric field (0.7 [MV/m]) at which switching initiates under pure electrical loading, is referred to as the 'coercive field'. Once the applied load exceeds the coercive field, then switching starts. This causes a change in spontaneous polarization and strain which is obviously renders the macroscopic response of the electric displacement and the total strain nonlinear. The applied load then reaches to 2.0 [MV/m], which corresponds to a saturated level. At this stage, almost all the domains are closely aligned with the direction of the applied field. Further increase of the electric field will cause linear response, since all the elements are already switched. When the field is unloaded, the electric displacement and strain vary linearly and at zero electric field, it exhibits some electric displacement and strain, which are so called remanent polarization and strain. Upon the application of negative electric field, the switching processes commence again in a significant reverse poling then taking place. During the forward loading the elements are oriented in a particular direction, but in the reverse loading the oriented elements should be reoriented and aligned along the same axis but in the opposite direction, which makes the element go for 180^0 switching. Concerning the strain response, during the reverse loading, the strain decreases nonlinearly to minimum level and then strain starts increasing as the elements undergo reverse poling. Since the elements undergo two consecutive 90^0 switches to align with the negative loading direction, the strain response decreases initially for first 90^0 switching and subsequently the second 90^0 switching occurs which makes the strain to increase. The computations, highlighted in Fig.1, nicely match with experimental results reported in Zhou et al. [18].

In the absence of the electric field, i.e., in the ferroelastic case, the mechanical depolarization response is predicted in terms of stress S versus strain e and versus dielectric displacement D as shown in Fig.2. The simulation is carried out by considering the test specimen from the unpoled virgin state, in which the initial random orientations give rise to net zero strain and dielectric displacement. The domain switching process is initiated, once the incrementally applied compressive stress reaches nearer to the coercive level, say for instance -75 [MPa]. Saturation is reached at a loading level of -400 [MPa] so that further increase in the applied stresses results in entirely linear response. Considering the ferroelastic case, the elements will go only four possible

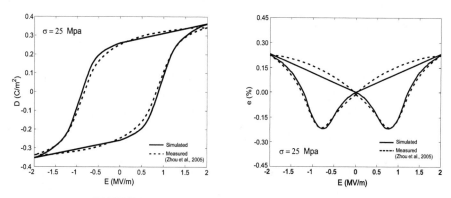

FIGURE 3. hysteresis and butterfly curves at $\sigma = 25$ MPa

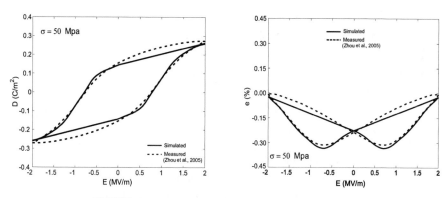

FIGURE 4. hysteresis and butterfly curves at $\sigma = 50$ MPa

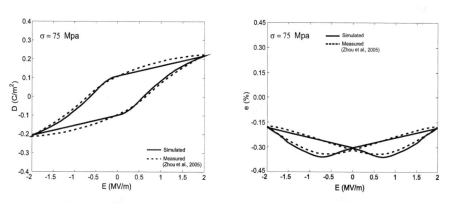

FIGURE 5. hysteresis and butterfly curves at $\sigma = 75$ MPa

90^0 switching. Upon reversing the loading direction, dissipative effects as reflected by the irreversible behavior are clearly displayed. Moreover, one observes different slopes in, for example, the S versus e response which stems from aligning the local material properties along the switching directions.

233

Subsequently, the macroscopic hysteresis and butterfly curves under constant axial compressive stresses are simulated. The applied stress magnitude is thereby chosen as 25, 50, and 75 MPa. Fig.3 displays an electric displacement versus electric field hysteresis curve as well as a strain versus electric field butterfly curve at a compressive stress of 25 MPa. In particular, an enlarged switching range, a reduced macroscopic remanent polarization value as well as decreased saturation polarization magnitudes are observed. The simulated butterfly curve perfectly approximates the characteristic remanent strain value. Computations under higher mechanical loading conditions are highlighted in Fig.4 and Fig.5 wherein the comparison with experimental data is additionally visualised. It can, for instance, clearly be seen that increasing compressive stresses result in enlarged domain switching ranges and that the hysteresis and butterfly curves flatten at higher mechanical loading levels. These effects reflect that the effective critical level for domain switching raises with increasing mechanical loading. As the compressive stresses are applied, the macroscopic total strain shifts along the axial loading direction resulting in a negative strain with reference to unpoled state.

SUMMARY

A three dimensional micromechanical model has been developed to simulate the nonlinear behavior of poly-crystalline ferroelectrics. Since the virgin state of these materials is commonly unpoled, initial orientations of polarization vectors have been randomly generated. The onset of domain switching was introduced via a firm thermodynamically consistent framework by means of a dissipation threshold. The adopted model for ferroelectric materials has been embedded into a finite element formulation, whereby each finite element was assumed to represent one individual grain. As a key aspect, grain boundary effects were incorporated by means of a micro-macromechanically motivated modeling aspect according to which the change in Gibbs free energy of the neighborhood or surrounding domains was employed. The material constants such as elastic, dielectric and piezoelectric constants are incorporated as a consequence of switching processes. Numerical tests show that the simulated hysteresis and butterfly loops are in qualitative agreement with experimental observations. In the future, different grain sizes such as the impact of size effects in the modeling aspects could be studied in detail, together with the effect of non-linearity due to micro cracking in the test specimen.

REFERENCES

1. R. C. Smith. *Smart material systems – Model development.* Society of Industrial and Applied Mathematics, Philadelphia, 2005.
2. M. Kamlah. Ferroelectric and ferroelastic piezoceramics-modeling of electromechanical hysteresis phenomena. *Continuum Mech. Thermodyn.*, 13(4):219–268, 2001.
3. C. M. Landis. Non-linear constitutive modeling of ferroelectrics. *Current opinion solid state Mat. Sc.*, 8:59–69, 2004.
4. J. E. Huber. Micromechanical modeling of ferroelectrics. *Current opinion solid state Mat. Sc.*, 9:100–106, 2005.
5. M. Kamlah and Q. Jiang. A constitutive model for ferroelectric pzt ceramics under uniaxial loading. *Smart Mat. Struct.*, 8:441–459, 1999.

6. C. M. Landis and R. M. McMeeking. A phenomenological constitutive law for ferroelastic switching and a resulting asymptotic crack tip solution. *J. Intelligent Mat. Sys. Struct.*, 10:155–163, 2000.
7. M. Kamlah and U. Böhle. Finite element analysis of piezoceramic components taking into account ferroelectric hysteresis behaviour. *Int. J. Solids Struct.*, 38:605–633, 2001.
8. J. Shieh, J. E. Huber, and N. A. Fleck. An evaluation of switching criteria for ferroelectrics under stress and electric field. *Acta Mater.*, 51:6123–6137, 2003.
9. R. M. McMeeking and C. M. Landis. A phenomenological multiaxial constitutive law for switching in polycrystalline ferroelectric ceramics. *Int. J. Engng. Sci.*, 40:1553–1577, 2002.
10. J. Schröder and D. Gross. Invariant formulation of the electromechanical enthalpy function of transversely isotropic piezoelectric materials. *Arch. Appl. Mech.*, 73:533–552, 2004.
11. M. Elhadrouz, T.B. Zineb, and E. Patoor. Constitutive law for ferroelectric and ferroelastic single crystals: a micromechanical approach. *Comput. Mater. Sci.*, 32:355–359, 2005.
12. S. C. Hwang, C. S. Lynch, and R. M. McMeeking. Ferroelectric/ferroelastic interactions and a polarization switching model. *Acta Metall. Mater.*, 43:2073–2084, 1995.
13. W. Lu, D. N. Fang, C. Q. Li, and K. C. Hwang. Nonlinear electric-mechanical behaviour and micromechanics modeling of ferroelectric domain evolution. *Acta Mater.*, 47:2913–2926, 1999.
14. A. Arockiarajan, A. Menzel, B. Delibas, and W. Seemann. Micromechanical modeling of switching effects in piezoelectric materials - a robust coupled finite element approach. *Euro. J. Mech. A/Solids*, 25:950–964, 2006.
15. M. G. Shaikh, S. Phanish, and S. M. Sivakumar. Domain switching criteria for ferroelectrics. *Comput. Mater. Sci.*, 37:178–186, 2006.
16. W. Chen and C. S. Lynch. A micro-electro-mechanical model for polarization switching of ferroelectric materials. *Acta Mater.*, 46:5303–5311, 1998.
17. J. E. Huber, N. A. Fleck, C. M. Landis, and R. M. McMeeking. A constitutive model for ferroelectric polycrystals. *J. Mech. Phys. Solids*, 47:1663–1697, 1999.
18. D. Zhou, M. Kamlah, and D. Munz. Effects of uniaxial prestress on the ferroelectric hysteretic response of soft pzt. *J. Euro. Ceramic Soc.*, 25:425–443, 2005.

SESSION D2

FERROELECTRIC MATERIALS—
PREPARATION AND CHARACTERIZATION

Accurate Characterization And Modeling of Piezoelectric And Electrostrictive Ceramics And Single Crystals

Binu Mukherjee

Department of Physics, Royal Military College of Canada, Kingston, Ontario K7K 7B4, Canada.

Abstract. The continuing need for large actuation and a continually increasing variety of applications has seen the use of piezoelectric materials under a growing range of conditions. These include large applied AC fields, DC bias fields, applied stresses as well as a wider range of frequencies and temperatures. Under these conditions the behaviour of the materials becomes non-linear and is not described well by the small signal parameters supplied by material manufacturers. It is therefore necessary to know and understand the behaviour of piezoelectric materials under these conditions. This presentation reviews the work that has been carried out at the Laboratory for Ferroelectric Materials in the Royal Military College of Canada where we have been studying the non-linear properties of piezoelectric and electrostrictive materials by observing their strain, dielectric and elastic behaviour under the various conditions mentioned above.

Keywords: Piezoelectric, Electrostrictive, Non-linearities, Strain, Frequency, Temperature
PACS: 77, 77.65-j, 77.84.-s.

INTRODUCTION

Piezoelectric and electrostrictive materials are important constituents of electromechanical sensors, actuators and smart structures. Piezoelectric materials produce a strain, S, under the influence of an external electric field, E, or become electrically polarised under the influence of an external stress, T. The property of piezoelectricity is closely related to the phenomenon of ferroelectricity, which describes the spontaneous polarisation in a crystal that can be changed between two or more distinct directions with respect to the crystal axes through the application of an external electric field. This ability of ferroelectric materials to switch polarisation under an external electric field from a random orientation to a preferred direction is used in a variety of polycrystalline ferroelectric ceramics to produce a polycrystalline piezoelectric ceramic with a preferred direction of net polarisation. This process is known as "poling". In an unpoled ceramic the individual domains in the ceramic have randomly oriented directions of polarisation so that the ceramic as a whole shows little or no

CP1029, *Smart Devices: Modeling of Material Systems, An International Workshop*
edited by S. M. Sivakumar, V. Buravalla, and A. R. Srinivasa
©2008 American Institute of Physics 978-0-7354-0553-0/08/$23.00

polarisation. The partial alignment of domains during poling creates a net spontaneous polarisation in the poling direction and the ceramic shows a C_∞ symmetry around that direction.

Piezoelectricity can be mathematically derived from a phenomenological model derived from thermodynamical potentials. The derivations are not unique and the set of equations describing the direct and converse piezoelectric effect depend on the choice of potential and the independent variables used.[1,2]. For example, one set of such linear constitutive relations is

$$S_p = s_{pq}^E T_q + d_{mp} E_m \quad \text{and} \quad D_m = \varepsilon_{mn}^T E_n + d_{mp} T_p, \tag{1}$$

where D is the electric displacement, s is the elastic compliance, d is the piezoelectric constant and ε is the dielectric permittivity. The superscripts of the constants designate the independent variable that is held constant when defining the material coefficient and the subscripts define directions that take account of the anisotropic nature of the material. It has been shown that the piezoelectric constant d_{mp} that occurs in the two equations is the same.[3]. The material constants form a 9 x 9 matrix with 1, 2, 3 designating the orthonormal directions and 4, 5, 6 designating the shear directions. For the commonly used piezoelectric ceramics with C_∞ symmetry, such as lead zirconate titanate or PZT, there are ten non-zero, independent matrix elements consisting of 5 independent elastic coefficients ($s_{11}^E, s_{12}^E, s_{13}^E, s_{33}^E, s_{55}^E$), 3 independent piezoelectric coefficients (d_{31}, d_{15}, d_{33}) and 2 independent dielectric constants ($\varepsilon_{11}^T, \varepsilon_{33}^T$) [4]. While the linear constitutive relations can be written in ways other than that shown in (1), there are only 10 independent constants. The full, reduced, matrix for these materials can be found in the IEEE standard [4], which also contains the equations that allow one to convert from one set of constitutive equations/matrix to another.

Ideally, under small fields and stresses and for materials with low losses within a limited frequency range, these 10 constants contain all the information required to predict the behaviour of the material when a stress, strain or electric field is applied to it. However, the continuing need for large actuation and a continually increasing variety of applications has seen the use of piezoelectric materials under a growing range of conditions. These include the application of large fields and stresses as well as a wider range of frequencies and temperatures. In practice most piezoelectric materials display dispersion and non-linearities, have measurable losses and their properties are temperature dependent. The last two decades has also seen the development of finite element packages to model piezoelectric materials, such as PZFlex [5], and some of these are sufficiently precise so as to require very accurate material constants. All of the above have pointed to the need for better characterisation of piezoelectric materials: both for the traditional ceramics and polymers as well as for the new high strain single crystals in the lead zinc niobate – lead titanate and lead magnesium niobate – lead titanate families. The Laboratory for Ferroelectric Materials at the Royal Military College of Canada has a continuing programme for improving the characterisation of these materials and this paper presents a summary of the work of this group since its establishment in 1988. Fuller details regarding the work can be obtained from the references mentioned in the paper.

IMPEDANCE SPECTROSCOPY OR RESONANCE METHODS

The most widely used technique for determining all the 10 relevant constants for piezoelectric ceramics is the resonance technique outlined in the IEEE standard on piezoelectricity. [4] A piezoelectric sample of specific geometry is excited with an AC signal and an impedance analyser is used to determine the complex impedance and admittance as a function of frequency. The impedance spectra contain resonances that result from ultrasonic standing waves in the material. Several particular frequencies may be defined from these spectra: the parallel resonance frequency, f_p, is the frequency at which the resistance R is a maximum, the sideband frequencies, $f_{+\frac{1}{2}}$ and $f_{-\frac{1}{2}}$, correspond to the maximum and minimum in the reactance X, the series resonance frequency f_s is the frequency at which the conductance G is a maximum and the sideband frequencies, $f_{+\frac{1}{2}s}$ and $f_{-\frac{1}{2}s}$, correspond to the maximum and minimum of the susceptance B. Five resonance modes are commonly used to find the full set of material coefficients for a piezoelectric ceramic and each mode requires samples with specific aspect ratios [4] to ensure that the sample is excited in a mode where the one-dimensional approximation is valid and coupling between the modes is invalid. It should be noted that the aspect ratios may be relaxed for materials with low mechanical Q whereas they need to be more stringent for materials with a high mechanical Q and high electromechanical coupling. The impedance equations that govern the various resonance spectra have been derived from the phenomenological theory of piezoelectricity [6,7] for the case of real material constants assuming lossless materials. These equations express the impedance as a function of the appropriate material constants and of the particular frequencies mentioned above. Holland [8] suggested that the losses in a piezoelectric material may be taken into account by representing the material constants as complex coefficients. Sherrit [9] has re-derived the impedance equations for the various resonance geometries using complex material constants so as to include the dielectric, mechanical and piezoelectric losses in the material. The complex material constants can be found by comparing the impedance equations with the appropriate experimental impedance curve around the fundamental resonance and finding the best fit. Numerous techniques have been proposed for this purpose [10,11,12,13,14,15,16], including some that have been developed in our laboratory. Some of the advantages and disadvantages of the various techniques have been discussed by Mukherjee and Sherrit [17], by Kwok et al [18] and by Pardo [19]. Commercial software is now available for carrying out the analysis of the experimental resonance curves [20]. The physical significance of the complex material constants has been discussed by Mukherjee and Sherrit [21] and a discussion may also be found on the PRAP website [20]. An example of a complete characterisation is shown in Table 1 [22] and it shows the full set of measured and calculated complex material constants for the Motorola 3203HD piezoceramic. The table shows that some

TABLE 1: The reduced matrix of Motorola 3203HD PZT ceramic including the electromechanical coupling determined at fundamental resonance of each mode [22]. The values shown are averages over 4 specimens and the spread of values is indicated by the standard deviations over the 4 specimens.

Material Constant	Mode	Frequency kHz	Value		Standard deviation %	
			Real	Imaginary	Real	Imaginary
s_{11}^E (m²/N) x 10¹¹	LTE	71.5	1.56	-0.030	0.63	5.2
s_{11}^E (m²/N) x 10¹¹	RAD	150.9	1.55	-0.032	0.45	2.8
s_{11}^E (m²/N) x 10¹¹	Average		1.56	-0.031		
s_{12}^E (m²/N) x 10¹¹	RAD	150.9	-0.420	0.012	3.9	4.9
s_{13}^E (m²/N) x 10¹¹	Calculated	Smits' formula	-0.821	0.034	N/A	N/A
s_{33}^E (m²/N) x 10¹¹	LE	199	1.89	-0.034	1.0	0.78
s_{55}^E (m²/N) x 10¹¹	TS	2730	3.92	-0.13	2.9	4.3
s_{66}^E (m²/N) x 10¹¹	Calculated	IEEE formula	3.96	-0.086	N/A	N/A
c_{33}^D (m²/N) x 10¹¹	TE	6390	1.77	0.023	2.0	11
d_{31} (C/N) x 10⁻¹²	LTE	71.5	-297	9.7	0.7	7.1
d_{31} (C/N) x 10⁻¹²	RAD	150.9	-293	10	0.68	5.8
d_{31} (C/N) x 10⁻¹²	Average		-295	9.9		
d_{33} (C/N) x 10⁻¹²	LE	199	564	-15	3.1	17
d_{15} (C/N) x 10⁻¹²	TS	2730	560	-30	4.6	11
ε_{11}^T (F/m) x 10⁻⁸	TS	2730	2.14	-0.13	0.44	6.8
ε_{33}^T (F/m) x 10⁻⁸	RAD	150.9	3.06	-0.11	1.1	6.5
ε_{33}^T (F/m) x 10⁻⁸	LT	71.5	2.83	-0.061	1.9	9.4
ε_{33}^T (F/m) x 10⁻⁸	Average		2.95	-0.083		
ε_{33}^S (F/m) x 10⁻⁸	TE	6390	1.06	-0.053	2.0	4.2
k_{33}	LE	199	0.763	-0.0029	0.52	45
k_{31}	LTE	71.5	0.447	-0.0054	0.90	16
k_{15}	TS	2730	0.611	-0.0034	3.1	37
k_p	RAD	150.9	0.706	-0.0062	0.45	6.1
k_t	TE	6390	0.536	-0.0050	0.46	12

redundancy is built in to the method, as the same coefficient may be determined from more than one resonance measurement. These may be used to find an average value but it should be noted that the measurements are not all at the same frequency and in the presence of significant dispersion it is important to take note of the frequency at which the constant has been determined. A more detailed understanding of dispersion in a piezoceramic can be obtained by carrying out the resonance analysis at the fundamental resonance and at higher order resonances [22]. While the dispersion in piezoceramics is not normally very significant, it becomes important in some other piezoelectric materials, as for example in polyvinylidene fluoride - tetrafluoroethylene copolymers, and in such cases the real and complex parts of the material constants may be expressed as polynomials in frequency [23] and the applications engineer can then calculate the value of the constant at the frequency of interest.

EQUIVALENT CIRCUITS

In designing devices it is sometimes useful to have an equivalent electrical circuit to represent the material. The most commonly used equivalent circuit to represent a piezoelectric vibrator has been the Van Dyke circuit, which uses four real circuit parameters. Sherrit et al have shown that the Van Dyke circuit cannot take account of all the losses in the piezoelectric material and they have proposed an alternative model [24,25] with three complex circuit parameters for each resonator geometry. The circuit parameters can be calculated from the material constants and vice versa.

TEMPERATURE DEPENDENCE OF THE MATERIAL CONSTANTS

The resonance experiments outlined above have been carried out with the ceramic samples placed in a Thermotron programmable temperature chamber that allows the temperature to be controlled between -170°C and 200°C. The variation of the dielectric, elastic and piezoelectric constants as a function of temperature have been measured for the hard EC-69 and soft EC-65 ceramics made by Edo Ceramics Inc. In Fig. 1, we show the variation of the piezoelectric constants of the two types of ceramic as a function of temperature. Full details, including the variation of the dielectric, elastic and electromechanical coupling coefficients can be found elsewhere [26]

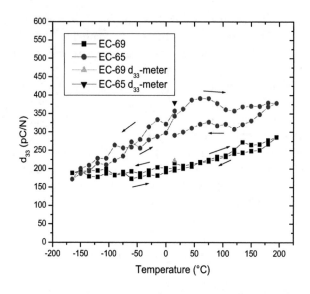

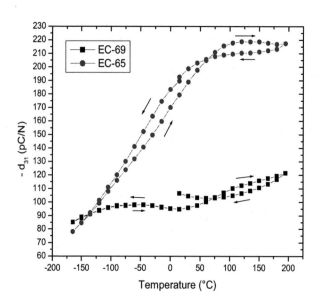

FIGURE 1. Temperature dependence of the piezoelectric coefficients d$_{33}$ (length extensional mode), -d$_{31}$ (radial mode) of soft (EC-65) and hard PZT (EC-69) samples.

ELECTRIC FIELD DEPENDENCE OF THE PIEZOELECTRIC RESPONSE

Although equation (1) suggests that the piezoelectric strain is a linear function of the applied field, this is only true when small fields are applied. As larger electric fields are applied in an effort to obtain larger strains, the response of the ceramic becomes non-linear. The phenomenological equations (1) can be uncoupled thereby allowing the strain and dielectric displacement to be monitored as a function of electric field, which facilitates the determination of the piezoelectric coefficients [27].

The variation of the piezoelectric constant d and the permittivity ε as a function of the applied electric field can be studied under two different experimental conditions: (a) quasistatic experiments that give the value of the constants under near DC conditions, and (b) resonance experiments that provide the material constants at the appropriate resonance frequency. The difference in the significance of these two types of experiments can be understood with reference to Fig. 2. The main curve in the figure shows a typical curve that is obtained when a varying electric field, below the coercive field, is applied to a piezoelectric material, and the resultant strain is measured. A measurable hysteresis is caused by irreversible domain wall motions that are characteristic of ferroelectric materials. Zhang et al [28] studied the onset of irreversibility in such materials and found that each material had a plateau region where the permittivity and the piezoelectric constant were independent of field and they attributed this field independence to reversible domain motions that characterise the intrinsic piezoelectric response. Under the application of a large electric field, the domain walls move to maintain a minimum in the domain energy and the some of the domains engulf other domains or change shape irreversibly, which contributes to the net strain and polarisation and constitutes the extrinsic piezoelectric response. The total piezoelectric response is the sum of the reversible and irreversible contributions. Figure 2 shows that the slopes of the two types of measurement are different and therefore the material constants found from these measurements cannot be the same. When a large low-frequency electric or stress field is applied and the corresponding hysteresis loop is measured, an <u>average</u> value of the dielectric or piezoelectric constant can be found from the average slope of the hysteresis curve. However, when a large DC electric or stress field is applied and a small AC field is used to measure the response under that DC field the slope of the AC response can be said to provide a <u>dynamic</u> value of the constant. This discussion underlines the importance of determining material constants under conditions appropriate for any particular application.

As will be clear from the above discussion, when small electric or stress fields are applied to a piezoceramic, no hysteresis is observed and a well defined slope yields values of either the piezoelectric constant or the permittivity resulting from intrinsic contributions. As the electric field is increased, domain effects begin to occur,

245

producing some hysteresis and the average slope increases as both intrinsic and extrinsic effects contribute to the piezoelectric and dielectric responses [29]. Such an average slope and average values of the material constants can only be defined when the hysteresis is not too pronounced and we have found that, in this regime, the average values of the piezoelectric and dielectric constants increase linearly with applied electric field [29].

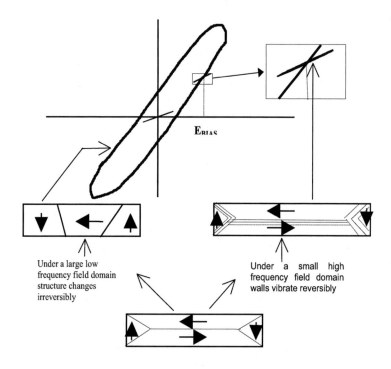

FIGURE 2. The relationship between the reversible and irreversible domain motion and the resultant strain S or electric displacement D as a function of applied stress T or electric field E.

Strain Response under Large Applied Electric Fields

When very high fields are applied and significant hysteresis is present, it becomes difficult to define an average slope. However the total strain/displacement can be measured as a function of applied electric field/stress and their ratio can be used to define an effective average material coefficient. Alternatively a large DC bias field may be applied and a small AC field can be used to find the response and to define a dynamic constant. The determination of the piezoelectric response under these conditions requires direct measurement of the strain. In our laboratory we have different experimental methods of measuring strains including the relatively simple optical lever [30], the differential variable reluctance transducer (DVRT) and laser Doppler interferometry. The latter is the most versatile of the direct strain measuring instruments as it allows the measurement of AC strains [31]. Our measurement system uses a Zygo ZMI 2000 heterodyne laser Doppler interferometer [32] with a resolution of 0.62 nm and it has been described elsewhere [33] The system may be used to directly measure linear strains as well as shear strains [33] and thus it allows us to determine all the piezoelectric d coefficients. Figure 3 shows the variation of the average values of the d_{33}, d_{31} and d_{15} as a function of applied electric fields up to about 1 MV/m for two types of lead zirconate titanate (PZT) ceramics: the fairly soft EC-65 and the hard EC-69 ceramics made by EDO Ceramics Inc. Soft PZT is characterised by a high piezoelectric constant and mobile domain boundaries [34] so that extrinsic contributions are inherently more important and their effect is increased at higher fields.[35] The observed increase in the d coefficients in the soft PZT, EC-65, is likely due to the larger extrinsic contribution resulting from increased domain switching under the influence of large fields. The nonlinearities observed are stronger than what would be predicted by the Rayleigh law as can be seen from the following polynomial fits to the curves for the EC-65 PZT:

$$d_{15} = 560 + 1170E - 205E^2 + 142E^3,$$
$$d_{33} = 455 + 529E - 153E^2 + 43E^3,$$
$$d_{31} = -192 - 312E + 126E^2 - 12E^3.$$

Although Fig. 3 does not show any significant non-linearity in the d coefficients of the hard EC-69 ceramic, this is only because the applied fields here were not high enough, as will be seen later, in Fig. 5. The acceptor doped hard PZT suppresses domain wall response so that higher fields are required for the same effects to be observed.

Strain Response under the Application of a DC Bias Field

Our laser Doppler interferometer system has also been used to study the DC bias dependence of the piezoelectric strain. Figure 4 shows the DC field dependence of the dynamic d_{33} for both types of piezoceramics at various frequencies. When a positive DC bias is applied to the well poled hard PZT, EC-69, a very small decrease in d_{33} is observed. However, when a negative DC bias is applied to this ceramic, the d_{33} increases significantly as a function of the bias field. For a well-poled ceramic, the bias field does not significantly increase the poling but rather helps to pin the domains leading to the observed small decrease in d_{33}. However, a negative bias causes de-pinning (de-ageing) [36] of the domain walls and this increases the extrinsic contribution to the piezoelectric strain and leads to higher values of d_{33}. In the case of the soft PZT, EC-65, a negative bias field initially causes a rise in the d_{33} due to similar de-pinning effects. However the soft PZT has low values of the coercive fields and when the negative fields reach this value the PZT depoles causing a sharp drop in the d_{33}. If the negative field is further increased the material now re-poles in the direction of the

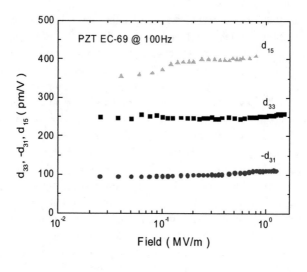

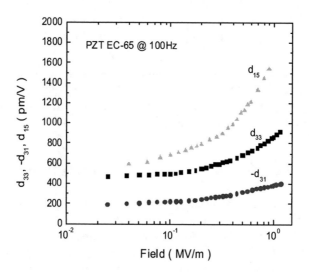

FIGURE 3. Piezoelectric d coefficients of hard EC-69 and soft EC-65 PZT ceramics as a function of an applied 100 Hz AC electric field.

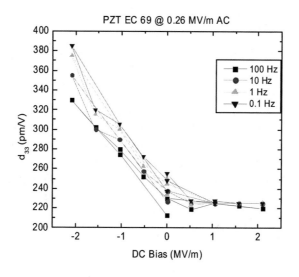

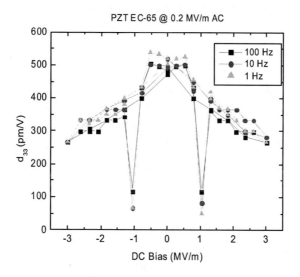

FIGURE 4: DC bias field dependence of the two types of PZT at various frequencies.

applied field, which now becomes a positive bias and contributes to domain pinning that reduces the d_{33}. Since extrinsic effects contribute more to the strain in soft PZT, these effects are seen to be larger in the case of the soft PZT.

Figure 5 shows the AC measuring field dependence of the dynamic d33 in the presence of various positive DC bias field levels. This figure shows measurements made up to AC fields of 5 MV/m. The hard EC-69 ceramic shows little dependence on the AC field or the DC bias field up to 1 MVm. However for higher AC fields the material shows a significant non-linear increase in d33 caused by the de-ageing and depoling processes that increase the extrinsic domain contributions but the non-linear increase is suppressed by the application of a positive DC bias, which stabilises the domain structure and inhibits de-ageing. In the case of the soft PZT, EC-65, also the non-linear response is seen to be suppressed by DC bias fields. The general result that a DC bias reduces the non-linear response in a piezoceramic is in agreement with earlier observations by Masuda and Baba [37].

High Frequency Properties under DC Bias

The methods of impedance analysis of piezoelectric resonators described earlier have been used to study the radial mode resonance of three types of PZT samples, EC-69, EC-65 and the very soft EC-76, when the samples were placed under DC bias fields that ranged up to ± 3 kV. PRAP software [20] was used to analyse the impedance curves and find the values of the elastic compliances s_{11}^E and s_{12}^E, the dielectric constant ε_{33}^T, the piezoelectric constant d_{31} and the planar electromechanical coupling constant k_p, all as a function of the DC bias field. Our results are shown in Fig. 6. It should be noted that the EC-76 ceramic is so soft that no meaningful measurements could be made with negative DC fields as very small negative fields were sufficient to depole these ceramics. The EC-65 ceramic depoled at negative fields of about –0.8 MV/m. The hard PZT, EC-69 showed no depoling up to the maximum DC fields applied. Similarly to what was observed in our low frequency interferometric measurements, the piezoelectric constant d31 decreases when a positive DC bias is applied to the specimen and it increases when a negative bias field is applied until depoling occurs. The constants s_{11}^E, ε_{33}^T and k_p also showed a similar behaviour. However, the elastic compliance s_{12}^E increased with the application of a positive DC bias and decreased with a negative bias before depoling occurred.

In summary, it can be said that a DC bias in the poling direction clamps the domains and reduces the strain. However as the AC field is increased the strain increases due to depinning and depoling effects. On the other hand a negative DC bias causes field-induced de-ageing (unclamping of domain walls) that results in an increase in the

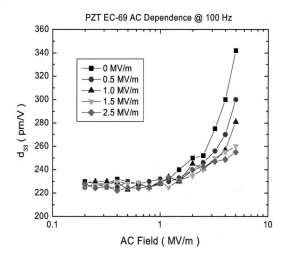

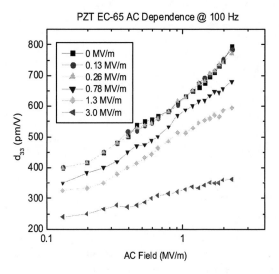

FIGURE 5. AC field dependence of the d_{33} for the two types of PZT under positive DC bias fields (in the same direction as the poling).

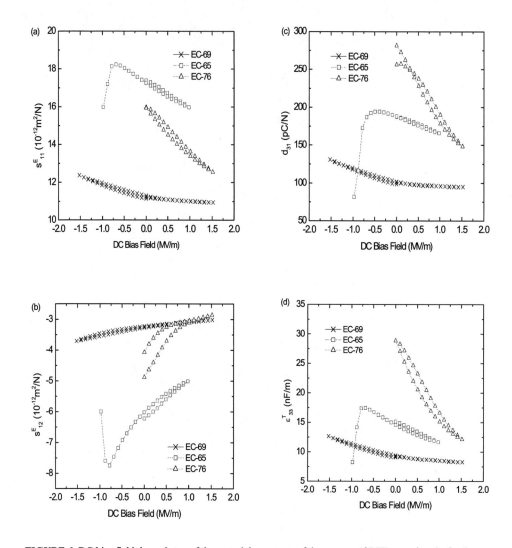

FIGURE 6. DC bias field dependence of the material constants of three types of PZT ceramics obtained from radial mode resonance measurements.

piezoelectric constant, the dielectric constant and the electromechanical coupling constant. Fuller details of our measurements are available elsewhere [33].

STRESS DEPENDENCE OF THE PIEZOELECTRIC RESPONSE

It has been shown that mechanical stress can cause re-orientation of polarisation and depoling in piezoelectric ceramics [38,39,40] Equation 1 shows that if the electric field is kept constant or is set to zero (short circuit condition), the dielectric displacement and the strain can be measured and the piezoelectric, dielectric and elasticconstants can be determined as a function of stress.[27,41]. Our measurements were carried out on the soft EC-65 and the EC-69 ceramics mentioned above. The test samples were 5 mm x 5 mm x 8 mm. In order to ensure that the stress was uniaxial and to reduce any lateral clamping effects [3,36], two additional 6mm. long pieces of the same material were stacked on either side of the sample. The basic experimental set-up used in our laboratory for making these measurements has been described earlier [27]. A Monsanto tensometer is used in the compression mode and exerts a uniaxial stress on the sample assembly mounted between two insulators. A semi-circular base plate ensures that the stress is normal to the specimen. A Keithley 617 electrometer measures the current that is generated when a stress is applied to the specimen. The stress on the sample is determined by using the tensometer force head, whose output is read by a Keithley 199 digital multimeter and all data are fed into a computer. Under short circuit conditions, E = 0, and the electric displacement is given by the integral of the current with respect to time:

$$D_3 = d_{33}T_3 = \frac{1}{A}\int I dt \text{ so that } d_{33}(T_3) = \frac{\int I dt}{AT_3}, \tag{2}$$

where A is the electroded cross-sectional area of the sample. As discussed above, this will give the average value of d_{33} up to the stress level T_3. A differential d_{33} can be found from the slope of the curve giving the electric displacement as a function of the applied stress. Since the piezoelectric response takes a finite time, as will be discussed in the next section, and exhibits hysteresis, the values of the piezoelectric constant determined will depend on the ramping rate of the stress and on the previous stress level experienced by the sample.

The experimental set-up used to to measure the dynamic response to a small AC stress superimposed on a DC pre-stress was based on the method described by Audigier et al.[42] A piezoelectric actuator and a piezoelectric dynamic force sensor were inserted between the sample and one of the insulators. The actuator was used to produce a small AC force and this force was detected by the force sensor. The electric displacement was determined as described above. The dielectric constant was measured at 1 kHz by using an HP 4284 LCR meter. The strain response to the AC stress was measured by a strain gauge and the elastic constant could thus be determined. All signals were measured by lock-in amplifiers to improve the signal to noise ratio. The dynamic values were measured with a 0.2 Hz AC stress with an amplitude of 1.5 MPa. The DC pre-stress was increased in steps of 10 MPa up to 160 MPa and the AC measurement was taken five minutes after each step increase in the pre-stress to give time for the full piezoelectric response to develop (see the following section).

Due to reasons of space only our dynamic measurements are discussed here. Figure 7 shows the experimental variation of the piezoelectric constant d_{33}, the elastic constant s_{33}^{E}, the dielectric constant ε_{33}^{T} and the electromechanical coupling coefficient k_{33}, all as a function of the pre-stress for the hard EC-69 and the soft EC-65 PZT ceramics. When a stress is applied to the hard EC-69 ceramic, domain switching can lead to the production of new non-180° domain walls.[43,44] Also, the stress can cause de-ageing (de-pinning of domain walls) and both effects lead to an increase in d_{33} as seen in Fig. 7. This increase in the extrinsic contribution also causes the dielectric constant and the loss to increase, as observed earlier [36'45] and shown in Fig. 7. As the stress is further increased the domain walls become progressively clamped by the stress and causes a decrease in the d_{33}. The figure also shows that when the stress is released the value of d_{33} recovers and indeed after a full stress cycle, the EC-69 ceramic shows a net increase in d_{33}. This confirms that there has been no depoling and the decrease in d_{33} under high stress is due to stress-induced clamping of domain walls. When the stress is removed the effects of de-ageing remain while no more clamping occurs thus causing a net increase in the response. If the specimen is left to age the zero stress response reduces to the well-aged value. Just like EC-69, the soft EC-65 also shows an initial increase in d_{33} as the pre-stress is increased but the d_{33} begins to decrease at a fairly low stress level and there is little recovery of the d_{33} when the stress is released. It is clear that in this case the stress has caused significant de-poling. This is the dominant effect at high stress, although a small recovery of d_{33} when the stress is being reduced suggests that some reversible stress-induced clamping of domain walls has also occurred. But it is clearly stress-induced de-poling that causes the dramatic reduction in d_{33} in EC-65 ceramic after a stress cycle. The significant drop in s_{33}^{E} after a full cycle is further evidence of the depoling of the specimen. On the other hand EC-69 shows little change in s_{33}^{E} but it shows a strong increase in its dielectric constant ε_{33}^{T} that is normally an indication of de-ageing in the material.[36] Our measurements of the average and differential values of d_{33} as a function of stress have been described elsewhere [27].

TIME DEPENDENCE OF THE PIEZOELECTRIC RESPONSE AND ACTIVATION ENERGIES OF PZT CERAMICS

As far back as 1959 it had been reported [46] that the piezoelectric response of a material was dependent on the rate of application of the electric field. Using an experimental method similar to that described in the previous section, we have found

254

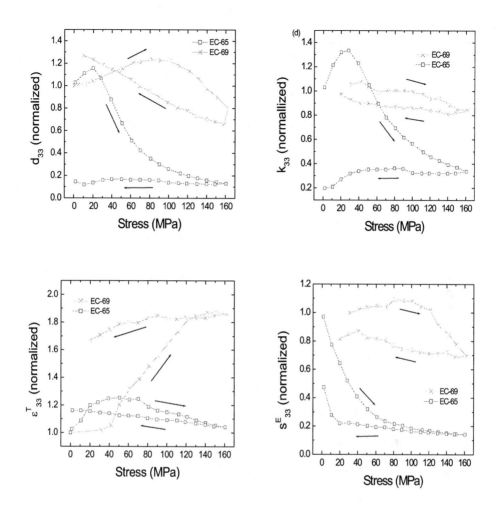

FIGURE 7. The stress dependence of the electromechanical properties of EC-69 and EC-65 PZT ceramics.

the time dependence of the piezoelectric response of PZT ceramics by applying a delta function stress step and measuring the polarisation produced by the piezoelectric response as a function of time. As we have reported earlier [47], the piezoelectric response to the delta function stress step is made up of two parts: a fast response that corresponds to the intrinsic, 180° domain changes, and a slow increase that corresponds to the extrinsic, non-180° domain changes. The extrinsic contribution continues to increase several hundred seconds after the application of the step stress. A plot of the extrinsic polarisation response as a function of the logarithm of time yields a straight line whose slope, m, is found to be a function of temperature, T. This suggests that the domain wall motions that contribute to the extrinsic response are thermally activated. Clearly a ceramic material would have a distribution of activation energies. Arrhenius plots of $\ln(mT)$ vs. $1/T$ have allowed us to determine the average activation energies of various PZT ceramics. Typical values of the average activation energies obtained are 0.29 eV for a Navy Type I PZT ceramic and 0.62 eV for a Navy Type III PZT ceramic.[47] The time dependence of the piezoelectric response is responsible for the frequency dependence of the piezoelectric constant. A DC measurement can allow the piezoelectric response to develop fully. However as the frequency increases, smaller numbers of domains can change in the time available and the piezoelectric constant decreases.

CHARACTERISATION OF THE NEW, HIGH STRAIN, PZN-PT AND PMN-PT SINGLE CRYSTALS

<001> oriented single crystals of the relaxor ferroelectric solid solution $Pb(Zn_{1/3}Nb_{2/3})O_3$-$PbTiO_3$ (PZN-PT) are known to exhibit high piezoelectric constants ($d_{33} > 2000$ pC/N), high electromechanical coupling ($k_{33} > 0.9$) and electric-field-induced strains up to 1.7% [48'49] and are therefore very promising for transducer and actuator applications. The very high strain values of these crystals are partly due to an applied electric-field-induced rhombohedral to tetragonal hysteretic phase transition.[49,50] We have investigated the strain response of two compositions of these PZN-PT crystals with PT contents of 4.5% and 8%, designated respectively as PZN-4.5%PT and PZN-8%PT. The crystals were grown by TRS Ceramics Inc. using a flux method and the samples provided to us had the dimensions of 5 mm x 5 mm x 1 mm and sputtered gold electrodes had been applied to them.

In a first experiment the strains produced by the crystals in the poling direction were measured as a function of a 0.2 Hz unipolar electric field and of temperature. A differential variable reluctance transducer was used to measure the strains while the samples were immersed in silicon oil that could be heated from 25° C to 105° C. Figure 8 shows the strain response of the <001> oriented PZN-8%PT crystals as a function of a unipolar electric field at various temperatures. At room temperatures the poled crystals are in a stable multi-domain rhombohedral state with <111> polar directions, called engineered domain configuration [49], and little hysteresis is observed in the strain response. As the applied field is increased, hysteresis loops appear indicating the occurrence of an electric-field-induced phase transition from the rhombohedral to a tetragonal single domain state with a <001> polar direction. The phase transition occurs at an electric field that decreases linearly with temperature [51] and at temperatures higher than 75°C PZN-8%PT crystals are already in the tetragonal phase without the need for any applied field. Similar results have been obtained for PZN-4.5%PT crystals [51]. The change in longitudinal strain, ΔS_{33}, due to the phase transition is almost independent of temperature in the temperature range investigated, although the absolute strain value and the phase transition fields decrease as the temperature increases. We have found $\Delta S_{33} \sim 0.4\%$ for the PZN-8%PT crystals in agreement with values obtained by Viehland [52] from bipolar field measurements. Figure 9 shows the temperature dependence of the piezoelectric constant d_{33} in the rhombohedral phase and in the field-induced tetragonal phase of PZN-4.5%PT, as calculated from our direct strain measurements. In the rhombohedral phase, d_{33} increases with temperature in both crystals [51] and in the case of PZN-4.5%PT, the rate of increase itself increases significantly as the temperature rises above 60° C as shown in Fig.9. In the tetragonal phase the d_{33} does not change very significantly as a function of temperature.

Given that the rhombohedral phase provides the greater strain, the strain response of this phase has been further investigated as a function of AC applied fields and also as a function of DC bias fields. In this case the strains were measured directly with the Zygo laser Doppler interferometer that has been mentioned earlier and the d_{33} coefficient was derived from the strain measurements. Figure 10 shows the variation of the d_{33} of the two types of crystal as a function of applied field. At low applied fields the strain is non-hysteretic and varies linearly with electric field. As the electric field is increased beyond a threshold value E_t, hysteresis appears and the slope of the strain curve increases. This non-linear behaviour appears to be related to domain wall motion, as is the case in piezoelectric ceramics [53], and the hysteresis is an indication of losses due to domain wall motions. E_t has been defined as the field at which $\Delta d_{33}/d_{33}$ begins to exceed 2.5%, as was done by Li et al.[53]. The E_t values obtained, 1.58 kV/cm for PZN-8%PT and 1.33 kV/cm for PZN-4.5%PT, were much higher than for soft PZT ceramics [53], which indicates the existence of a metastable domain configuration in the single crystal PZN-8%PT shows greater non-linearity than PZN-4.5%PT and this is likely due to the fact that PZN-8%PT is closer to the morphotropic phase boundary (MPB) [48] so that domain walls move more easily and more significantly in response to the external field. It is also possible that a newly discovered phase [54] around the MPB may also play an important role in the non-linear behaviour of the PZN-8%PT crystals. The dielectric constant of the crystals, as determined from polarisation measurements varies as a function of the AC field in a manner similar to the variation of d_{33}.[55]. Figure 10 shows that the d_{33} measured with increasing fields is somewhat greater than the values

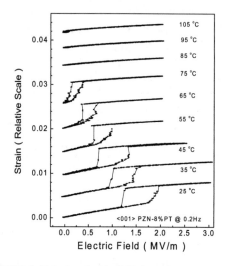 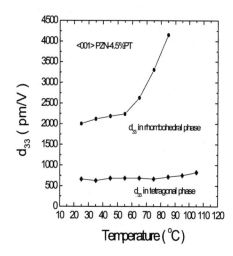

FIGURE 8. Unipolar electric field dependence of the strain response of PZN-8%PT single crystals at various temperatures.

FIGURE 9. Temperature dependence of d_{33} in the rhombohedral phase and the tetragonal phase respectively in PZN-4.5%PT single crystals.

found when the field is being decreased. This clearly indicates that irreversible domain motions have occurred and that the crystals are partially depoled even though the maximum field applied may be lower than the coercive fields of the crystals. The possibility of depoling severely limits the high power performance of the piezocrystals. We have investigated the possibility of using a DC bias to stabilise the domain configuration and prevent the depoling caused by the application of large AC fields. Figure 11 shows the d_{33} coefficients of the two types of crystals, calculated from the strain responses, as a function of DC bias field. Clearly the region of linear response increases with DC bias and Fig. 12 shows that the E_t increases linearly with DC bias. Our measurements show that a positive DC bias can effectively stabilise the domain configuration and prevent the depoling caused by the application of large AC fields. Figure 11 shows the d_{33} coefficients of the two types of crystals, calculated from the strain responses, as a function of DC bias field. Clearly the region of linear response increases with DC bias and Fig. 12 shows that the E_t increases linearly with DC bias. Our measurements show that a positive DC bias can effectively stabilise the domain configuration in PZN-PT crystals and enhance the field interval over which their piezoelectric response is linear without significantly reducing the piezoelectric response of the crystal as happens in the case of piezoelectric ceramic which their piezoelectric response is linear without significantly reducing the piezoelectric response of the crystal as happens in the case of piezoelectric ceramics [53]. Although the PZN-PT piezocrystals poled along the <001> direction have large values of the longitudinal piezoelectric constant d_{33}, they have relatively poor shear strains with $d_{15} \sim$ 140 pC/N [56]. But Zhang et al [57] have reported that PZN-PT crystals poled along the <111> direction show large shear strains but the values obtained by them, using resonance measurements, depended on approximate calculations since the elastic compliances required for a proper calculation were not available. Our Zygo laser Doppler interferometer system has been used to directly measure the shear strains in these materials for a 1 Hz AC excitation. From these strain measurements we have the following, very high, values of the piezoelectric shear coefficient: $d_{15} \sim 5500$ pC/N for the PZN-4.5%PT crystal and $d_{15} \sim 7500$ pC/N for the PZN-8%PT crystal. The effects of DC bias field and AC frequency on the shear piezoelectric coefficient of <111> oriented PZN-PT crystals have also been studied. Figure 13 shows that the d_{15} coefficient decreased as the DC bias field increased. The d_{15} was also found to decrease as the AC frequency increased but the frequency dependence itself decreased as the DC bias was increased. These observations suggested that an increasing DC bias caused the crystal to approach a single domain state. Further details of our observations have been published elsewhere [58].

Similar measurements have been made on two compositions of single crystals in the lead magnesium niobate – lead titanate family [59].

FINITE ELEMENT MODELING OF PIEZOELECTRIC MATERIALS AND DEVICES

Many efforts have been made to develop finite element modeling of piezoelectric materials and devices for virtual prototyping. Of these the most successful is perhaps PZFlex, the FEM tool developed by Weidlinger Associates with funding from the US Navy [60]. This programme has the advantages of being of moderate size, of being able to deal with elastic and acoustic composite materials including a small amount of active piezoelectric materials, of being able to model transient behaviour with input and output signals occurring naturally in the time domain and the potential to include non-linear and thermal effects. By use of Fourier analysis the time domain solutions contain all necessary frequency information It has been extensively used in the development of underwater sonar, medical imaging and NDE transducers. Virtual prototyping using PZFlex speeds up design cycles and improves understanding and functionality. As an example of the static modeling possible, Fig. 14 shows the modeling of piezocomposite impedance for a 1-3 polymer-ceramic composite. Note that the modeling accurately picks up even small resonances.

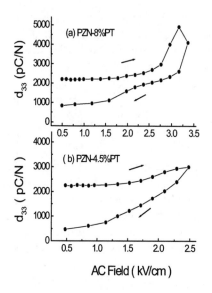

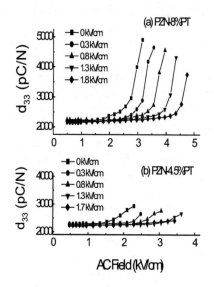

FIGURE 10. AC field dependence of the piezoelectric d_{33} for (a) PZN-8%PT and (b) PZN-4.5%PT, first with increasing field and then with decreasing field.

FIGURE 11. Effective piezoelectric coefficient d_{33} of (a) PZN-8%PT and (b) PZN-4.5%PT crystals, as a function of a 1 Hz electric field under various DC bias fields. The bias fields increase from left to right.

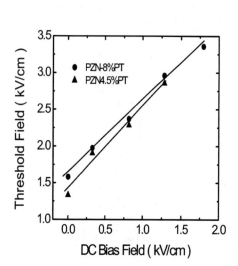

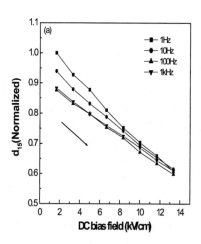

FIGURE 12. Threshold AC fields for the onset of non-linearity in PZN-8%PT and PZN-4.5%PT crystals as a function of DC bias. as a function of DC bias.

FIGURE 13: .DC bias field dependence of d_{15} for <111> oriented PZN-8%PT single crystals at four different frequencies (1 Hz, 10 Hz, 100 Hz, and 1 kHz) as DC bias is increased.

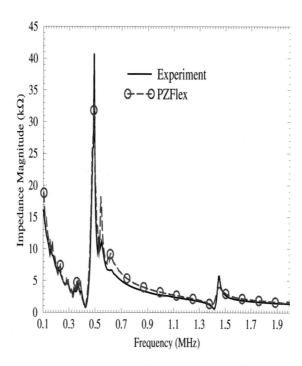

FIGURE 14. Comparison between PZ Flex modelling prediction of the impedance around resonance for a piezoelectric resonator compared with experimental measurements.

CONCLUSIONS

This paper has presented experimental observations to show that, except when very small excitation signals are applied, the piezoelectric, dielectric and elastic responses of a piezoelectric ceramic are essentially non-linear. The material constants have to be expressed as complex coefficients in order to take account of all the losses in the material and they have also been shown to change as a function of temperature and frequency. The high strain PZN-PT single crystals have also been found to have non-linear properties. It is clear that, from the point of view of an applications/design engineer, it would be appropriate to have the material characterised under the same conditions as the material would be exposed to during the application. For reasons of space, we have only summarised the work carried out in our laboratory. A more general survey of some of the non-linear properties has been given by Gonnard [61] and by Damjanovic and Robert.[62].

ACKNOWLEDGEMENTS

The author thanks Weidlinger Associates and Dr. Paul Reynolds for furnishing figure 14. He thankfully acknowledges the contributions of his research associates and students and he gratefully acknowledges funding support from the US Office of Naval Research, the Canadian Department of National Defence and the National Science and Engineering Research Council of Canada.

REFERENCES

1. A.F. Devonshire, *Phil. Mag. Supp.*, **3**, 85 (1952).
2. W.P. Mason, *Physical Acoustics and the Properties of Solids*, 1958, Princeton, N.J., D. Van Nostrand Co. Inc.
3. Q.M. Zhang, J. Zhao, K. Uchino and J. Zheng, *J. Mater. Res.*, **12**, 226 (1997).
4. IEEE Standard on Piezoelectricity (1987), ANSI/IEEE Standard 176-1987.
5. *PZFlex*, from Flex Support, Weidlinger Associates, Inc., 4410 El Camino Real, Suite 110, Los Altos, CA 94022, USA, see http://www.wai.com/AppliedScience/Software/Pzflex/index-pz.html or contact Dr. Paul Reynolds at reynolds@wai,com
6. D.A. Berlincourt, D.R. Curran and H. Jaffe, Chapter 3 in *Physical Acoustics I Part A*, edited by W.P. Mason, pp.169-270, Academic Press: 1964,. pp.169-270.
7. A.H. Meitzler, H.M. O'Bryan and H.F. Tiersten, *IEEE Trans. on Sonics and Ultrasonics*, **SU-20**, 233 (1973).
8. R. Holland, *IEEE Trans. on Sonics and Ultrasonics*, **SU-14**, 18 (1967).
9. S. Sherrit, Chapter 15 in *Electroactive Polymer (EAP) Actuators as Artificial Muscles: Reality, Potential and Challenges*, edited by Y. Bar-Cohen, Bellingham, WA, SPIE Press, 2001..
10. R. Holland and E.P. Eernisse, *IEEE Trans. on Sonics and Ultrrasonics*, **SU-16**, 173 (1969).
11. H. Ohigashi, T. Itoh, K. Kimura, T. Nakanishi and M. Suzuki, *Jap. J. Appl. Phys.*, **27**, 354 (1988).
12. T. Tsurumi, T. Ichihara, K. Asaga and M. Daimon, *J. Amer. Cer. Soc.*, **73(5)**, 1330 (1990).
13. J.G. Smits, *IEEE Trans. on Sonics and Ultrrasonics*, **SU-23**, 393 (1976).
14. S. Sherrit, H.D. Wiederick and B.K. Mukherjee, *Ferroelectrics*, **134**, 111 (1992).
15. S. Sherrit, N. Gauthier, H. D. Wiederick and B. K. Mukherjee, *Ferroelectrics*, **119**, 17 (1991).
16. C. Alemany, A.M. Gonzalez, L. Pardo, B. Jimenez, F. Carmona and J. Mendiola, *J. Phys. D: Appl. Phys.*, **28**, 945 (1995).
17. B.K. Mukherjee and S. Sherrit, *Proc. of the Fifth Int. Congress on Sound and Vibration*, **Vol.1**, 385 (1997), The International Institute of Acoustics and Vibration, Dept. of Mech. Eng., Auburn University, AL 36849-5341, USA.
18. K.W. Kwok, H.L.W. Chan and C.L. Choy, *IEEE Trans. on Ultrasonics, Ferroelectrics, and Frequency Control*, **44,** 733 (1997).
19. L. Pardo, M. Alguero and K. Brebol, *Advances in Science and Technology*, **45**, 2448-2558 (2006).
20. PRAP (Piezoelectric Resonance Analysis Programme) software available from TASI Technical Software Inc. See www.tasitechnical.com
21. S. Sherrit and B.K. Mukherjee, *Proc. of 1998 IEEE International Ultrasonics Symposium*, IEEE, Piscataway, NJ ,633 (1998).
22. S. Sherrit, H.D. Wiederick and B.K. Mukherjee, *Medical Imaging1997: Ultrasonic Transducer Engineering (SPIE Proceedings* **Volume 3037***)*, 158 (1997). SPIE Press, Bellingham, WA., USA.
23. S. Sherrit, J.E. Haysom, H.D. Wiederick, B.K. Mukherjee and M. Sayer, *Proc. of the Tenth Int. Symp. on the Applications of Ferroelectrics, ISAF'96*, IEEE, Piscataway, N.J., USA., 959 (1996).
24. S. Sherrit, H.D. Wiederick, B.K. Mukherjee and M. Sayer, *J. Phys.D: Appl. Phys.*, **30**, 2354 (1997).
25. S. Sherrit, H.D. Wiederick and B.K. Mukherjee, *Proc. of 1997 IEEE Ultrasonics Symposium*, IEEE, Piscataway, NJ (1997), **Vol.2**, 931 (1997).
26. R.G. Sabat and B.K. Mukherjee, *J. Appl. Phys.* **101**, 064111: 1-7, (2007)..
27. G. Yang, S-F. Liu, W. Ren and B.K. Mukherjee, *Smart Structures and Materials 2000: Smart Materials Technologies (SPIE Proceedings* **Volume 3992**), SPIE Press, Bellingham, WA., USA. 103 (2000).
28. Q. M. Zhang, W.Y. Pan, S.J. Jang and L.E. Cross, *J. Appl. Phys.* **64**, 6445 (1988).
29. S. Sherrit, H.D. Wiederick, B.K. Mukherjee and M. Sayer, *Smart Structures and Materials 1997: Smart Materials Technologies (SPIE Proceedings* **Volume 3040***)*, SPIE Press, Bellingham, WA., USA, 99 (1997).

30 .H.D. Wiederick, S. Sherrit, R.B. Stimpson and B.K. Mukherjee, *Ferroelectrics*, **186**, 25 (1996).

31. K.M. Rittenmyer and P.S. Dubbelday, *J. Acoust. Soc. Am.*, 91, 2254 (1992).

32. Zygo Corp., Laurel Brook Rd., Middlefield, CT 06455-0448, USA.

33. A.J. Masys, W. Ren, G. Yang and B.K. Mukherjee, *J. Appl. Phys.*, **94**, 1155 (2003).

34. Q.M. Zhang, W.Y. Pan and L.E. Cross, *J. Appl. Phys.*, **63**, 2492 (1988).

35. Q.M. Zhang, H. Wang, N. Kim and L.E. Cross, *J. Appl. Phys.*, **75**, 454 (1994).

36. Q.M. Zhang and J. Zhao, *IEEE Trans. on Ultrasonics, Ferroelectrics, and Frequency Control*, **46**, 1518 (1999).

37. Y. Masuda and A. Baba, *Jpn. J. Appl. Phys.*, **24 (Suppl 3)**, 113 (1985).

38. A.G. Luchaninov, A.V. Shil'nikov, L.A. Shuvalov and V.A. Malyshev, *Ferroelectrics*, **145**, 235 (1993).

39. H. Cao and A. Evans, *J. Amer. Cer. Soc.*, **76**, 890 (1993).

40. S.C. Hwang, C.S. Lynch and R.M. McMeeking, *Acta Metall. Mater.*, **43**, 2073 (1995).

41. D. Guyomar, D. Audigier and L. Eyraud, *Proc. of the Eleventh Int. Symp. on the Applications of Ferroelectrics, ISAF'98*, 307 (1998). IEEE, Piscataway, N.J., USA.

42. D.Audigier, C. Richard, C. Descamps, M. Troccaz and L. Eyraud, *Ferroelectrics*, **154**, 219 (1994).

43. A. Schaufele and K.H. Hardtl, *J. Amer. Cer. Soc.*, **79**, 2637 (1996).

44. C.S. Lynch, *Acta Mater.*, **44**, 4137 (1996).

45. H.H.A. Krueger, *J. Acoust. Soc. Am.*,**42**, 636 (1967).

46. D. Berlincourt and H.H.A. Krueger, *J. Appl. Phys.*, **30**, 1804 (1959).

47. S. Sherrit, D.B. Van Nice, J.T. Graham, B.K. Mukherjee and H.D. Wiederick, *Proc. of the Eighth Int. Symp. on the Applications of Ferroelectrics, ISAF'92*, 1005 (1992). IEEE, Piscataway, N.J., USA.

48. J. Kuwata, K. Uchino and S. Nomura, *Ferroelectrics*, **37**, 579 (1981).

49. S.E. Park and T.R. Shrout, *J. Appl. Phys.*, **82**, 1804 (1997).

50. S-F. Liu, S.E. Park, T.R. Shrout and L.E. Cross, *J. Appl. Phys.*, **85**, 2810 (1999).

51. W. Ren, S-F. Liu and B.K. Mukherjee, *Appl. Phys. Letts.*, **80**, 3174 (2002).

52. D. Viehland, *J. Appl. Phys.*, **88**, 4794 (2000).

53. S. Li, W. Cao and L.E. Cross, *J. Appl. Phys.*, **69**, 7219 (1991).

54. B. Noheda, D.E. Cox, G. Shirane, S.E. Park, L.E. Cross and Z. Zhong, *Phys. Rev. Lett.*, **86**, 3891 (2001).

55. W. Ren, S-F. Liu and B.K. Mukherjee, *Appl. Phys. Lett.*, **83**, 5268 (2003).

56. J.H. Yin, B. Jing and W. Cao, *IEEE Trans. on Ultrasonics, Ferroelectrics, and Frequency Control*, **47**, 285 (2000).

57. S.J. Zhang, L. Laurent, S-F. Liu, S. Rhee, C.A. Randall and T.R. Shrout, *Jpn. J. Appl. Phys.*, **41**, L1099 (2002).

58. S-F. Liu, W. Ren, B.K. Mukherjee, S.J. Zhang, T.R. Shrout, P.W. Rehrig and W.S. Hackenberger, *Appl. Phys. Lett.*, **83**, 2886 (2003).

59. W. Ren, A.J. Masys, G. Yang and B.K. Mukherjee, *J. Phys. D: Appl. Phys.*, **35**, 1550-1554 (2002).

60.. Weidlinger Associates Inc., 399 W., El Camino Real, #200, Mountain View, CA 94022, USA.: www.wai.com (e-mail: reynolds@wai.com)

61. P. Gonnard, Chapter 16 in *Piezoelectric Materials and Devices*, Ed.: N. Setter, Ceramics Laboratory, EPFL Swiss Federal Institute of Technology, Lausanne 1015, Switzerland, (2002).

62. D. Damjanovic and G. Robert, Chapter 17 in *Piezoelectric Materials and Devices*, Ed.: N. Setter, Ceramics Laboratory, EPFL Swiss Federal Institute of Technology, Lausanne 1015, Switzerland, (2002).

Models and Characterization of Piezoelectric Ceramics

D. D. Ebenezer

[a]Naval Physical and Oceanographic Laboratory
Kochi 682021, INDIA
Phone +91 484 2571703; Fax: +91 484 2424858; email: tsonpol@vsnl.com

Abstract. Selected literature on the use of linearized constitutive equations and complex piezoelectric coefficients to develop models and characterize materials is presented in brief. One-dimensional models of simple piezoelectric ceramic components are derived using linearized constitutive equations and other fundamental relations to illustrate the procedure used to develop multi-dimensional models. Then, the procedure for building models of multi-layer devices using models of components is illustrated. Models are also used to characterize materials. A method to determine the errors when the ANSI/IEEE Standard method is used to determine real piezoelectric coefficients and methods to characterize non-Standard shapes often used in devices are also presented

Keywords: piezoelectric ceramics, constitutive models, linearized models

INTRODUCTION

Models of polarized piezoelectric ceramic components are used for modeling devices and characterization of materials. They are obtained using constitutive models of the material, strain-displacement relations, equilibrium equations, boundary conditions, and initial conditions or excitations. Approximations can be used to develop the model if the characteristics important for the application of interest can be explained by using the model. For example, linearized constitutive equations are sufficient if the strain and electric field are small and complex piezoelectric coefficients can be used if only steady-state conditions are of interest.

The following are presented here: various forms of the piezoelectric equations of state and the relationship between the various coefficients; conditions satisfied by the imaginary parts of the complex coefficients; procedures for developing analytical models of piezoelectric components of simple shape and devices with several components; software packages for numerical models; errors in determining the real piezoelectric coefficients by using the ANSI/IEEE Standard method; and methods to determine the complex coefficients of some shapes that are not included in the Standard methods.

CP1029, *Smart Devices: Modeling of Material Systems, An International Workshop*
edited by S. M. Sivakumar, V. Buravalla, and A. R. Srinivasa
©2008 American Institute of Physics 978-0-7354-0553-0/08/$23.00

PIEZOELECTRIC EQUATIONS OF STATE

Linearized Equations

The linearized constitutive equations of state of elastic solids [1] can be expressed as

$$\begin{bmatrix} T_{xx} & T_{yy} & T_{zz} & T_{xy} & T_{yz} & T_{zx} \end{bmatrix}_t = [C^e] \begin{bmatrix} S_{xx} & S_{yy} & S_{zz} & S_{xy} & S_{yz} & S_{zx} \end{bmatrix}_t \qquad 1.$$

where T_{ij} and S_{ij} are the components of stress and strain respectively and $[\ ]_t$ denotes the transpose of a matrix. $[C^e]$ is a 6x6 symmetric matrix of coefficients where the superscript e is used for elastic materials. Therefore, in the most general case, elastic solids are defined by 21 coefficients. In the case of isotropic solids, 2 coefficients completely define the material and the non-zero elements of the $[C^e]$ matrix are

$$c_{12} = c_{13} = c_{21} = c_{23} = c_{31} = c_{32} = \lambda,$$
$$c_{44} = c_{55} = c_{66} = \mu, \qquad 2.$$
$$c_{11} = c_{22} = c_{33} = \lambda + 2\mu$$

where λ and μ are Lame's constants. Other pairs of coefficients are in common use: for example, Young's modulus, Y, and Poisson's ratio, σ where

$$[Y, \ \sigma] = \left[\frac{\mu(3\lambda + 2\mu)}{\lambda + \mu} \quad \frac{\lambda}{2(\lambda + \mu)} \right]. \qquad 3.$$

Similarly, the linearized constitutive equations of state [2] for piezoelectric materials can be expressed as

$$\begin{bmatrix} T_{xx} & T_{yy} & T_{zz} & T_{yz} & T_{zx} & T_{xy} & D_x & D_y & D_z \end{bmatrix}_t =$$
$$[C^p] \begin{bmatrix} S_{xx} & S_{yy} & S_{zz} & S_{yz} & S_{zx} & S_{xy} & E_x & E_y & E_z \end{bmatrix}_t \qquad 4.$$

where D_x, D_y and D_z are the components of charge density and E_x, E_y and E_z are components of the electric field. $[C^p]$ is a 9x9 symmetric matrix where the superscript p is used for piezoelectric materials. It is expressed as

$$[C^p] = \begin{bmatrix}
c_{11}^E & c_{12}^E & c_{13}^E & 0 & 0 & 0 & 0 & 0 & -e_{31} \\
c_{12}^E & c_{11}^E & c_{13}^E & 0 & 0 & 0 & 0 & 0 & -e_{31} \\
c_{13}^E & c_{13}^E & c_{33}^E & 0 & 0 & 0 & 0 & 0 & -e_{33} \\
0 & 0 & 0 & c_{44}^E & 0 & 0 & 0 & 0 & 0 \\
0 & 0 & 0 & 0 & c_{44}^E & 0 & 0 & -e_{15} & 0 \\
0 & 0 & 0 & 0 & 0 & 0.5(c_{11}^E - c_{12}^E) & -e_{15} & 0 & 0 \\
0 & 0 & 0 & 0 & e_{15} & 0 & \varepsilon_{11}^S & 0 & 0 \\
0 & 0 & 0 & e_{15} & 0 & 0 & 0 & \varepsilon_{11}^S & 0 \\
e_{31} & e_{32} & e_{33} & 0 & 0 & 0 & 0 & 0 & \varepsilon_{33}^S
\end{bmatrix} \qquad 5.$$

where $c_{11}^E, c_{12}^E, c_{13}^E, c_{33}^E$ and c_{44}^E are the elastic stiffness coefficients, e_{31} and e_{33} are the piezoelectric stress coefficients, ε_{33}^S is the dielectric permittivity coefficient. 10 coefficients are necessary to completely define a polarized piezoelectric ceramic.

It is convenient to write the above equations of state as a pair of equations in the form

$$T = c^E S - e_t E$$
$$D = eS + \varepsilon^S E$$

6.

where the matrices c and ε are used with the superscripts E and S, respectively because strain and electric field are used as independent variables. This is the most convenient form of the equations of state when addressing problems in which none of the components of the field variables can be assumed to be zero because it can be directly used in the dynamic equilibrium equations and the Gauss electrostatic condition after expressing the strain in terms of displacement.

The equation of state is also expressed in other forms that are sometimes convenient:

$$T = c^D S - h_t D$$
$$E = -hS + \beta^S D$$

7a.

$$S = s^E T + d_t E$$
$$D = dT + \varepsilon^T E$$

7b.

and

$$S = s^D T + g_t D$$
$$E = -gT + \beta^T D$$

7c.

where the notation is apparent from the context. A particular form is useful if one of the components of the field variables on the right hand side is zero everywhere. Eqs. 6 and 7a are convenient in plane-strain conditions whereas Eqs. 7b and 7c are useful in plane-stress conditions. Eqs. 7a and 7b have been used in the analysis of radially polarized cylindrical shells while Eqs. 6 and 7c have been used in the analysis of axially polarized cylinders and cylindrical shells, respectively.

The coefficients in the various matrices are related to each other. For example,

$$d = \varepsilon^T g = es^E,$$

8a.

$$g = \beta^T d = hs^D,$$

8b.

$$e = \varepsilon^S h = dc^E,$$

8c.

$$h = \beta^S e = gc^D,$$

8d.

$$\beta^T = [\varepsilon^T]^{-1}, \quad \beta^S = [\varepsilon^S]^{-1},$$

8e.

$$[c^E]=[s^E]^{-1}, \quad [c^D]=[s^D]^{-1},$$

8f.

$$\varepsilon^S = \varepsilon^T - es^E e_t,$$

8g.

and

$$s^D = s^E - g_t \varepsilon^T g.$$

8h.

These relations are often needed for comparing different models.

Internal Losses

Internal losses give rise to hysteresis loops and internal heating. The effect of internal losses can be modeled using complex coefficients and the linearized governing equations under steady-state conditions. Each coefficient is expressed as the sum of real and imaginary parts. For example,

$$s_{11}^E = s_{11}^{E'} + j s_{11}^{E''}.$$

9.

where ' denotes the real part and '' denotes the imaginary part. However, when transients are of interest, this approach is not suitable and nonlinear analysis is essential.

Holland [3] derived conditions that are satisfied by the imaginary parts of complex coefficients:

$$-s_{11}^{E''} \geq 0 \quad -s_{33}^{E''} \geq 0 \quad -s_{44}^{E''} \geq 0 \quad -\varepsilon_{11}^{T''} \geq 0 \quad -\varepsilon_{33}^{T''} \geq 0 \quad -s_{11}^{E''} \geq \left| s_{12}^{E''} \right|$$

$$s_{11}^{E''} s_{33}^{E''} \geq \left| s_{13}^{E''} \right|^2 \quad s_{11}^{E''} \varepsilon_{33}^{T''} \geq \left| d_{31}^{''} \right|^2 \quad s_{33}^{E''} \varepsilon_{33}^{T''} \geq \left| d_{33}^{''} \right|^2 \quad s_{44}^{E''} \varepsilon_{11}^{T''} \geq \left| d_{15}^{''} \right|^2$$

10.

$$s_{33}^{E''}\left(s_{11}^{E''} + s_{12}^{E''} \right) \geq \left| s_{13}^{E''} \right|^2 \quad \varepsilon_{33}^{T''}\left(s_{11}^{E''} + s_{12}^{E''} \right) \geq \left| d_{31}^{''} \right|^2$$

When the equations of state in one of the other forms in Eq. 7 are used, it is convenient to determine the above coefficients and ensure that the conditions are satisfied.

MODELS OF PIEZOELECTRIC CERAMIC COMPONENTS AND DEVICES

Models of piezoelectric ceramic components of different shapes are related to the corresponding models for elastic bodies. The development of the piezoelectric model has followed the development of the elastic model and the sophistication is at the most equal to that in the elastic model. Interest in elastic bodies is much wider and absence of piezoelectric counterparts to some elastic models is, therefore, not surprising.

Analytical and numerical models of piezoelectric ceramics are widely used. Each type has its own advantages as in the case of elastic structures. Popular numerical packages have included the ability to correctly model piezoelectric materials only in recent years though special packages have been available for over a decade.

ANALYTICAL MODELS

Models of simple piezoelectric shapes with boundary conditions that are easily achieved in practice are of interest for characterization. Therefore, for example, there is interest in slabs with the electric field perpendicular to the direction of vibration and rods with electric field along the direction of vibration with zero stress at the ends.

Models of devices with piezoelectric components are of ultimate interest. Therefore, analytical models of piezoelectric components with general boundary conditions that can be used as building blocks are of interest. Slabs, rods, axially polarized rings, and radially polarized tubes are generally used in devices and models of these components with specified stress or velocity at the boundaries are of interest.

One example of each type of analytical model is illustrated here: a slab with zero stress at both ends, a rod with specified stress or velocity at the ends, and a multilayer Tonpilz transducer.

Stress-free Slab

Consider the piezoelectric slab [2] of length L and width W shown in Fig. 1. The distance between the electrodes is h. The length is much greater than the width and thickness. A voltage, Φ, is applied across the electrodes. The input electrical admittance and the vibration of the slab are of interest. Stress is zero everywhere on the surface of the slab. Slabs that satisfy the conditions on the dimensions are used to characterize new materials. They are also used in segmented ring underwater transducers.

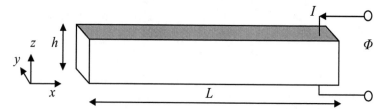

FIGURE 1. A length expander piezoelectric slab with perpendicular field.

By convention, polarization is along the z direction. It is assumed that the electric field is only in the z direction. It is also assumed that the only non-zero stress inside the rod is T_{xx} because $L \gg W$ and h. Therefore, stress and electric field are chosen as independent variables and the most convenient form of the equation of state is in Eq. 7b. Making use of the assumptions yields

$$\begin{Bmatrix} S_{xx} \\ D_z \end{Bmatrix} = \begin{bmatrix} s_{11}^E & d_{31} \\ d_{31}^T & \varepsilon_{33}^T \end{bmatrix} \begin{Bmatrix} T_{xx} \\ E_z \end{Bmatrix}. \qquad 11.$$

It is noted by comparing the above equation with the equation of state for a long thin elastic rod that s_{11}^E is analogous to $1/Y$. The assumptions are reasonable at frequencies not much greater than the lowest resonance frequency.

The dynamic equilibrium equation for the slab is obtained by considering a short length dx of the rod and is expressed as

$$\rho\omega^2 A U \, dx = A\frac{\partial T_{xx}}{\partial x} dx \qquad\qquad 12.$$

where ρ is the density, A is the cross-sectional area and U is the displacement in the x direction. It is noted that the local acceleration depends on the stress-gradient and that the direction of predominant vibration is perpendicular to the direction of the applied electric field.

After defining the strain-displacement relationship for the rod:

$$S_{xx} = \frac{\partial U}{\partial x}, \qquad\qquad 13.$$

combining Eqs. 11 – 13 yields the wave-equation for the slab:

$$\frac{\partial^2 U}{\partial x^2} + \frac{\omega^2}{c^2}U = 0 \qquad\qquad 14.$$

where $c = (\rho s_{11}^E)^{-0.5}$ is the speed of longitudinal waves in the slab. The solution to Eq. 14 is expressed as

$$U = C\sin(kx) + D\cos(kx) \qquad\qquad 15.$$

where $k = \omega/c$ and U is in upper-case to denote that it is frequency-dependent. C and D can be determined by using the boundary conditions. There are only 2 unknown coefficients and only 1 boundary condition can, therefore, be specified at each end. Consider for example, a rod that is fixed at both ends. The boundary conditions, $U = 0$ at $x=0$ and L, can be used to show that $D = 0$ and $C\sin(kL) = 0$.

Expressions for the input electrical admittance and other functions of interest can now be easily derived and are illustrated next for a rod.

Rod with General Boundary Conditions

Consider a piezoelectric ceramic rod [2] shown in Figure 2. The length, L, and the lateral dimensions, W and h, are shown in the figure. The distance between the electrodes is L. The length of the long slender rod is much greater than the width and thickness. A voltage, Φ, is applied across the electrodes. Non-zero stresses or velocities are specified at the ends.

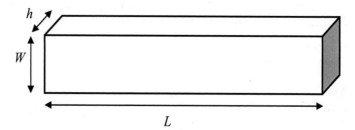

FIGURE 2. A long, slender, piezoceramic rod. The distance between the electrodes is L.

The stresses on the lateral surfaces are zero and are assumed to be zero inside the rod also and the displacement along the length of the rod is assumed to be uniform across the cross-section. It is not assumed that the electric field is constant as in the case of the slab. However, the electrodes have a small area and it is assumed that the charge density is uniform across the cross-section. The equations of state consistent with these assumptions are

$$\begin{Bmatrix} S_{zz} \\ E_z \end{Bmatrix} = \begin{bmatrix} s_{33}^D & g_{33} \\ -g_{33} & \varepsilon_{33}^T \end{bmatrix} \begin{Bmatrix} T_{zz} \\ D_z \end{Bmatrix} \qquad 16.$$

where, as noted earlier, the direction of polarization is the third or z direction.

The equilibrium equation for the rod is expressed as

$$\rho\omega^2 U = \frac{\partial T_{zz}}{\partial z} \qquad 17.$$

where ρ is the density and U is now the displacement in the z direction. The strain-displacement relationship for the rod is

$$S_{zz} = \frac{\partial U}{\partial z}. \qquad 18.$$

Combining Eqs. 16 – 18 and solving the differential equation for the displacement yields

$$U = C\sin(kz) + D\cos(kz). \qquad 19.$$

The current, I, is defined as

$$I = j\omega A D_z \qquad 20.$$

where A is the surface area of each electrode and the voltage across the electrodes is

$$\Phi = -\int_0^L E_z dz. \qquad 21.$$

Then, substituting the expressions for strain and charge density in the equation of state yields

$$T_{zz} = \{k[C\cos(kz) - D\sin(kz)] - g_{33}I/(j\omega A)\}/s_{33}^D \qquad 22.$$

270

The relationship between the stresses and velocities at the ends, the applied voltage, and the induced current is of interest. It is convenient to determine this after expressing the coefficients C and D in terms of the velocities at the ends of the layers as:

$$C = \frac{V_i^+ - V_i^- \cos(kL)}{j\omega \sin(kL)}$$

23a.

and

$$D = \frac{1}{j\omega} V_i^-$$

23b.

where V_i^- and V_i^+ are the velocities at $z = 0$ and L, respectively. The subscript i is used here for convenience and is used later to indicate that the rod is the ith layer in a multilayer device.

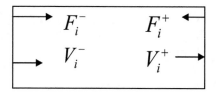

FIGURE 3. Schematic of the ith layer of a multilayer device showing forces and velocities at the ends.

After defining F_i^- and F_i^+ as the forces at $x = 0$ and L, respectively as shown in Figure 3, it is seen after some algebra that

$$\begin{Bmatrix} F_i^+ \\ V_i^+ \\ I_i \end{Bmatrix} = \begin{bmatrix} 1 + \dfrac{Z_1}{Z_2} & -Z_1\left(2 + \dfrac{Z_1}{Z_2}\right) & -\kappa \dfrac{Z_1}{Z_2} \\ -\dfrac{1}{Z_2} & 1 + \dfrac{Z_1}{Z_2} & \kappa \dfrac{1}{Z_2} \\ -\kappa \dfrac{1}{Z_2} & -\kappa \dfrac{Z_1}{Z_2} & \dfrac{\rho c A \omega C_0}{Z_2 \sin(kL)} \end{bmatrix} \begin{Bmatrix} F_i^- \\ V_i^- \\ \Phi \end{Bmatrix}$$

24.

where $Z_1 = -j\rho c A \tan(kL/2)$ and $Z_2 = \dfrac{j\rho c A}{\sin(kL)} - \dfrac{jk^2}{\omega C_0}$. Eq. 24 is convenient to use in the model of multi-layer devices because the force and velocity at one end of the layer and the current are expressed in terms of the force and velocity at the other end and the applied voltage.

A similar procedure is used in the analysis of axially polarized cylinders of arbitrary length to radius ratio using exact equations.

Multilayer Device

Consider, next, a device with many piezoelectric ceramic and elastic layers. An example is a Tonpilz transducer used in many underwater applications [4]. A schematic of the transducer is shown in Figure 4.

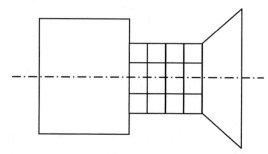

FIGURE 4. Schematic of an axisymmetric Tonpilz transducer with a cylindrical elastic tail mass, four piezoelectric ceramic rings, and a head mass in the shape of a frustum of a cone.

A model of the device is built by using models of each layer [5] and the first such model was developed about 40 years ago. A schematic of the ith layer is shown in Figure 3. First, a model of each layer is used to relate the force and velocity at one end of the layer to the force and velocity at the other end – as done above for the piezoelectric rod. Then, the continuity conditions at the interfaces between the layers

$$\begin{Bmatrix} F_i^+ \\ V_i^+ \end{Bmatrix} = \begin{Bmatrix} F_{i+1}^- \\ V_{i+1}^- \end{Bmatrix} \qquad\qquad 25.$$

are used to develop a model of the device. The results of the model are in fairly good agreement with experimental results even when there is a discontinuity in the areas of adjacent layers as between the rings and the ends masses as shown in Fig. 4.

The procedure for using the continuity conditions at the interface between the layers to build a model of the multilayer device is now illustrated by combining the models of two layers.

The equations for the two layers in Eq. 24 are expressed in the form

$$\begin{Bmatrix} F_i^+ \\ V_i^+ \end{Bmatrix} = [T_i]\begin{Bmatrix} F_i^- \\ V_i^- \end{Bmatrix} + \{t_i\}\Phi, \ i = 1, 2 \qquad\qquad 26a.$$

and

$$I_i = [Q_i]\begin{Bmatrix} F_i^- \\ V_i^- \end{Bmatrix} + \{q_i\}\Phi \ \ i = 1, 2. \qquad\qquad 26b.$$

Then, using the continuity conditions at the interface yields

$$\begin{Bmatrix} F_2^+ \\ V_2^+ \end{Bmatrix} = [T_2][T_1]\begin{Bmatrix} F_1^- \\ V_1^- \end{Bmatrix} + [[T_2]\{t_1\} + \{t_2\}]\Phi \qquad\qquad 27.$$

where Φ is known and the two equations are solved by using the boundary conditions at the two ends.

Various types of boundary conditions occur in common use. The force at the end is zero when the device is air-backed and the velocity is zero when the device is rigidly mounted. The radiation impedance

$$Z_{rad} = \frac{F_2^+}{V_2^+}$$
28.

is complex and determined independently when the device radiates into water. Thus in all cases, in addition to the two equations in Eq. 27, there are two boundary conditions that can be used to solve the four unknowns: F_1^-, V_1^-, F_2^+, and V_2^+. Then, the stress at the interface and the current in each layer are determined using Eqs. 26. Finally, the stresses and displacement within each layer are then determined. This completes the model of the multilayer device.

Numerical Models

The first paper on finite element analysis of piezoelectric structures was probably by Allik and Hughes [7] in 1970. In the early years the emphasis was on application to sonar transducers and the effect of fluid loading on radiation from simple structures [8]. Then several types of underwater transducers were modeled [5, 9, 10]. Meanwhile, several new piezoelectric finite elements were presented and are reviewed by Benjeddou [11]. Later, there was interest in sensing and control of structures [12].

Now commercial packages are available. ATILA [13] is developed by ISEN, France and is a special package for the analysis of sonar transducers. A recent version has boundary elements also. ANSYS and PAFEC can also be used to model piezoelectric structures. A bibliography "Guide to the Literature of Piezoelectricity and Pyroelectricity" is periodically published in the journal Ferroelectrics [14].

CHARACTERIZATION OF PIEZOELECTRIC CERAMICS

Piezoelectric ceramics are characterized when new materials are being developed and for quality control when they are produced in large numbers in a particular shape and size. In the former case, methods recommended by the ANSI / IEEE Standard on Piezoelectricity [15] are usually used to determine real piezoelectric coefficients using linearized equations. The size of the sample used satisfies the recommendations in the Standard. The Standard equations have been adapted to determine complex coefficients [16, 17]. Several of the earlier methods have been compared by Kwok et al [18]. When quality control during production is of interest, the shape and size of the piezoelectric components seldom satisfy the recommendations in the Standard. Therefore, the Standard method cannot be used and other methods to determine the coefficients using linearized equations are presented here.

The input electrical admittance of a piezoelectric ceramic is easily and accurately measured using, for example, an impedance analyzer. It is, therefore, convenient to determine the piezoelectric coefficients using the measured admittance. However, it should be ensured that the strain – especially at resonance – is small and only linearly dependent on the applied electric field.

273

In devices, such as hydrophones, where the piezoelectric is acoustically excited the electric field is very low and linear characterization is sufficient. In some devices, such as transducers used as underwater projectors, the electric field is high but not so high as to make non-linear analysis essential. Appropriate models of the material and device are necessary when they are deliberately driven non-linearly.

Standard Methods

The ANSI / IEEE Standard is used to determine the piezoelectric coefficients of slabs, rods, and discs with large aspect ratios using one-dimensional models. A slab and a rod are shown in Figure 1 and 2, respectively. One-dimensional models of slabs and rods are in good agreement with experiment when the length is much greater than the width and thickness. For slabs, it is also required that the length be greater than three times the width. Discs should have a radius greater than 20 and 40 times the thickness in order to determine s_{11}^E and s_{12}^E, respectively. Explicit expressions are given in the Standard for the coefficients of slabs and rods. However, for discs, numerical or approximate methods are used to solve equations and determine the coefficients. Slabs are used to find ε_{33}^T, s_{11}^E and d_{31}; rods are used to find ε_{33}^T, s_{33}^D and d_{33}; and discs are used to find ε_{33}^T, s_{11}^E, s_{12}^E and d_{31}. The coefficients are also used to find the piezoelectric coupling coefficient k_{31} in slabs and discs and k_{33} in rods.

The Standard does not give any indication of the error caused by using a one-dimensional model to determine the piezoelectric coefficients. Even though the error is systematic when the size of the sample is fixed, different sizes that satisfy the recommendations on size are likely to be used. Further, knowledge of the error when using a particular size is useful.

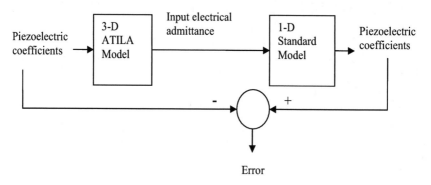

FIGURE 5. Flow chart for finding the error in determining piezoelectric coefficients.

The error [19] in determining the coefficients using the Standard method because of the use of one-dimensional models is determined as shown in Fig. 5. The input electrical admittance of the structure is computed using a set of 10 piezoelectric coefficients and the 3-D finite element program ATILA. The computed admittances and the approximate 1-D Standard equations are then used to calculate some piezoelectric coefficients. The differences between the actual coefficients and those obtained using the Standard equations are the errors for that particular case. Errors exist because the Standard equations are based on 1D models. Normally, measured values are used in place of the computed admittances and the actual coefficients are not known and the errors cannot be determined.

For example, consider the case where the 10 piezoelectric coefficients input to ATILA are s_{11}^E=12.3x10^{-12} m^2/N, s_{12}^E=-4.05x10^{-12} m^2/N, s_{13}^E= -5.31x10^{-12} m^2/N s_{33}^E=15.5x10^{-12} m^2/N, s_{44}^E=39x10^{-12} m^2/N, d_{31}=-123 x10^{-12} C/N, d_{33}=289x10^{-12} C/N, d_{15}=496x10^{-12} C/N, ε_{11}^S, and ε_{33}^S. The values of ε_{11}^S and ε_{33}^S are computed using the values of the other 8 coefficients, $\varepsilon_{11}^T / \varepsilon_0$=1475, and $\varepsilon_{33}^T / \varepsilon_0$=1300. The coefficients in the set input to ATILA, and those calculated using the relationships in Eqs. 8 are the *actual* values.

The errors in coefficients of slabs determined using the Standard method are shown in Table 1 for a few cases. More results are shown in Ref. [19]

TABLE 1. Coefficients of slabs obtained using Standard equations.

No.	L (mm)	W (mm)	t (mm)	$\varepsilon_{33}^T / \varepsilon_0$	s_{11}^E (pm^2/N)	d_{31} (pC/N)	k_{31}
1.	100	30	10	1.30e3	12.4	-119	-0.31
2.	100	25	10	1.30e3	12.4	-120	-0.32
3.	100	12	5	1.30e3	12.3	-123	-0.33
4.	40	20	5	1.30e3	12.7	-108	-0.28
5.	40	15	3	1.30e3	12.5	-116	-0.30
6.	40	10	5	1.30e3	12.4	-121	-0.32
7.	40	10	3	1.30e3	12.4	-120	-0.32
8.	40	8	3	1.30e3	12.4	-121	-0.32
Actual Values⟶				1.300e3	12.30	-123.0	-0.327

OTHER SHAPES AND METHODS

Axially polarized rings and radially polarized tubes are examples of commonly used shapes that cannot be characterized using the Standard methods. Therefore, models of piezoceramics in these shapes and methods to characterize them are of particular interest.

An axially polarized ring is shown in Figure 6. The distance between the electrodes, h, is usually approximately equal to the wall thickness, b-a. It has been modeled by Martin [20], Ebenezer and Abraham [21], and Ramesh and Ebenezer [6] using membrane theory, thin shell theory, and exact equations respectively. Ramesh and Ebenezer showed that the resonance and anti-resonance frequencies computed using their model is in good agreement with finite element results and can be used to determine piezoelectric coefficients. Martin [22] presented a method to determine the piezoelectric coefficients of a stack of rings by using the measured admittance. These stacks are used in Tonpilz transducers as shown in Fig. 4.

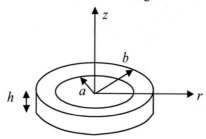

FIGURE 6. An axially polarized ring with electrodes on the flat surfaces. The inner and outer radii are a and b, respectively. The distance between the electrodes is h.

Radially polarized tubes are modeled using membrane theory by Haskins and Walsh [23] and Ebenezer [24]. Ebenezer and Sujatha used this model to characterize them [25]. Drumheller and Kalnins [26], Roghaceva [27], and Ebenezer and Abraham [28] developed thin shell models of the tubes and solved those using numerical, approximate, and analytical methods. The effect of bending strain is included in the thin shell models and they are, therefore, more useful than membrane models when tubes with thicker walls are of interest.

Ebenezer [29] used an iterative method to determine four complex piezoelectric coefficients of radially polarized piezoelectric tubes: ε_{33}^T, s_{11}^E, s_{12}^E, d_{31}. The complex value of ε_{33}^T is determined by using the measured value of the admittance at a low frequency. The complex values of the remaining three coefficients are determined by iteratively refining them until the measured values of 3 frequencies and 3 bandwidths are very nearly equal to computed values. The frequencies are: f_s^L and f_s^U in the lower and upper branches, respectively, at which the conductance, G, reaches a local maximum; the frequency f_p^U in the upper branch at which the resistance, R, reaches a local maximum. The three bandwidths are: $f_{1/2s}^L - f_{-1/2s}^L$, $f_{1/2s}^U - f_{-1/2s}^U$, and $f_{1/2p}^U - f_{-1/2p}^U$ where $f_{1/2s}^L$ and $f_{-1/2s}^L$ are the frequencies in the lower branch at which B reaches a local minimum and maximum, respectively, $f_{1/2s}^U$ and $f_{-1/2s}^U$ are the frequencies in the upper branch at which B reaches a local minimum and maximum, respectively, and $f_{1/2p}^U$ and $f_{-1/2p}^U$ are the frequencies in the upper branch at which the reactance, X, reaches a local minimum and maximum, respectively.

The errors in the real parts of the coefficients determined using a cylindrical shell of length 20 mm, average radius 15 mm, wall thickness 1 mm are comparable to the errors in the coefficients determined using a thin axially polarized disc of radius 30 mm and thickness 1 mm and the Standard method, i.e., no errors in $\mathrm{Re}(s_{11}^E)$ and $\mathrm{Re}(d_{31})$ to 3 significant digits, and about 1/3 % error in $\mathrm{Re}(s_{12}^E)$, even though the shell is not as slender as the disc. This occurs because the effect of the finite dimensions of the shell and bending strain are included in the analytical model of the shell whereas the plane-strain assumption is made in the Standard model.

CONCLUSIONS

The use of linearized piezoelectric equations of state and complex piezoelectric coefficients is illustrated through examples. It is appropriate to use these approximations whenever models developed using them can be used to explain experimentally observed behavior. Modeling and designing using these approximations is much easier than dealing with non-linear equations of state and other means of representing hysteretic effects. In many cases, the device of interest will go through an initial transient phase during which it is not possible to precisely explain its behavior using these approximations. However, when the excitation is harmonic and not large enough to induce non-linear behavior, nearly steady-state conditions will be reached within a few cycles and the approximations will be valid. This is usually the case when the piezoelectric ceramic is a part of device or structure in which other components also contribute to damping.

ACKNOWLEDGMENT

Facilities and encouragement provided by Director, NPOL are gratefully acknowledged.

REFERENCES

1. Kolsky, *Stress Waves in Solid* (Dover, New York, 1963).
2. D. A. Berlincourt, D. R. Curran, and H. Jaffe, ''Piezoelectric and piezomagnetic materials and their function in transducers,'' in *Physical Acoustics* Vol. I, Pt. A edited by W. P. Mason (Academic, New York, 1964).
3. R. Holland, "Representations of the Dielectric, Elastic, and Piezoelectric Losses by Complex Coefficients", IEEE Trans.Sonics Ultrasonics, **SU-14**, 18-20 (1967).
4. D. Stansfield *Underwater Electroacoustic Transducers* (Bath, U. K. 1990).
5. J.-N. Decarpigny, J.-C. Debus, B. Tocquet, and D. Boucher, "In-air analysis of piezoelectric Tonpilz transducers in a wide frequency band using a mixed finite element-plane wave method," *J. Acoust. Soc. Amer.*, vol. 78, pp. 1499–1507, (1985).
6. R. Ramesh and D. D. Ebenezer, "Analysis of axially polarized piezoelectric ceramic rings," *Ferroelectrics* Vol. 323 pp 17 - 23 (2005).
7. H. Allik and T. J. R. Hughes, "Finite element method for piezoelectric vibration," *Int. J. Num. Meth. Engineering* 2, 151 – 157 (1970).
8. R. R. Smith, I. T. Hunt, and D. Barach, "Finite element analysis of acoustically radiating structures with applications to sonar transducers," *J. Acoust. Soc. Amer.* 54, 1277- 1288 (1977).
9. R. Lerch, "Simulation of piezoelectric devices by two and three dimensional finite elements," *IEEE Ultrasonics, Ferro. Freq. Cont.* 37, 233 – 247 (1990).
10. K. D. Rolt, "History of flextensional electroacoustic transducer," *J. Acoust. Soc. Amer.* 87, 1340–1349 (1990).
11. A. Benjeddou, "Advances in piezoelectric finite element modeling of adaptive structural elements: a survey," *Computers and Structures* 76, 347 – 363 (2000).
12. S. S. Rao and M. Sunar, "Piezoelectricity and its use in the disturbance sensing and control of flexible structures: A survey," *Applied Mechanics Review* 47, 113 – 123 (1994).

13. *ATILA Users Manual*, ISEN, Lilee CEDEX, France.
14. S. B. Lang, "Guide to literature of piezoelectricity and pyroelectricity. 21," *Ferroelectrics*.
15. IEEE Standard on Piezoelectricity, ANSI/IEEE Std. 176-1987. *IEEE Trans. Ultrason., Ferroelect., Freq. Cont.* UFFC-43 717-772 (1996).
16. S. Sherrit, H. D. Wiederick, B. K. Mukherjee, "A complete characterization of the piezoelectric, dielectric, and elastic properties of Motorola PZT 3203 HD including losses and dispersion," in *Ultrasonic Transducer Engineering:* Proc. SPIE **3037**, pp. 158-169, SPIE, Bellingham, WA 98227, USA, 1997.
17. M. Alguero, C. Alemany, L. Pardo, and A. M. González, "Method for obtaining the full set of linear electric, mechanical, and electromechanical coefficients and all related losses of a piezoelectric ceramic," *J. Amer. Cer. Soc.* 87, 209 – 215 (2004).
18. K. W. Kwok, H. L. W. Chan, and C. L. Choy, "Evaluation of the material parameters of piezoelectric materials by various methods," IEEE Trans. Ultrason., Ferroelect., Freq. Contr., vol. 44, 733–742 (1997).
19. D. D. Ebenezer and A. J. Sujatha, "Errors in the characterization of piezoelectric ceramics," Proc. ISSS-SPIE Intl. Conf. Smart Materials Structures Systems 131 – 138 (Bangalore, India, 2002).
20. G. E. Martin, "Vibrations of longitudinally polarized ferroelectric cylindrical tubes," *J. Acoust. Soc. Amer.* 35, 510-520 (1963).
21. D. D. Ebenezer and P. Abraham, "Analysis of axially polarized piezoelectric ceramic cylindrical shells of finite length with internal losses," *J. Acoust. Soc. Amer.*, Vol. 112 No. 5 pp 1953-1960 (2002).
22. G. E. Martin, "New standard for measurements of certain piezoelectric ceramics," *J. Acoust.Soc. Amer.* 35, 925 (1963).
23. J. F. Haskins and J. L. Walsh, "Vibrations of ferroelectric cylindrical shells with transverse isotropy. I. Radially polarized case," *J. Acoust. Soc. Amer.* 29, 729–734 (1957).
24. D. D. Ebenezer, "Three-port parameters and equivalent circuit of radially polarized piezoelectric ceramic cylinders of finite length," *J. Acoust. Soc. Amer.* 99, 2908-2912 (1996).
25. D. D. Ebenezer and A. J. Sujatha, "New methods to characterize radially polarized piezoelectric ceramic cylindrical shells of finite length," *J. Acoust. Soc. Amer.* 102, 1540-1548 (1997).
26. D. S. Drumheller and A. Kalnins, "Dynamic shell theory for ferroelectric ceramics," *J. Acoust. Soc. Amer.* **47**, 1343–1353 (1970).
27. N. N. Rogacheva, *The Theory of Piezoelectric Shells and Plates*, (CRC, London, 1993).
28. D. D. Ebenezer and Pushpa Abraham, "Piezoelectric thin shell theoretical model and eigenfunction analysis of radially polarized ceramic cylinders," *J. Acoust. Soc. Amer.* 105, 154-163 (1999).
29. D. D. Ebenezer, "Determination of complex coefficients of radially polarized piezoelectric ceramic cylindrical shells using thin shell theory," *IEEE Trans. on Ultrasonics, Ferroelec., Freq. Cont.* 51, 1209-1215 (2004).

Preparation and Characterization of PZT Thin Films

A. Bose[a], M. Sreemany[a], S. K. Halder[b], D. K. Bhattacharyya[a], and Suchitra Sen[a*]

[a]Central Glass and Ceramic Research Institute, CSIR, 196 Raja S.C.Mullick Road, Kolkata-700032, India
[b]National Physical Laboratory, Dr. K. S. Krishnan Road, New Delhi-110012, India
*E-mail: suchitra@cgcri.res.in

Abstract. In analogy with Piezoelectric Wafer Active Sensors (PWAS), Lead Zirconate Titanate (PZT) thin films also seem to be promising for Structural Health Monitoring (SHM) due to a number of reasons. Firstly, PZT thin films with well oriented domains show enhanced piezoelectric response. Secondly, PWAS requires comparatively large voltage leading to a demand for thin PZT films ($<<$ μm in thickness) for low voltage operation at ≤ 10V. This work focuses on two different aspects: (a) growing oriented PZT thin films in ferroelectric perovskite phase in the range of (80 – 150) nm thickness on epitaxial Si/Pt without a seed layer and (b) synthesizing perovskite phase in PZT thin films on Corning glass 1737 using a seed layer of TiO_x (TiO_x thickness ranging between 30 nm to 500 nm).

Keywords: Thin Films, R F Sputtering, Piezoelectric, Perovskite, X-ray diffraction, Field Emission Scanning Electron Microscopy, Residual Stress

PACS: 68.37, 68.55, 61.05

INTRODUCTION

Multicomponent ferroelectric oxide thin films, in particular perovskites - the oxygen octahedral class of materials such as lead zirconate titanate (PZT) exhibit high permittivity, large spontaneous polarization and pyroelectric, electro-optic and ferroelectric effects These properties make these materials attractive for a wide range of device applications such as non-volatile memory, DRAM, sensors and actuators, accelerometers and optical switches.

Thin PZT wafers (PWAS), surface-mounted or embedded in the structure can act as active sensors and detect structural damage through Lamb waves. In analogy with PWAS, PZT thin film also seems to be a futuristic material for structural health monitoring (SHM) by Lamb wave propagation [1]. As embedded sensor material, they are highly attractive due to a number of reasons. Firstly, the PWAS has to be bonded or embedded to the structure under interrogation and this bonding is the weak link between the sensor element and the structure. On the other hand, thin films can be directly deposited on the structural element to be interrogated and so the bonding problem can be easily overcome. Secondly, PZT thin films when grown epitaxially can have well oriented domains and show much improved piezoelectric effect.

CP1029, *Smart Devices: Modeling of Material Systems, An International Workshop*
edited by S. M. Sivakumar, V. Buravalla, and A. R. Srinivasa
©2008 American Institute of Physics 978-0-7354-0553-0/08/$23.00

Thirdly, thin PZT films with thickness $<< 0.1$ µm are ideal for low voltage operation ($\leq 10V$) compared to PWAS [2].

The properties of PZT thin films are most dependent on the method of synthesis, structure and crystalline phases. It is important to synthesize epitaxial, single-crystalline films with well-defined interfaces for reproducible ferroelectric properties for miniaturizing the conventional ferroelectric/piezoelectric devices. The orientation and the ferroelectric properties of these films can be controlled by i) controlling the growth and substrate parameters during synthesis ii) controlling the post annealing conditions during recrystallization and iii) introducing a buffer/seed layer between substrate and film. The main objectives of the present work are to develop PZT films with pure perovskite (ferroelectric) phase with epitaxial growth by Off-axis Radio Frequency (RF) Magnetron Sputtering under optimized deposition conditions and finally, formation of Pt/PZT/Pt structure, keeping their applications in mind. Ferroelectric PZT thin films offer potential applicability in high-density non-volatile memory devices, actuators and microelectromechanical systems. PZT film sandwiched between two electrodes acts as a capacitor and this electrode/PZT/electrode structure is the basic building block for such a device.

This work focuses on two different aspects:

1. To grow epitaxial/well-oriented PZT thin films in ferroelectric perovskite phase in the range of thickness (80-150)nm on epitaxial substrates without a seed layer
2. To synthesize polycrystalline/oriented pure perovskite PZT thin films in the same range of thickness on Corning glass using a seed layer of TiO_x of varying thickness

It is important to mention here that other than processing conditions, the parameters that affect epitaxial or oriented growth are i) orientation of the substrate ii) lattice mismatch of film and substrate which introduces misfit strain and iii) thermal expansion of film and substrate and the resulting residual stress. All these points will be discussed in the paper.

EXPERIMENTAL

This study focuses on synthesis of epitaxial/oriented PZT films from single oxide Pb-rich target controlling the growth during deposition and post-deposition annealing conditions on epitaxial platinised Si by Off-axis RF Magnetron Sputtering in a 3-target Sputtering System (Maker: Hind Hi Vac, Bangalore, India). Films have been prepared in the thickness range of (80-150) nm from a target of composition $[Pb_{1.1}(Zr_{0.52}Ti_{0.48})O_3]$ near morphotropic phase boundary. The commercially available Si/Pt with layered structure i.e. Si/SiO$_2$/TiO$_2$/Pt(111) and Si/SiO$_2$/TiO$_2$/Pt(200) are used as substrates where the thickness of Pt is ~150 nm and that of TiO$_2$ is ~20nm . Thickness and phase formation in PZT films are optimized by varying the deposition parameters i.e. substrate temperature (T_s), gas pressure during deposition (P_p), sputtering gas composition (P_c) and RF power density (W_d). Post-deposition annealing of the films are carried out systematically under varied conditions like annealing temperature (600^0- 700^0)C, annealing time (1-3) mins, ramp rate (10 - $20)^0$C/s and annealing environment (air and O$_2$) in Rapid Thermal Annealer (RTA).

The phase evolution by X-ray Diffraction (XRD), microstructural study by Field Emission Scanning Electron Microscope (Supra 35 VP FESEM) with local chemical analysis by Energy Dispersive X-ray Analysis (EDX, Link ISIS) and surface roughness by Stylus profilometry (Talysurf I20) are done on as-deposited and annealed PZT films for complete structural characterization. Residual stress measurement is done by Double Crystal X-ray Diffractometer by measuring the radius of curvature of the substrates [3]. In a separate experiment, studies are performed on optimization of TiO_x seed layer and synthesis of PZT films with this layer on polished 1737 corning glass. The objectives of this study are to grow pure perovskite PZT thin films with well-oriented grains on epitaxial and amorphous substrates with and without seed layer.

OBSERVATIONS

Synthesis of PZT thin films on epitaxial Si/Pt without a seed layer

The deposition variables used in our study are given in table 1.

TABLE 1. Deposition parameters used for synthesis of PZT thin films on Si/Pt by Off-axis R F Magnetron Sputtering

Target composition	Substrates	Substrate temperature	R F Power density (W_d) $(W.cm^{-2})$	Sp. gas composi-tion $(Ar:O_2)$	Pr. during deposition (mtorr)
$Pb_{1.1}(Zr_{0.52}Ti_{0.48})O_3$	Si/SiO$_2$/TiO$_2$/ Pt(111)	25 ^{0}C (RT), 400^{0}C	2.96	80:20 – 50:50	15.0 - 36.7
Do	Si/SiO$_2$/TiO$_2$/ Pt(200)	Do	Do	Do	Do

The as-deposited films are weakly crystalline or amorphous depending on the orientation of the substrate (Fig 1). On Pt(111), presence of undesirable phases are noticeable which do not match with JCPDS data of oxygen deficient (26-142) or Pb-deficient (45-533) pyrochlore phase. However, these could be either α- PbO (35-1482)or ZrO$_2$ (37-1484). Higher substrate temperature of 400^{0}C results in better growth of ferroelectric perovskite phase compared to those deposited at room temperature. Optimum deposition conditions, in our case, are found to be: T$_s$= 400^{0}C, W$_d$ = 2.96 W.cm^{-2}, P$_c$ = Ar:O$_2$ :: 80:20, P$_p$=36.7 mortor Post-deposition annealing of the films are done at different temperatures ~ (600^0–700^0)C, varying the ramp rate and soaking time. Ambience during annealing is also very important in this growth process. So the furnace is flushed with oxygen once before annealing.

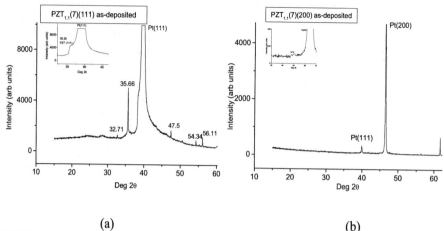

(a)   (b)

FIGURE 1: XRD pattern of as-deposited PZT film synthesized at substrate temperature of 400^0C (a) on Si/Pt (111) (b) Si/Pt (200)

Phase evolution by XRD

Effect of deposition variables

Optimization of the deposition variables are performed with a set of films deposited with varied parameters and annealed at 6500C for 3 mins. The ramp rate is 100C/s for (111) and $20^0C/s$ for (200).

Phase evolution study of the films after post annealing shows that higher substrate temperature of 400^0C during deposition facilitates the formation of the perovskite phase (Fig 2). The films are fine-grained and denser. This is due to the higher mobility of the adatoms at higher deposition temperature enabling better rearrangement, thus helping nucleating the desired phase and finally resulting in a denser microstructure. The lattice constant 'a' is calculated from (111) peak assuming a pseudo-cubic structure and is found to be 0.4091 nm for PZT (111) deposited at 400^0C and is little higher than that reported by Khaaenamkaew et al [3].

The gas composition $Ar:O_2$ during synthesis is another important variable in sputtering. In our experiment, gas composition of 80:20 to 60:40 is used to see its effect on growth behavior of perovskite phase. Considering the phase purity, 80:20 is found to be optimum since films synthesized by sputtering with this gas composition at a pressure of 36.7 torr facilitate perovskite phase formation as the major one after annealing. However the growth rate is small (Fig 3). At 70:30, growth of perovskite phase is much faster, but presence of undesirable phases is observed. The occurrence of peaks at $\sim36^0$ and $\sim61.5^0$ 2θ in case of $Ar:O_2$ ratio of 70:30 and 60:40 indicates presence of second phases like ZrO_2/PbO and pyrochlore.

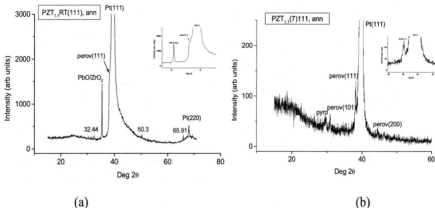

(a)   (b)

FIGURE 2 XRD patterns of annealed PZT films on Si/Pt (111) deposited at (a) RT and (b) 4000C

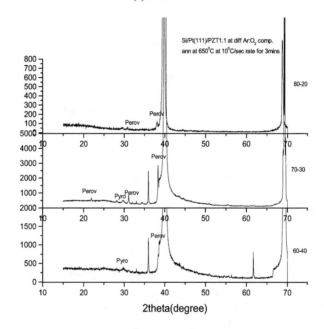

FIGURE 3: XRD patterns of PZT films deposited on Si/Pt(111) at 4000C under different sputtering gas composition and post-annealed

Effect of post annealing conditions

To optimize the post-annealing conditions, films prepared under optimized deposition conditions mentioned above are subjected to annealing at temperatures T_a from 600-700^{0}C with ramp rates of 10^{0}C and 20^{0}C/sec in RTA. The annealing time is varied between (1- 3) mins. This optimization leads to the following observations:

1. The annealing temperature and time play important roles in the development of PZT perovskite phase. Annealing at 650^0C for 3 mins is found to be most suitable for the formation of perovskite phase for films on both (111) & (200) Si/Pt substrates.

2. The interesting phenomena is that ramp rate during annealing plays a key role in developing the oriented growth. 10^0C/sec is favourable for growing oriented perovskite structure on Si/Pt(111). (111) PZT is grown preferentially in this case (JCPDS 33-784) (Fig 2b). But for films on Si/Pt(200), the perovskite phase is more favourably formed with ramp rate of 20^0C/sec and is more random (Fig 4). Degree of orientation f(hkl)is calculated by the eqn [4]: f(hkl)= I(hkl) / ΣI(hkl) and is found to be ~70% for (111) and ~ 51% for (200). So, the substrate orientation and the ramp rate control the orientation of the PZT phase developed on these epitaxial substrates.

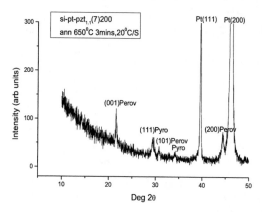

FIGURE 4: . XRD pattern of PZT films deposited at 400^0C and annealed at 650^0C for 3 mins at ramp rate of 20^0C/s on Si/Pt (200)

Thickness and surface roughness

Thickness and average surface roughness are measured in Talysurf I 20 Surface Profilometer. An important observation is that both on Si/Pt(111) and Si/Pt(200), the thickness of the films increases after annealing which may be due to the volume Change after phase transition. Thickness is further confirmed by cross-sectional FESEM study (Fig 5a). A systematic surface roughness study indicates better surface quality for (111) i.e.(Ra= 8.8nm for (111) and 12.3nm for (200)).

Microstructural study

Fig 5. shows a series of microstructure of PZT films by FESEM. A fine-grained and well-crystallized growth of PZT on both Si/Pt(111) and Si/Pt(200) deposited under optimized conditions after annealing, supports the evidence from XRD study. Local chemical analysis by EDX shows an increase of Pb/Ti ratio after annealing. This could be due to diffusion of Ti from the whole system consisting of the multilayers of PZT and the bottom electrode, towards the Si wafer. The films deposited at 400^0C and post-annealed under optimized conditions have finer grain size and a denser microstructure compared to those deposited at RT (Fig 5 b, c, d).

Gas composition has a noticeable effect on the microstructure of the films. Poor structural conformity in films deposited at 60:40 indicate poor crystallinity as evidenced in XRD compared to those deposited at 80:20. At 70:30, the grain structure is much coarser than at 80:20.

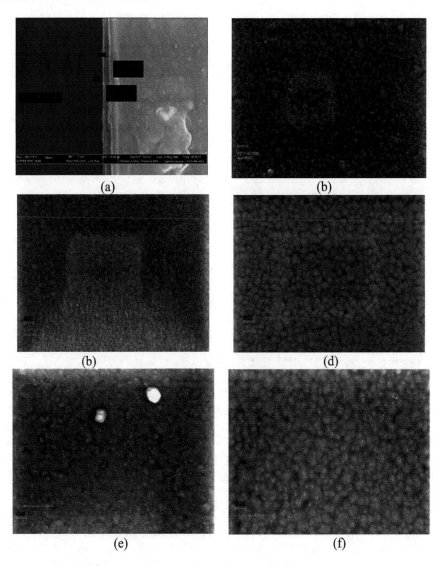

FIGURE 5. FESEM micrographs of annealed PZT films on Si/Pt substrates:
(a) Cross-sectional FESEM for thickness measurement (b) Annealed PZT film (deposited at 400^0C) of thickness ~110nm on Si/Pt(200), Mag 100k (c) Ann PZT film of thickness ~ 96 nm (deposited at 400^0C) on Si/Pt(111), Mag 100k (d) Ann PZT film of thickness ~ 120 nm (deposited at RT) on Si/Pt(111), Mag 100k (e) Ann PZT film deposited on Si/Pt(111) at gas composition 60:40 (f) Ann PZT film, deposited on Si/Pt(111) at gas composition 70:30

RESIDUAL STRESS MEASUREMENT

An important physical parameter in thin films is residual stress which affect many of the properties which are directly related to device life [5]. It is reported that residual stress significantly affect the piezoelectric property [6], dielectric constant etc. of PZT thin films. In this work, bi-axial stress is calculated from the radius of curvature of the substrates by x-ray diffraction method using CuKα radiation in a Double Crystal X-ray Diffractometer. The negative radii of curvature shows that the bi-axial stress is compressive in nature and the stress is high (~1.3 GP_a for PZT on Si/Pt(111) and ~0.8 GP_a for (200)) in case of annealed samples which are comparative with those reported in the literature [7].

Synthesis of PZT thin films on corning 1737 with a seed layer

It is observed that even under optimized deposition and post-deposition annealing conditions, a trace of pyrochlore phase exists in PZT films. For this purpose a study is performed on the development of PZT films using a seed layer of TiO_x with an aim to synthesize phase-pure perovskite films.

Effect of film thickness and phase of TiO_x ($0 \leq x \leq 2$) seed layers

Using a thin TiO_x ($0 \leq x \leq 2$) seeding layer, one can control both the crystallization property and texture of a lead zirconate titanate (PZT) film [8]. In this regard, phase and thickness of the TiO_x seeding layer are two important factors. Dielectric and ferroelectric properties of PZT films are strongly dependent on their crystalline orientation.

TiO_x ($0 \leq x \leq 2$) seed films suitable for perovskite PZT phase formation, are grown on corning-1737 glass substrates by RF sputtering process using various deposition parameters and then post-annealing these films in air at different temperatures in a conventional furnace. PZT films are then deposited on these substrates pre-coated with a thin TiO_x seed layer. Deposition and air annealing conditions for PZT films are kept identical for all the films. Processing parameters for both TiO_x seed films and PZT films are listed in table-2. To get information regarding the phase and crystalline orientation of the films, XRD study is carried out. Surface morphology and film thickness are obtained from FESEM studies.

Phase evolution and microstructure

TiO_x ($0 \leq x \leq 2$) films having thickness ~50 nm transform into anatase phase at an annealing temperature, T_a, of ~ 350°C and the films remain in the same phase up to T_a ~ 650°C. TiO_x seed films having intermediate thickness of about ~100 nm exist in TiO phase up to T_a ~ 350°C and start to transform into rutile at T_a ~ 550°C. Phase evolution with T_a for TiO_x films having thickness of ~ 500 nm closely follow the same trend as that of ~100 nm.

TABLE 2: Sputtering and annealing conditions of TiO_x ($0 \leq x \leq 2$) seed films and PZT films

Parameters	TiO_x seed films	PZT films
Target used	Ti	$Pb_{1.1}(Ti_{0.48}Zr_{0.52})O_3$
Base pressure	$\sim 3.0 \times 10^{-3}$ mtorr	$\sim 3.0 \times 10^{-3}$ mtorr
Working pressure	$\sim(1.5 - 6.0)$ mtorr	~ 3.75 mtorr
Power density	$\sim(1.1 - 2.2)$ W cm^{-2}	~ 3.0 W cm^{-2}
Sputtering gas	Ar	Ar/O_2
Gas mixing ratio (Ar : O_2)	100: 0	$\sim 90 : 10$
Target to substrate distance	~11 cm	~ 11 cm
Substrate rotation	~3 r.p.m.	~ 3 r.p.m.
Substrate temperature (T_s)	32 - 250°C	250°C
Substrate	Corning-1737	TiO_x/C-1737

Effect of both the film thickness and phase of TiO_x ($0 \leq x \leq 2$) seed layers on perovskite PZT phase formation is shown in Fig.6. Thickness and annealing temperature of TiO_x seed films on C-1737 are shown as d_{seed} and $(T_a)_{seed}$ respectively in the figure. Deposition and air annealing conditions for all the PZT films studied are kept identical. Hence, the variation in crystalline orientation and phase of PZT films develops entirely from the difference in phase and thickness of TiO_x seed layers. Polycrystalline perovskite PZT [$Pb(Ti_{0.48}Zr_{0.52})O_3$] phase forms on those C-1737 substrates that are pre-coated with a ~50 nm thick TiO_2 seed films ($x = 2$) of anatase form. Partially oriented perovskite PZT [$Pb(Ti_{0.48}Zr_{0.52})O_3$] phase along (100) direction is formed on those C-1737 substrates that are pre-coated with a ~100 nm thick TiO_2 seed films of rutile form. But in case of those C-1737 substrates pre-coated with a ~500 nm thick TiO_2 seed films of rutile form, no clear evidence about the perovskite PZT phase formation can be observed from XRD results. From XRD results, it also appears that a TiO_x seed layer of thickness in the range $\sim$50-500 nm either in amorphous or in sub-oxide form, i.e. TiO does not help in forming perovskite PZT phase (Fig.6).

As shown in Fig.7(a), the microstructure of air-annealed (600°C) TiO_2 (poly-crystalline rutile) seed layer of thickness ~100 nm reveals the sub-micron size grains of rutile phase Fig.2(b) shows the surface morphology of perovskite PZT film formed on the TiO_2 (rutile) seed layer of thickness ~100 nm after air-annealing at 650°C. Surface coverage density of the rosettee structures in this PZT film is found to be high. Seed and PZT layers can be identified clearly from the cross-sectional FESEM image of PZT/ TiO_2 (~100 nm)/C-1737 (Fig.7(c)). Thickness of the above PZT film is about ~120 nm.

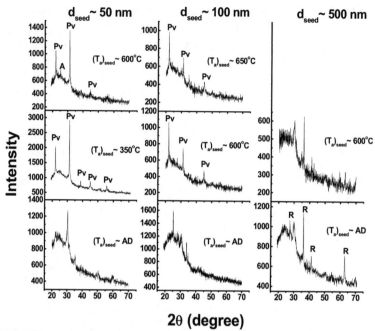

FIGURE 6: XRD patterns of PZT films on C-1737 pre-coated with ~50, 100, 500 nm thick TiO_x ($0 \leq x \leq 2$) seed films (Pv: Perovskite PZT; R: Rutile: A: Anatase)

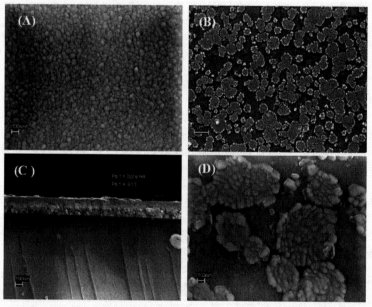

FIGURE 7: FESEM images of TiOx seed layer and PZT film surface microstructure of TiO2 (rutile) seed layer (~100 nm) on C-1737   surface microstructure of perovskite PZT film on (A) at low magnification cross-sectional image of (B) (D) surface microstructure of perovskite PZT film on (A) at high magnification

CONCLUSIONS

1. Oriented PZT thin films in perovskite phase in the thickness range of (80-150) nm could be grown on both epitaxial Si/Pt(111) and Si/Pt(200) substrates

2. PZT (111) is grown on Si/Pt(111) under optimized deposition and annealing conditions. On Si/Pt(200), PZT(200) grows preferentially, but degree of orientation is less compared to that on (111).

3. Substrate orientation, substrate temperature during deposition and ramp rate during post-annealing control the orientation of the grown films.

4. Ar:O_2 gas ratio during sputtering influences the phase purity of the PZT films.

5. On Corning 1737, a thin titania (TiO_2) seed layer helps to form perovskite PZT phase.

6. Polycrystalline perovskite PZT film develops when the titania (TiO_2) seed film exists in anatase form.

7. Oriented perovskite PZT film along (100) direction develops when the thickness of the titania (TiO_2) rutile seed film is ~100 nm.

8. Perovskite PZT phase does not form on a titanium or on other sub-oxides of titanium seed films of any thickness.

9. Thickness of the seed titania film is also an important parameter for perovskite phase formation in PZT films. If the seed film thickess becomes thicker (d $\geq$500 nm), perovskite PZT phase does not form easily even with rutile phase of the seed film.

ACKNOWLEDGEMENTS

The authors wish to acknowledge with thanks, the help of Dr. P. K. Das, Sri A. K. Mondal, Sm. T. Maity, Sri N. Dey of CGCRI. They are grateful to Dr. H. S. Maiti, Director, CGCRI for his constant encouragement during the course of the work and for permission to present the paper. Finally, the financial support from CSIR, Govt. of India in the form of Network project (COR022) is gratefully acknowledged

REFERENCES

1. V Giurgiutiu, SPIE 10th International Synposium on Smart Structures and Materials, March 2002
2. Victor Giurgiutiu, Timir Datta, Michael L. Myrick : Technology Review on Nano Piezo Sensors for Structural Health Monitoring, Damage Detection and Failure Prevention, May 2004 (Web article)
3. Z.B. Zhao , J. Hershberger, S.M. Yalisove, J.C. Bilello, Thin Solid Films 415 (2002) 21–31
4. P. Khaenamkaew, S. Muensit, I. K. Bdikin, A. L. Kholkin, Mat Chem & Phys 102 (2007) 159-164
5. F. K. Lotgerling, J. Inorg. Nucl. Chem. 9 (1959) 113
6. R. Ali. D Roy Mahapatra, S. Gopalkrishnan, Sensors and Actuators A, 116 (2004) 424
7. M. Alquero, M.L. Calzada, L. Pardo, E. Snoeck, J. Mater. Res. 14 (1999) 4570–4580.
8. F. Yang, W.D. Fei, Z.M. Gao, J.Q. Jiang, Surface & Coatings Technology 202 (2007) 121–125
9. B. Valquin et al, Appl. Surf. Sci. 195 (2002) 63.

Preparation and Characterization of PZT Wafers

A. Seal[*], B.S.S.Chandra Rao[#], S. V. Kamath[#], A. Sen[*], and H. S. Maiti[*]

[*]*Central Glass and Ceramic Research Institute, CSIR, Kolkata, India.*
[#]*Defence Metallurgical Research Laboratory, DRDO, Hyderabad, India*

Abstract. Piezoelectric materials have recently attracted a lot of attention for ultrasonic structural health monitoring (shm) in aerospace, defence and civilian sectors, where they can act as both actuators and sensors. Incidentally, piezoelectric materials in the form of wafers (pwas-piezoelectric wafer active sensor, approx. 5-10 mm square and 0.2-0.3 mm thickness) are inexpensive, non intrusive and non-resonant wide band devices that can be surface-mounted on existing structures, inserted between the layers of lap joints or embedded inside composite materials. The material of choice for piezoelectric wafers is lead zirconate titanate (PZT) of composition close to morphotropic phase boundary [pb ($zr_{0.52}$ $ti_{0.48}$) o_3]. However, an excess pbo is normally added to pzt as a densification aid and also to make up for the lead loss during high temperature sintering. Hence, it is of paramount importance to know how the shift of the lead content from the morphotropic composition affects the piezoelectric and mechanical properties of the sintered wafers, keeping in view the importance of mechanical properties of wafers in shm. In the present study, we observed that with the increase in the lead content of the sintered wafers, the dielectric and piezoelectric constants decreased. However, the elastic modulus, hardness and fracture toughness of the wafers increased with increasing lead content in the composition. Hence, the lead content in the sintered wafers should be optimized to get acceptable piezoelectric and mechanical

Keywords: PWAS, mechanical, piezoelectric, dielectric.
PACS: 77.22.Ch, 77.22.Gm, 62.20.de, 43.40.+s

INTRODUCTION

Piezoelectric materials are currently used in a variety of applications as sensors and actuators and they play a pivotal role in the arena of smart materials and structures. Though wafers or substrates of alumina are common items in electronics industry [1-2], piezoelectric wafers are seldom used in real-life applications. However, recently, PZT based piezoelectric wafers (approx. 5-10 mm square and 0.2-0.3 mm thickness) have opened new avenues for ultrasonic testing of structures. Piezoelectric wafer active sensors (PWAS) can act as both sensors and actuators. Several investigators [3-5] have explored the generation of Lamb waves with PWAS. Piezoelectric wafers are non-intrusive, nominally invasive and non-resonant wide-band devices with surface pinching in-plain strain and can be surface-mounted on existing structures or inserted between the layers of lap joints or inside composite materials. PWAS can act as the generator as well as detector of Lamb wave. Incidentally, guided waves (like Lamb waves in thin plates) have certain advantages for NDE of structures e.g., they travel

CP1029, *Smart Devices: Modeling of Material Systems, An International Workshop*
edited by S. M. Sivakumar, V. Buravalla, and A. R. Srinivasa
©2008 American Institute of Physics 978-0-7354-0553-0/08/$23.00

long distances and follow the contour of the structure in which they are propagating and these modes allow inspection in regions that are inaccessible, such as buried structures.

Tape casting or doctor-blade process [1-6] is generally employed to produce thin flat ceramic sheets (substrates/wafers). Tape-casting is advantageous for preparing relatively large-area uniform thin sheets with high density [1], which can be cut to small pieces according to the requirements. Tape casting, basically, consists of preparing a suspension of a ceramic powder in a solvent with addition of dispersants, binders and plasticizers. The suspension is cast onto a stationary or moving surface. After evaporation of the solvent, the dried green tape is stripped from the surface followed by cutting to the appropriate size and shape. The green tapes are sintered after removal of the organic components at a low temperature. Apparently, this simple process of making substrates/wafers turns out to be a not-so-easy process when one tries to prepare dense, flat, warpage-free PZT wafers. Keeping in mind the requirements of indigenous source of low-cost PZT wafers for NDE of thin wall structures like aircraft shells, dams, bridges, pressure vessels, industrial tubes, pipes etc, it is the need of the hour to address the technological challenges of making thin, dense PZT wafers in a cost-effective way.

Generally, during the fabrication of substrates and wafers, it needs optimization of tape casting slurry characteristics, binder burnout schedule, sintering schedule, weight of the applied load to avoid warpage without hampering densification, selection of the right kind of setter plate to avoid sticking of the wafer etc. Other than the aforesaid factors, one of the key issues in the synthesis of PZT wafers is the volatilization of PbO during sintering. The loss of PbO results in the change in composition, variation of the microstructure and poor densification of PZT ceramics [7, 8]. The conventional method of tackling lead volatilization is to sinter the material in a closed crucible surrounded with lead containing 'atmosphere powder' [9]. Other efforts are focused on decreasing sintering temperature by adding different liquid-phase-sintering aids among which PbO plays an important role [10]. Incidentally, PbO volatilization is a pressing issue for PZT wafers because of their high surface area to volume ratio. It needs an utmost control of the parameters like the heating schedule, the amount of surrounding 'atmosphere powder' with respect to the crucible volume and the amount of excess lead oxide in the batch composition so that dense and flat PZT wafers with desired composition, which do not stick to the setter plates, can be obtained. As an excess PbO is added to facilitate sintering and to adjust the lead loss during sintering, it is of paramount importance to know how the PbO content in the fired PZT wafers affects the dielectric as well as mechanical properties. While the effect of PbO content on the dielectric properties of bulk PZT is well-known [11, 12], similar information is lacking for wafers. Also very few reports are available on the effect of PbO content on the mechanical properties of conventionally processed bulk PZT ceramics [13, 14]. As the mechanical properties of PZT wafers are of great importance for SHM, the present study deals with the effect of PbO content (excess or deficient with respect to the morphotropic composition) on the mechanical as well as the dielectric and piezoelectric properties of PZT wafers.

EXPERIMENTAL

PZT powder of composition Pb $(Zr_{0.52}Ti_{0.48})O_3$, which is close to the morphotropic phase boundary and shows the optimum piezoelectric properties, was prepared by a mixed route comprising citrate-nitrate gel method followed by solid state mixing and calcination [7]. The raw materials used for the preparation of PZT were reagent grade $Pb(NO_3)_2$, citric acid, TiO_2, $ZrOCl_2.8H_2O$, and ethylene diamine. Initially $ZrOCl_2.8H_2O$ was dissolved in distilled water to get a clear solution. The solution was treated with NH_4OH solution followed by 1:1 nitric acid to get $ZrO(NO_3)_2$. The strength of the solution was determined gravimetrically. The $ZrO(NO_3)_2$ solution was taken in a beaker. The required amount of $Pb(NO_3)_2$ was added to the $ZrO(NO_3)_2$ solution to get Zr to Pb ratio equal to 0.52 : 1. Anhydrous citric acid was added to this solution with molar ratio of citrate to nitrate equal to 0.37. The above solution was concentrated by heating and stirring continuously on a hot plate and during this period ethylene diamine was added as a complexing agent. The solution set to a viscous gel due to continuous heating and the gel began to foam and swell, and burned on its own (autoignition) due to strong exothermic reaction between the citrate and nitrate species. The ash thus produced was calcined at $750°$ C for 2h to get a fine mixture of $PbZrO_3(PZ)$ and PbO. In the homogeneous fine mixture of PZ and PbO, TiO_2 was mixed in the right proportion (to form PZT) in an agate mortar and a pestle under acetone. The mixture was calcined at $800°$ C for 4h to get the desired PZT powder.

Stable suspension of PZT was prepared using reagent grade solvent consisting of an azeotropic mixture of methyl ethyl ketone (MEK) and ethanol (66:34 by volume). PZT and the solvent in the weight ratio of 1:15 along with 1-3 wt% phosphate ester (Emphos PS21-A, Witco Chemicals, USA) were ball milled for 20 h using zirconia balls to get the desired suspension. A typical PZT tape casting slurry was made as per the composition given in Table 1. An excess PbO (3wt%) was used as a sintering aid [12, 17] in the batch composition. To prepare the tape casting slurry, PZT powder along with the excess PbO and dispersant was ball-milled in the solvent for 20 h using zirconia balls. The binder, plasticizer and homogenizer were added to the slurry followed by further milling for 2 h. The slurry was tape-cast using a moving doctor-blade at a speed in the range of 20-25 mm/sec and was allowed to dry for 12 h. The dried tapes were cut into square shapes (10-15 mm square) and fired in air at a slow heating rate ($5°C/h$) from room temperature to $600°C$ so as to remove the organics. The fired tapes were sintered at a temperature in the range of $1150°C$-$1200°C$ with heating rates between 150-$400°C/h$ under controlled atmosphere (created by using a mixture of lead zirconate and lead oxide powders inside a closed crucible).

By manipulating the processing parameters i.e., the amount of surrounding atmospheric powder and the sintering schedule three varieties of wafers were fabricated. Wafers-A were lead deficient (-1.52%, i.e., the lead content in the sintered wafers was 1.52% less than that of the morphotropic composition), wafer-B were near morphotropic (-0.5%, i.e., the lead content in the sintered wafers was only 0.5% less than that of the morphotropic composition) and wafer-C were lead excess (+1.26%, i.e., the lead content in the sintered wafers was 1.26% more than that of the morphotropic composition). Table 2 gives a glimpse of the firing schedule to get the three varieties of wafers.

TABLE 1.

TABLE 1. Composition of PZT tape casting slurry

Ingredients	Function	Wt%
PZT [Pb(Zr$_{0.52}$Ti$_{0.48}$)O$_3$]	Ceramic	75.00
PbO	Sintering additive	2.25
Phosphate Ester	Dispersant	1.93 (Equivalent to 2.5 wt% of PZT)
Methyl Ethyl Ketone + Ethanol	Solvent	14.56
Polyvinyl Butyral	Binder	2.47
Polyethylene Glycol	Plasticizer	2.66
Butyl Benzyl Phthalate	Plasticizer	0.74
Cyclohexanone	Homogenizer/ skin inhibitant	0.39

TABLE 2. Characteristics of sintered wafers under varying processing conditions

Sorrounding Powder, [Batch wt / Powder wt]	Firing Schedule	PbO (wt%) excess or deficient (with respect to PZT) after firing.
PZT:PbO (1:1) [0.213]	Heated @ 300°C/h to 950° C (1h), then heated @300°C/h to 1175°C (10 min) and cooled @ 300°C/h	-1.4 to -1.6
PZT:PbO (1:1) [0.213]	Heated @ 400°C/h to 950° C (1h), then heated @400°C/h to 1175°C (10 min) and cooled @ 400°C/h	+0.37 to -0.5
PZT:PbO (1:1) [0.085]	Heated @ 400°C/h to 950° C (1h), then heated @400°C/h to 1175°C (10 min) and cooled @ 400°C/h	+1.08 to 1.51

The bulk density of the fired wafers (around 8mm x 8mm x 250µm) was calculated geometrically and the fracture surfaces were viewed in a SEM (Leo430i). After polishing the wafers with ceria powder (0.2 µm) the hardness and Young's modulus of the wafers were measured on a CSM-make instrumented microindentation system, where one can record the load versus indentation depth during the indentation process. A Vickers indenter was used for carrying out the microindentation tests. The fracture toughness was also determined using the CSM make instrumented microindentation system with appropriate selection of load to ensure well-defined cracks at the corners of the indent. The radial crack lengths and indentation half sizes were measured directly using Leicavm Ht Auto Optical Microscope with a calibrated eyepiece.

The Young's modulus was measured using the Oliver and Pharr method [15] from the slope of unloading portion of P-h (load – displacement) curves while the hardness values of these samples were calculated from the remnant depth of indentation from the P-h curves using the following relation [15]

$$H_V = 0.038 \frac{P}{h^2} \qquad\qquad 1.$$

where P is the maximum indentation load and h is the remnant depth of indentation. The mean fracture toughness (K_c) values were calculated from the equation suggested by Lawn et al. [17] as given below:

$$K_c = 0.016 \left(\frac{E}{H_V} \right)^{\frac{1}{2}} \frac{P}{c^{\frac{3}{2}}}$$

2.

where E, H_V, c and P are Young's modulus, Vickers hardness, radial crack length and indentation load, respectively.

The dielectric studies of the fired wafers were made using a Hioki 3532-50 LCR Hitester in the frequency range of 100-1MHz after electroding the as fired wafers with gold paste and curing at a temperature of 900° C for 15 min. The piezoelectric strain constant (d_{33}) was measured on a d_{33} meter (Pennbeker 8000 d_{33} tester) after poling the samples in silicone oil at a temperature of 120°C at 3kV/mm field for 30 min. The planer electrochemical coupling coefficient (k_p) was determined from the resonance (f_r) and antiresonance (f_a) frequencies using the formula

$$K_p = \sqrt{2.51 \frac{\Delta f}{f_r}}$$

3.

where Δf is equal to $(f_a - f_r)$. The ferroelectric hysteresis studies of the wafers were carried out on a loop tracer (Precision LC, Radiant Technologies Inc.)

RESULTS AND DISCUSSION

The bulk densities of the wafers were in the range of 95–97%. The SEM micrographs of the fracture surfaces for all the three varieties of PZT wafers are shown in Fig. 1. Equiaxed PZT grains with a grain size of 8-10 μm were observed in all the samples. Very small amounts of fine ZrO_2 grains were observed in wafers-A (PbO-deficient), whereas needles of $PbTiO_3$ were seen in wafer-C (PbO-excess). These observations were also confirmed from the XRD traces as well as EDS analysis. Figure 2 shows the representative load versus indentation depth (P-h) curves obtained from the instrumented microindentation experiments for the three varieties of PZT wafers. It can be observed that the loading paths for all the three varieties are similar till about 50 mN load. Beyond this load the indenter penetrates the maximum depth in wafers-A (PbO-defficient) and the minimum in wafers-C (PbO-excess).

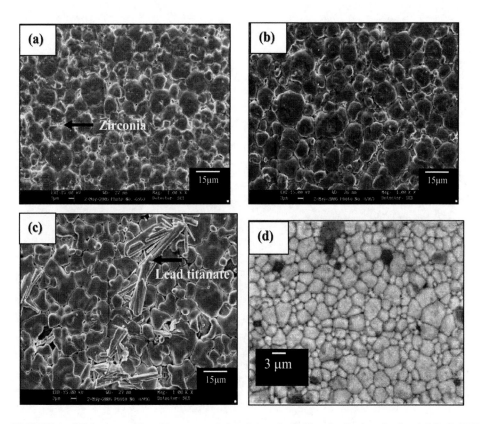

FIGURE 1. SEM micrographs of fracture surfaces of a) wafer-A, b) wafer-B c) wafer-C and d) Bulk PZT

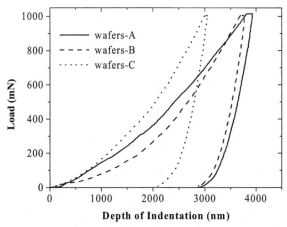

FIGURE 2. Load displacements curves for the three varieties of PZT wafers.

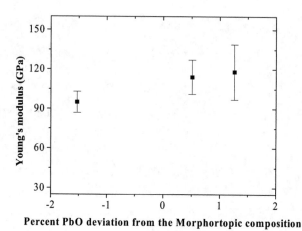

FIGURE 3. Variation of Young's modulus with percent PbO deviation from the morphotropic composition.

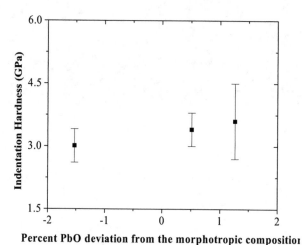

Percent PbO deviation from the morphotropic compositioɪ

FIGURE 4. Variation of hardness with percent PbO deviation from the morphotropic composition.

The Young's modulus values, as measured from the slopes of the unloading portion of P-h curves (Oliver and Pharr method [17]), are depicted in Fig. 3. It can be observed that the Young's modulus is the lowest for wafers-A (PbO deficient) and it monotonically increases from wafers-B (PbO near-morphotropic) to wafers-C (PbO excess). The hardness values are depicted in Fig. 4. It can be observed that the hardness also follows the same trend as that of Young's modulus and is the lowest for wafers-A and increases from wafers-B to wafers-C. The excess PbO, which helps in sintering, is expected to accumulate in the grain boundaries for wafers-A, while in case of the wafers-B and wafers-C the excess PbO volatilizes away thereby imparting higher Young's modulus and hardness values to the wafers. However, the little excess PbO present in the grain boundaries in the PbO-excess wafers may not be high enough

296

to cause appreciable weakening of the grain boundaries as has been reported in the literature [18,19].

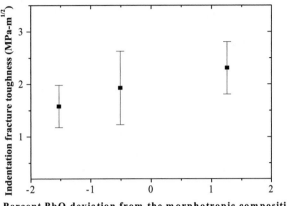

FIGURE 5. Variation of indentation fracture toughness with percent PbO deviation from stoichiometry

The K_c values calculated from equation 2 are depicted in Fig. 5. It can be observed from the figure that the fracture toughness follows the same trend as the Young's modulus and hardness, and is the lowest for wafers-A and it monotonically increases from wafers-B to wafers-C. The factors likely to affect the fracture toughness K_c of PZT ceramics are: mode of fracture, presence of second phase particles, grain size, grain boundary strength and the amount and nature of porosity [17]. In all the three varieties of wafers, the mode of fracture is intergranular and the grain size, nature and the volume fraction of porosity are also similar and hence the effect of PbO content on fracture toughness of the PZT wafers in the present case can be understood by taking into account the grain boundary strength and the presence of second phase particles (probably, PbO) in the grain boundaries. In the present case although ZrO_2 particles are present in wafers-A (PbO-defficient), they are extremely small in size (Fig. 3) and a few and far between and hence are not expected to have a significant influence in deflecting the crack and affect the fracture toughness of the samples. On the other hand, in wafers-C (PbO-excess), the $PbTiO_3$ particles are much larger in size (Fig. 3) and have a large enough volume fraction to cause the cracks to deflect significantly and consequently increase the fracture toughness of the sample.

Table 3 shows the comparison of the dielectric and piezoelectric properties of the three varieties of the PZT wafers with bulk samples. Figure 6 depicts the dielectric properties of the wafers at different frequencies. The nature of the polarization vs. electric field curve of the PZT wafers, as depicted in Fig.7, confirms the ferroelectric nature of the wafers. From Table 3 it is evident that the dielectric constant values of the wafers are comparable (especially, wafers-A) with those of the bulk samples.

TABLE 3: Dielectric and piezoelectric properties of PZT wafers and bulk samples

Sample Code	Excess / deficient PbO content (wt %)	Density (% of ρ_{th})	Dielectric constant (at 1 kHz)	Dissipation factor (at 1 kHz)	d_{33} (pC/N)	Coupling coefficient (k_p)
Wafers A	-1.523	97.03%	668.41	0.031	220-230	0.35
Wafers B	-0.50	96.19%	570.98	0.037	200-205	0.35
Wafers C	+1.26	95.35%	479.83	0.073	142-146	0.31
Bulk	Near morphotropic	96.91%	612.00	0.004	223	0.52

Among the three varieties of wafers, wafers-C (PbO-excess) show the lowest dielectric constant and the dielectric constant monotonically increases from wafers-B (PbO- near-morphotropic) to wafers-A (PbO-deficient). The decrease in dielectric constant of the wafers with the increase in PbO content may be attributed to the presence of non-ferroelectric PbO phase at the grain boundaries [17]. Relatively higher dissipation factors of the wafers may result from p-type conduction [20] and the dissipation factor decreases with excess PbO content probably because of the non-ferroelectric PbO phase at the grain boundaries. The d_{33} coefficient also follows the same trend. It is the highest for wafers-A (PbO-deficient) and decreases monotonically from wafers-B (PbO near-morphotropic) to wafers-C (PbO-excess). This may also be attributed to the presence of non-ferroelectric PbO phase at the grain boundary for higher PbO content wafers. The planar coupling co-efficient (k_p) values of the wafers remain more or less constant for the three varieties of the wafers, however their values are lower than those of the bulk samples.

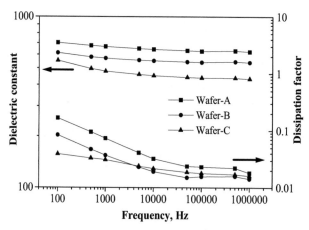

FIGURE 6. Variation of dielectric constant and dissipation factor of the wafers with frequency as a function of PbO content.

298

It is well-known that the k_p values vary with the grain size. From the SEM micrographs it is seen that the grain sizes of the three varieties of wafers are almost similar. The bulk samples have also grains of similar size like that of the wafers [Fig 1(d)]. Hence, the lower k_p values of the wafers probably arises from low domain wall mobility [21] due to domain wall stabilization by defect dipoles [9], the latter may be present in more numbers in wafers as evident from the higher DF values of the wafers. Further work should be carried out to probe this matter.

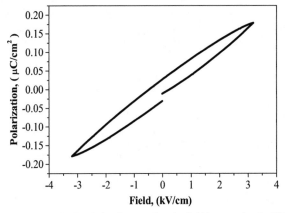

FIGURE 7: Typical polarization vs. electric field hysteresis of a PZT wafer. wafers.

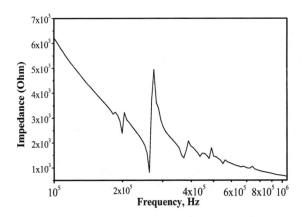

FIGURE 8. Typical nature of Impedance vs frequency plot of a PZT wafer.

CONCLUSIONS

From the present work it can be concluded that the hardness, Young's modulus and fracture toughness values of the PZT wafers are the lowest for wafers-A (PbO-deficient), intermediate for wafers-B (PbO near-morphotropic) and the highest for wafers-C (PbO-excess). This behaviour of the hardness and Young's modulus can be attributed to the

299

strengthening of the grain boundaries due to enhanced sintering in the presence of PbO liquid phase. For fracture toughness, the observation can be attributed to both the strengthening of the grain boundaries as well as the presence of needle shaped $PbTiO_3$ particles in wafers-C, which can deflect the cracks.

The dielectric constant and the d_{33} coefficient of PZT wafers were found to be the lowest for wafers-C (PbO-excess), intermediate for wafers-B (PbO near-morphotropic) and the highest for the wafers-A (PbO deficient). The dielectric constant values of the wafers-A and wafers B are comparable with, while the dielectric constant values of wafers-C are lower than those of the bulk samples. Similarly, the d_{33} values of the bulk sample are comparable to those of the wafers-A and wafers-B, whereas the d_{33} values of the wafers-C are lower than those of the bulk samples.

REFERENCES

1. E.P. Hyatt, *Ceram. Bull.* **65**, 637-638 (1986).
2. M.P. Albano and L.B. Garrido, *Ceram.Int.* **31**, 57-66 (2005).
3. V. Giurgiutiu, A. Zagrai and J. Bao, "Damage identification in aging aircraft structures with piezoelectric wafer active sensors", *12th International Conference on Adaptive Structures Technology*, 15-17 Oct, University of Maryland, 2001.
4. X.Lin, and F. G. Yuan, *Smart Materials and Structure.* **10**, 907-913 (2001).
5. V. Gurgiutiu, A.N. Zagrai and J. Bao, *ASME Journal of PresS. VeS. Tech.* **124**, 293-302, (2002).
6. C. Pagnoux, T. Chartier, M. D. Granja, F. Doreau, J. M. Ferreira and J.F. Baumard, *J. Eur. Ceram. Soc.* **18**, 242-247 (1998).
7. A.K. Sinha, D. Kumar, O. Parkash and H.S. Maiti," *Mat. Res. Bull.* **38**, 1165-1174 (2002).
8. S.S. Chiang, M.Nishiokar, R.M. Fulrath and J.A. Pask, *Am. Ceram. Soc. Bull.* **60**, 484-489, (1981).
9. A.J. Moulson and J.M. Herbert, Electroceramics: Materials, properties, applications, Chapman & Hall, London, pp. 280-284, 1990.
10. A. Seal, R. Mazumdar, A.Sen and H.S. Maiti, *Mat. Chem. Phys.* **97**, 14-18 (2005).
11. A.H. Webster, T.B. Weston and N.F.H. Bright, *J. Am. Ceram. Soc.* **50**, 490 – 491 (1967).
12. S. S. Chiang, M. Nishiokar, R. M. Fulrath and J. A. Pask, *Am. Ceram. Soc. Bull.* **60**, 484-489 1981).
13. L.Z. Storz, R.H. Dungan, *J. Am. Ceram. Soc.* **68**, 530–533 (1985).
14. S.W. Freiman, *Ferroelectrics* **102**, 381–390 (1990).
15. W.C.Oliver and G.M.Pharr, *J.Mater. Res.* **7**, 1564-1583 (1992).
16. B.R. Lawn, A.G. Evans and D.B. Marshall, *J. Am. Ceram. Soc.* **63**, 574-581 (1980).
17. A. Garg and D.C. Agarwal, *Mater. Sci. Eng. B* **56**, 46-50 (1999).
18. J.J. Kim and D.Y. Kim, *J. Am. Ceram. Soc.* **71**, 228–229 (1988).
19. B.M. Song, D.Y. Kim, S.I. Shirasaki, H. Yamamura, *J. Am. Ceram. Soc.* **72**, 833–836, (1989).
20. H.Hu and S.B. Krupanidhi, *J. Mater. Res.* **9**, 1484-1498 (1994).
21. C.A. Randal, N. Kim, J.P. Kucera, W. Cao and T.R, Shrout, *J. Am.Ceram.Soc.* **81**, 679-688 (1998).

POSTER SESSION

STRUCTURAL HEALTH MONITORING OF FRAME STRUCTURES USING PIEZO-TRANSDUCER

R.Shanker[a*] S.Bhalla[a] and A. Gupta[a]

[a]Department of civil Engineering, Indian Institute of Technology Delhi, Hauz Khas, New Delhi 110 016
[*]amashankeriitd@yahoo.com)

Abstract: Monitoring of civil structures is crucial for their proper functioning. Any crack in a structure changes its static and dynamic behaviours. To detect the damage/crack at the initiating time itself is challenging task in modern time. This paper describes an experimental study to extract the dynamic characteristics of a frame structure using piezo-electric ceramic (PZT) transducers. Tests are conducted on steel frame to extract the natural frequencies and the experimental mode shapes. Free vibration response is first acquired in the time domain and then transformed into frequency domain using Fast Fourier Transforms (FFT) analyser. Only single PZT patch is sufficient to extract the first nine modes shape of the steel frame .By using numerical model, mode shapes are extracted corresponding to each identified natural frequency. After determining natural frequencies and experimental mode shape, damages can be located by method of Naidu and Soh (2004). This approach can be used for damage/crack detection at very earlier stage.
Keywords: structural health monitoring, piezoelectric transducers, structural dynamics

INTRODUCTION

Civil structures are backbone of the economy of any nation. It has been emphasized that failure of civil-infrastructure to perform well could have an adverse effect on the gross domestic product [1]. It is most important to maintain these structures after completing construction because the overall performance depend upon post construction monitoring as well initial strength of structure. It is not possible to maintain the structure by only providing greater initial strength. Even lower strength structure performs satisfactorily if proper maintenance/monitoring are insured. It concludes that monitoring of any structure is must. During the last two decades, several research attempts have been made to develop techniques and methodologies to automatically monitor the health of structures. Global dynamic techniques have gained significant popularity during the last one and half decade. These are based on two parameters, the natural frequencies and the mode shape of the structure. Lower frequencies of structure can be determined easily but it is generally observed that at the 60 to 75 % of their ultimate failure load, the measured first frequency hardly shift by 1-2% only [2]. Pandey and Biswas [3] a that 50% reduction in the Young's modulus of elasticity, over the central 3% length of a 2.44m long beam, only resulted in about 3% reduction in the observed first natural frequency .In addition most low

CP1029, *Smart Devices: Modeling of Material Systems, An International Workshop*
edited by S. M. Sivakumar, V. Buravalla, and A. R. Srinivasa
©2008 American Institute of Physics 978-0-7354-0553-0/08/$23.00

frequency vibration based structural health monitoring (SHM)/ non destructive evaluation (NDE) methods on real-life structures are likely to encounter the presence of noise. The noise could be (a) mechanical noise, caused by sources such as vehicle movement or wind; (b) electrical noise, generated by variations in the power supply; or (c) electromagnetic noise, caused by communication waves, which affect the signal acquisition and transmission through cables and other susceptible circuitry [4]. The contamination by electrical interference and mechanical ambient noise degrades the quality of the emission signals [5, 6]. It can therefore be concluded that to capture the incipient damage, higher frequencies and mode shapes are mandatory [7].

This paper covers the determination of higher mode shapes and the associated frequencies using single (PZT) sensor of steel frame structure.

PZT SENSOR AND BASIC CONCEPT

Mode shape of any structure can be defined as the deflected shape of structure when the structure is vibrating with certain frequency. Also, by Maxwell's law of reciprocal deflection, the deflection of point n due to force P at point m is numerically equal to the deflection of the point m due to applied at point n. Hence deflection profile of the entire structure can be determined by a single fixed sensor at a point and by applying load of same magnitude at any other point where deflection is to be determined [10].

In this study, only one type of sensor namely PZT patch was evaluated on frame structures. Piezoelectric materials, which are commercially available as ceramics, such as lead zirconate titanate (PZT), are a type of 'smart' materials. They exhibit two related phenomena, (i) on application of mechanical stress, they undergo the development of surface charges (**direct effect**), and (ii) on application of electric field, they undergo mechanical strain (**converse effect**). The following equation describes the direct effect, which is used in sensor applications

$$D_3 = \overline{\varepsilon_{33}^T}E_3 + d_{31}T_1 \qquad\qquad 1.$$

where D_3 is the surface charge density, E_3 the electric field, T_1 the mechanical stress, $\overline{\varepsilon_{33}^T}$ the complex electric permittivity at constant stress and d_{31} the piezoelectric strain coefficient. In the absence of electric field, $E_3 = 0$. Further,

$$T_1 = \overline{Y^E}S_1 \qquad\qquad 2.$$

where $\overline{Y^E}$ is the complex Young's modulus of elasticity of the PZT material at constant electric field and S_1 the strain of the host structure (and also on the PZT patch). By using the theory of parallel place capacitors, we can derive

$$D_3 = \frac{\overline{\varepsilon_{33}^T}V}{h} \qquad\qquad 3.$$

where V is the potential difference across the terminals of the PZT patch and h is the thickness of the patch. Hence, from equations (1) to (3),

$$S_1 = \left(\frac{\overline{\varepsilon_{33}^T}}{d_{31}h\overline{Y^E}}\right)V = K_pV \qquad\qquad 4.$$

Thus, the output voltage across the terminals of the PZT patch is proportional to the strain of the host structure. The output voltage can be easily measured by oscilloscopes supported by conditioning circuit [8]. In this study, 2-channel FFT analyser has been used for this purpose.

EXPERIMENTAL PROCEDURE AND RESULTS

In this investigative study, steel frame made up of box section with column height 1.0 m, beam length 0.75 m and outer cross sectional dimension of 5.0 cm x 2.4 cm and thickness 1.3 mm was instrumented with single external PZT patch as shown in Fig.1. Randomly, nine nodes were created in frame at a distance 0.0, 0.30 cm, 0.75 cm, and 1.0 m from base in first column and at a distance 0.0, 0.25 cm, 0.50 cm, 0.75 cm and 1.0 m from base in second column. The structure was excited at these points by electric hammer of fixed magnitude of force, which was connected to channel 1 of the FFT analyzer. Free vibration response was measured after each excitation by PZT patch, which was connected to channel 2 of the FFT analyzer. The FFT analyzer recorded measurement from the PZT sensor after hammering at all different nine point one by one. Fig. 2 shows the measurement set up, consisting of the test structure, FFT analyzer and the hammer and the personal computer. Using the PZT patch, the instantaneous voltage was measured in the time domain. All the data recorded in the FFT analyzer was transferred to the PC via the USB interface .Using the ICATS software, the transferred data was automatically converted from time domain to frequency domain and hence typical frequency response function (FRF) were obtained. FRF is shown in Fig.3. The above frame structure was modeled in ICATS software considering the nine points as nodes and these FRF data were given as input resulting mode shape as output. The first nine Mode shapes are shown in Fig.4.

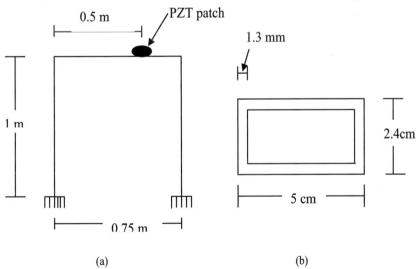

FIGURE 1. (a) Geometry of frame structure (b) Member cross-section

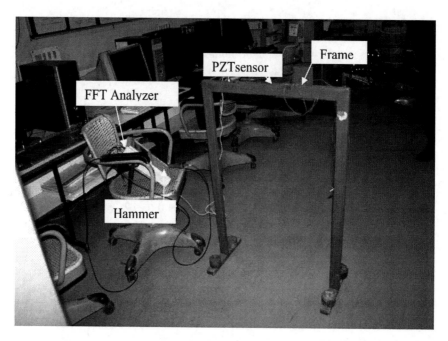

FIGURE 2: Experimental setup

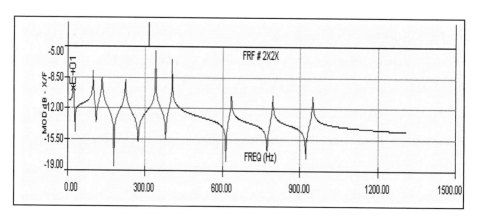

FIGURE 3: FRF of frame structure

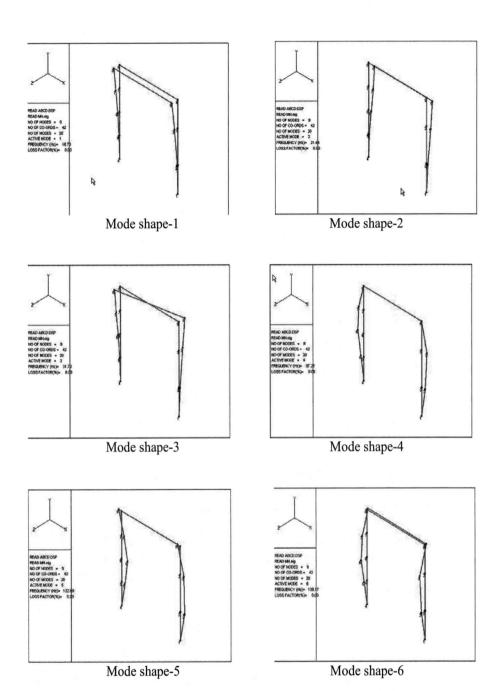

Mode shape-1

Mode shape-2

Mode shape-3

Mode shape-4

Mode shape-5

Mode shape-6

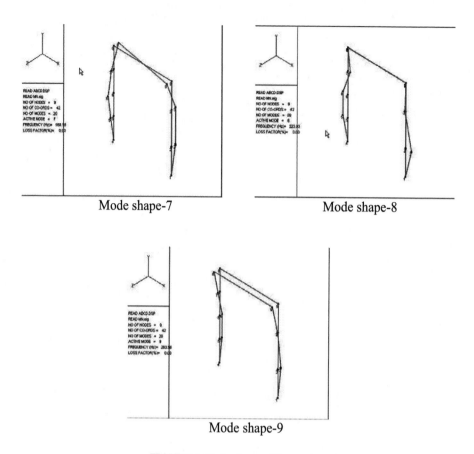

Mode shape-7

Mode shape-8

Mode shape-9

FIGURE 4. Mode shapes of frame structure

The first nine natural frequencies of the frame structure were identified as 18.70, 21.46, 31.70, 97.25, 132.69, 138.17, 168.16, 223.93 and 283.56 Hz. The frequencies and the mode shape of the structure are also determined numerically by ANSYS [11] which were found 19.40, 30.78, 46.52, 84.70, 134.68, 138.17, 184.33, 242.80 and 279.13 Hz. It can be found that the experimental frequencies and mode shape agree reasonably well with the numerical frequencies.

DISCUSSION

It can be observed that single PZT sensor has captured the first nine natural frequencies and the corresponding mode shape of the frame structure reasonably well. To detect the initial damage the higher natural frequencies and mode the shape are required [9]. Naidu and Soh (2004) used the numerical mode shape by ANSYS to detect the initial damage in the beam. More accurate result can be determined by using actual mode shape by the present method at the place of numerical mode shapes. Also, this technique can be used on any type of structure like beam, frame, truss and bridge due to small size of PZT to determine the experimental mode shape compared to accelerometer, which are conventionally used for modal data extraction; PZT patches

308

are extremely low cost. Since both low and higher frequency can be determined by single PZT patch, so this technique can be integrated with electro-mechanical impedance (EMI) technique to monitor the health of structures locally as well globally.

CONCLUSION

Results presented in this paper demonstrate that higher frequencies and mode shapes can be extracted by using single PZT sensor. So experimental mode shapes can be used at the place of numerical mode shapes which will give more accurate result to determine the initial damage of the structures. This technique can be combined with (EMI) technique, which will be more effective for structural health monitoring.

REFERENCES

1. Aktan, A. E., Helmicki, A. J. and Hunt, V. J. (1998), "Issues in Health Monitoring for Intelligent Infrastructure", Smart Materials and Structures, Vol. 7, No. 5, pp. 674-692.
2. Stubbs, N. and Osegueda,R.(1987) "Damage Evaluation in Solids-experimental Verification,"Mechanics and Material centre, Taxas A&M University MM-2424-22.
3. Pandey, A. K. and Biswas, M. (1994), "Damage Detection in Structures Using   Changes in Flexibility", Journal of Sound and Vibration, Vol. 169, No. 1, pp. 3-17.
4. Samman, M.M. and Biswas, M. (1994a), "Vibration Testing for Non-Destructive Evaluation of Bridges.I:Theory", Journal of Structural Engineering, ASCE, Vol. 120, No. 1, pp. 269-289.
5. Park, G., Cudney, H.H. and Inman, D. J. (2000a), "Impedance-based Health Monitoring of Civil Structure Components", Journal of Infrastructure System, ASCE, Vol. 6, No. 4, pp. 153-160.
6. Kawiecki, G.(2001), " Modal Damping Measurement for Damage Detection ", Smart Materials and Structures, Vol. 10, pp. 466-471.
7. Naidu, A. S. K. (2004), "Structural Damage Identification with Admittance Signatures of Smart PZT Transducers" Ph.D.Thesis,Nanyang Technological University, Singapore.
8. Sirohi, J. and Chopra, I. (2000b), "Fundamental Understanding of Piezoelectric Strain Sensors", Journal of Intelligent Material Systems and Structures, Vol. 11, No. 4, pp. 246-257.
9. Naidu, A. S. K. and Soh, C. K. (2004), "Identifying Damage Location with Admittance Signatures of Smart Piezo-Transducers", Journal of Intelligent Material Systems and Structures, Vol. 15, pp 627-642.
10. Reddy,C. S. (1996) Basic Structural Analysis, Tata McGraw-Hill Publication Company Limited, New Delhi
11. ANSIS Reference Manual; Release 9.0 (2004) www.ansys.com

AUTHOR INDEX

A

Arockiarajan, A., 183, 209, 221

B

Bandopadhya, D., 171
Bhalla, S., 303
Bhattacharya, B., 171
Bhattacharyya, D. K., 279
Bose, A., 279
Buravalla, V., 104

C

Chandra Rao, B. S. S., 290

D

Dai, H.-H., 77
Dayananda, G. N., 92
Deshpande, A. P., 161

E

Ebenezer, D. D., 264

G

Golden, M. A., 127
Gopalakrishnan, S., 140
Gosh, P., 58
Gupta, A., 303

H

Halder, S. K., 279

J

Jayabal, K., 221

K

Kamath, S. V., 290
Kamlah, M., 199
Khandelwal, A., 104
Kittipoomwong, D., 127
Klingenberg, D. J., 127

M

Maiti, H. S., 290
Mangalgiri, P. D., 13
Menzel, A., 209
Mohanraj, S., 50
Mukherjee, B., 239

N

Nakabdullah, N., 27
Namuduri, C. S., 127

P

Pandit, D., 161

R

Rajagopal, K. R., 3

S

Sansour, C., 183, 221
Seal, A., 290
Sen, A., 290
Sen, S., 279
Shanker, R., 303
Sivakumar, S. M., 161, 221

Skatulla, S., 183
Smith, A. L., 127
Sreemany, M., 279
Srinivasa, A. R., 58
Subba Rao, M., 92
Sun, Z., 127

T

Thamburaja, P., 27

U

Ulicny, J. C., 127
Utzinger, J., 209

V

Varughese, B., 92
Vedantam, S., 50

X

Xu, Z.-L., 77

Z

Zhou, D., 199

1-MONTH